W9-CUA-659

Gateway to Engineering

2nd Edition

George Rogers, Ed.D., DTE
Professor and Coordinator, Engineering/Technology Teacher Education
Purdue University

Michael Wright, Ed.D., DTE
Dean, College of Education
The University of Central Missouri

Ben Yates, DTE
Associate Affiliate Director for Missouri Project Lead the Way
Missouri University of Science and Technology

Australia • Brazil • Japan • Korea • Mexico • Singapore • Spain • United Kingdom • United States

Gateway to Engineering, Second Edition
George Rogers, Michael Wright, and Ben Yates

Vice President, Careers & Computing: Dave Garza

Director of Learning Solutions: Sandy Clark

Senior Acquisitions Editor: James DeVoe

Managing Editor: Larry Main

Senior Product Manager: Mary Clyne

Editorial Assistant: Aviva Ariel

Vice President, Marketing: Jennifer Ann Baker

Marketing Director: Deborah Yarnell

Senior Marketing Manager: Erin Brennan

Associate Marketing Manager: Jillian Borden

Senior Production Director: Wendy Troeger

Production Manager: Mark Bernard

Content Project Manager: David Barnes

Senior Art Director: Casey Kirchmayer

Media Editor: Deborah Bordeaux

Cover Image: ©Shutterstock.com/ Michael Drager

Library of Congress Control Number: 2012937278

ISBN-13: 978-1-1339-3564-3

ISBN-10: 1-1339-3564-8

Delmar
5 Maxwell Drive
Clifton Park, NY 12065-2919
USA

Cengage Learning is a leading provider of customized learning solutions with office locations around the globe, including Singapore, the United Kingdom, Australia, Mexico, Brazil, and Japan. Locate your local office at: **international.cengage.com/region**

Cengage Learning products are represented in Canada by Nelson Education, Ltd.

To learn more about Delmar, visit **www.cengage.com/delmar**

Purchase any of our products at your local college store or at our preferred online store **www.cengagebrain.com**

Notice to the Reader

Publisher does not warrant or guarantee any of the products described herein or perform any independent analysis in connection with any of the product information contained herein. Publisher does not assume, and expressly disclaims, any obligation to obtain and include information other than that provided to it by the manufacturer. The reader is expressly warned to consider and adopt all safety precautions that might be indicated by the activities described herein and to avoid all potential hazards. By following the instructions contained herein, the reader willingly assumes all risks in connection with such instructions. The publisher makes no representations or warranties of any kind, including but not limited to, the warranties of fitness for particular purpose or merchantability, nor are any such representations implied with respect to the material set forth herein, and the publisher takes no responsibility with respect to such material. The publisher shall not be liable for any special, consequential, or exemplary damages resulting, in whole or part, from the readers' use of, or reliance upon, this material.

Printed in Canada
1 2 3 4 5 6 7 16 15 14 13 12

Brief Contents

PART 5 ELECTRICITY AND ELECTRONICS

PART 6 PRODUCTION SYSTEMS

Table of Contents

PART 2 COMMUNICATING ENGINEERING IDEAS

PART 3 MODELING ENGINEERING DESIGNS

PART 4 THE USE OF ENERGY IN ENGINEERING

Chapter 13
Flight and Space 254

PART 5 ELECTRICITY AND ELECTRONICS

Chapter 14
Electrical Theory 272

Chapter 15
Electrical Circuits 290

Chapter 16
Electronics 304

PART 6 PRODUCTION SYSTEMS

Chapter 17

Chapter 18

Chapter 19

Chapter 20

Preface

Technology is as old as the human race. It began when the first humans learned to use such natural materials as rocks, sticks, and even animal parts to change the world around them. When people discovered how to attach a rock to a stick to make a hatchet or a spear, a technology was born.

Think about all the technologies you use during the day from the moment you wake up. Does an alarm clock wake you up? Where does the electricity for the alarm come from, and how does it get to your clock? Do you have hot running water in your home to take a shower? How does the water get to your home? How does the water move through your house? How is it heated? Do you use utensils and appliances to cook and eat food? Who designed them? How were they made?

COURTESY OF NASA

COURTESY OF AMERICAN HONDA MOTOR COMPANY, INC.

© BRIEDIS/SHUTTERSTOCK.COM

These examples might seem simple, but they were all designed and produced by engineers and engineering technologists. Imagine how your life would be without electricity, hot water, or simple forks and spoons. Engineers and engineering technologists design and manufacture everything from silverware to lifesaving medical devices. They design some robotic devices that are small and delicate enough to explore inside your body and others that are large and powerful enough to explore outer space. Your study of engineering and technology can help prepare you for a career in designing ways to make life better for other people.

HOW TO USE THIS BOOK

Just as products are designed and tested by engineers to meet people's needs, this book was designed and tested by experts to help you learn. Each chapter contains up-to-date information and special features that show how engineers tackle design problems in the real world. Look for the following features in every chapter:

Engineering in Action

S T
E M

As we know it today, the Ferris wheel came from the technical skills and engineering knowledge of a Pittsburgh, Pennsylvania, bridge builder named George Ferris. Ferris had graduated from Rensselaer Polytechnic Institute with a degree in civil engineering. He had worked for the railroads as a bridge builder and later founded a steel company in Pittsburgh. Ferris was approached at an engineering banquet in 1891 to design an American counterpart to the Eiffel Tower, the landmark of the 1889 Paris Exhibition. He sketched out a vertical merry-go-round on a napkin. A couple of years later, his Ferris wheel was unveiled at the 1893 Chicago World's Fair. The ride stood 264 feet high and, as you might expect, was constructed of steel. We need to thank the engineering education and technical skill of George Ferris for today's amusement ride.

- **Engineering in Action:** Each chapter begins with a special story that shows how engineers tackle design problems and create solutions for society.
- **Engineering Challenge:** Each chapter gives students and teachers a chance to improve their understanding of text ideas by doing hands-on problem-solving activities.

Engineering Challenge

ENGINEERING CHALLENGE 2

Design, build, and test an amusement ride that contains simple machines and mechanisms to move passengers (marbles) from a holding tank (waiting line) to a destination that is 24 inches higher and 24 inches to the side of the holding tank. Because this is an amusement ride, you should use as many simple machines and mechanisms as you can that will fit within the 24-inch-cube space.

- **Math in Engineering** and **Science in Engineering** use interesting facts to explain how science, technology, engineering, and mathematics are all connected.

Science in Engineering

S T
E M

It was Sir Isaac Newton's *Laws of Motion* (1687) that provided engineers with the scientific background to develop mathematical calculations related to levers. Newton's third law states, "For every action there is an equal and opposite reaction." When we force one side of the lever down, the opposite side moves up with an equal force. Engineers can, however, change the length of the lever arms in their design process to control the force that is required.

- **Career Spotlights** tell the stories of real engineers working in a variety of jobs. The inspiring engineers in these stories explain how they followed their educational pathway to exciting careers as engineers.

Career Spotlight

Name:
Oksana Wall

Title:
Consulting Structural Engineer, Celtic Engineering, Inc.

Job Description:
As a 13-year-old girl from Venezuela, Wall visited Disney World in Florida, and it immediately cast a spell on her. "It was such a happy and magical place," she says. Her father was an electrical engineer, and she was good at math and science, so she was already leaning toward an architecture or engineering career. But on that trip, she made up her mind to do engineering work for theme parks when she grew up.

After achieving her dream of working for Walt Disney World, Wall left for Celtic Engineering, Inc., which was started by her husband. Wall works with theme parks to create structures for rides and shows. For example, she helped design a multimedia show in Asia that includes a 360° movie experience. "The movie totally engulfs you, with several animated elements coming down from the ceiling," she says. "You feel like you're underwater."

Wall worked with other engineers on mechanisms that allow the show's elements to lower over the audience, perform their function, and then lift up again. The biggest element weighs about 25,000 pounds. "Behind the scenes there are at least 10 different mechanisms that work separately to bring the show together," she says.

Wall has a great time doing this kind of work. "It's very challenging doing structural engineering for entertainment because so much of what you do has never been done before in exactly that way," she says. "You have to adapt conventional engineering methods to something that is quite unique. It's fun never working on the same thing twice."

Education:
Wall received her bachelor's and master's degrees in civil engineering from the Florida Institute of Technology. She knew she wanted to work in theme parks, so she stayed focused on that goal in school. "I kept talking to all my teachers and to anybody else who would listen about what I wanted to do," she says. "Eventually, I spoke with other engineers to find out what I should be doing. Is this a good course to take? What is a good practical experience to have?"

Advice to Students:
Wall sees advantages in work that seems too difficult. "Don't get discouraged when something is hard," she says, "because, when things are really hard and you feel really dumb, that is when your brain is really stretching."

One difficulty Wall had to overcome is being a woman in engineering. "There are very few women in the field, but I never let that discourage me," she says. "When you have a hard time, think about your dream, and you will eventually get there if you keep on working at it."

- **Did You Know?** These miniarticles highlight interesting facts about engineering discoveries.

Did You Know?

Amusement rides, such as Ferris wheels and roller coasters, all use simple machines and mechanisms to change energy into motion. This motion provides excitement and entertainment for you and your friends. You could say that amusement rides are all work and all play.

- **Vocabulary:** We define and highlight important vocabulary words throughout the text. You will have a chance to test your understanding at the end of each chapter by writing definitions in your own words.
- **Stretch Your Knowledge:** You can test your understanding and learn more by completing the activities, problems, and projects at the end of each chapter.

FOR INSTRUCTORS

Gateway to Engineering introduces middle school students to the process of design, the importance of engineering graphics, and applications of electricity and electronics, mechanics, energy, communications, manufacturing processes, automation and robotics, and control systems. This textbook will help students build a solid foundation in technological literacy while they study engineering-related careers and educational pathways. Everyday examples show how engineers and their innovations affect the world around them. A clear, straightforward writing style complements the book's strong technical focus. Discussion of the social impacts of new technologies will allow students to explore the ramifications of engineering design. Finally, every chapter explores possible career pathways in engineering and engineering technology.

New to This Edition

The second edition of *Gateway to Engineering* has been fully updated to support the latest Project Lead the Way (PLTW) Gateway curriculum and to stay current with the pace of technology. The authors have updated information on current engineering careers and applications, including a revised chapter covering sustainable architecture.

An all-new CourseMate supplement for *Gateway to Engineering* has been developed to support the second edition. Students will find a rich array of online learning resources to help them succeed, while instructors will gain a suite of assessment and classroom management tools. Please see the full description that follows for instructions on how to access CourseMate through www.cengagebrain.com.

Gateway to Engineering is supplemented by a robust Instructor's Resource package that includes several powerful tools to help you spend more time teaching and less time preparing to teach:

- Instructor's Manual
- PowerPoint presentations
- Computerized test bank
- Correlation grid: ITEA Standards for Technological Literacy

COURSEMATE

The second edition of *Gateway to Engineering* includes the all-new Technology and Engineering CourseMate to help students make the grade.

The Technology and Engineering CourseMate for *Gateway to Engineering* includes:

- An interactive eBook, with highlighting, note taking, and search capabilities
- Interactive learning tools, including:
  - quizzes
  - flashcards
  - games
  - PowerPoint lecture slides
  - and more!

Instructors will be able to use CourseMate to access Instructor Resources and other classroom management tools.

To access these supplemental materials, please visit www.cengagebrain.com. At the cengagebrain.com homepage, search for the ISBN for your title (from the back cover of your book) using the search box at the top of the page. This will take you to the product page where these resources can be found.

Additional Reading Supplement

In addition to the tools included in the Instructor's Resource, teachers can benefit from the supplemental reader *Through the Gateway: Readings to Accompany Gateway to Engineering*. The reader contains a wealth of original fiction and nonfiction stories directly linked to the concepts in each chapter. Reviewed by technology teachers *and* teachers of English language arts, this mix of contemporary and historical vignettes and longer fiction tells the stories behind the innovations and inventions that shape our engineered world. Extended learning opportunities follow the readings and provide teachers with a relevant resource for integrating reading and writing across the curriculum.

GATEWAY TO ENGINEERING AND PROJECT LEAD THE WAY

This textbook resulted from a partnership forged with Project Lead the Way, Inc., in February 2006. As a nonprofit foundation that develops curriculum for engineering, Project Lead the Way provides students with the rigorous, relevant, reality-based knowledge that they need to pursue education in engineering or engineering technology.

The developers of the Project Lead the Way curriculum strive to make math and science relevant for students by building hands-on, real-world projects in each course. To support the project's curriculum goals and to support all teachers who want to develop projects and problem-based programs in engineering and engineering technology, Delmar Cengage Learning is developing a complete series of textbooks to complement all nine Project Lead the Way courses:

1. Gateway to Technology
2. Introduction to Engineering Design
3. Principles of Engineering
4. Digital Electronics
5. Aerospace Engineering
6. Biotechnical Engineering
7. Civil Engineering and Architecture
8. Computer Integrated Manufacturing
9. Engineering Design and Development

To learn more about Project Lead the Way's ongoing initiatives in middle school and high school, please visit www.pltw.org.

ACKNOWLEDGMENTS

The authors acknowledge the positive impact on the discipline of technology education provided by the vision and leadership of Richard Blais and Richard Liebich. As a secondary school teacher in upstate New York in 1996, Blais began the process of reviewing, developing, and evaluating a series of technology education courses that incorporated engineering concepts into the secondary school curriculum. These new "pre-engineering" courses were designed to provide students with the competencies to be successful in postsecondary school programs in engineering and engineering technology. This forward-thinking initiative was financially supported by the Charitable Venture Foundation founded by Liebich's family. This financial support has allowed the Project Lead the Way curriculum to become the nation's premier secondary school engineering-focused technology education program. The profession owes a debt of gratitude to the vision of Blais and the leadership of Liebich. The authors hope that *Gateway to Engineering,* in a small way, supports their vision and leadership.

The authors and publisher acknowledge the following reviewers, who contributed to the quality and accuracy of this text:

Todd Benz, *Pittsford-Mendon High School, Pittsford, NY*
Brent Blackburn, *Centennial Junior High School, Kaysville, UT*
Christine Calvo, *Cooley Middle School, Roseville, CA*

Casey Coon, *Oliver Middle School, Brockport, NY*
Bill Ganter, *Young Middle School, Tampa, FL*
Mike Gorman, *Woodside Middle School, Fort Wayne, Indiana*
Connie Hotze, *St. Joseph's Catholic School, Wichita, KS*
Jeff House, *Churchville-Chili Junior High School, Churchville, NY*
Robby Jacobson, *Lancaster Middle School, Lancaster, WI*
Carl Kramer, *Springfield Middle School, SC*
Barb Kubinski, *Nichols Junior High School, Arlington, TX*
Carrie McCune, *South Dearborn Middle School, Aurora, IN*
Ken Odell, *Churchville-Chili Junior High School, Churchville, NY*
Lauren Olson, *Biloxi Junior High School, Biloxi, MS*
Jeffrey Sullivan, *Menomonie Middle School, Menomonie, WI*
Bill Rae, *Lake Fenton Middle School, Fenton, MI*
Pamela Urbanek-Quagliana, *Clarence Middle School, Clarence, NY*
Matthew Wermuth, *Walden University*
Lola Whitworth, *Westminster Middle School, Westminster, SC*

Joanne Donnan, a curriculum developer for Project Lead the Way, reviewed portions of the book as a teacher at Galway Middle School, Galway, NY. Curriculum developers B. J. Brooks, Sam Cox, and Wes Terrell also reviewed the manuscript for Project Lead the Way.

ABOUT THE AUTHORS

Dr. George E. Rogers is a professor and coordinator of Engineering/Technology Teacher Education at Purdue University.

Dr. Michael D. Wright is the dean of the College of Education at the University of Central Missouri.

Ben Yates is the Associate Affiliate Director for Missouri Project Lead the Way at the PLTW affiliate Missouri University of Science and Technology. Yates is a former classroom teacher, school administrator, and university professor.

Gateway to Engineering

2nd Edition

CHAPTER 1
Engineering and Technology

Menu

Before You Begin

Think about these questions as you study the concepts in this chapter:

1. What is engineering?
2. What is the difference between an engineer and an engineering technologist?
3. Is there a difference between science and technology?

4. Are innovation and invention the same thing?

5. Why is the engineering profession important to our country's economic competitiveness and our standard of living?

6. What career fields are available in engineering?

7. What high school courses are available to learn more about engineering?

COURTESY OF NASA

FIGURE 1-1 Astronaut Buzz Aldrin poses for a photograph beside the United States flag during an Apollo 11 extravehicular activity (EVA) on the lunar surface.

"I believe that this nation should commit itself to achieving the goal, before this decade is out, of landing a man on the Moon and returning him safely to the Earth."

—President John F. Kennedy, 1961

Engineering in Action

S T
E M

In 1961, President John F. Kennedy offered the United States the greatest engineering challenge in human history. At the time, none of the technology even existed to fly to the moon, let alone keep a human being alive in space. Many people said it was impossible and pure science fiction to send a man to the moon (Figure 1-1).

In spite of these obstacles, the first lunar module, *Eagle,* landed on the surface of the moon on July 20, 1969. The next day, astronaut Neil Armstrong stepped out of the spacecraft and onto the moon. The words he spoke are now famous: "That's one small step for [a] man, one giant leap for mankind" (Figure 1-2).

COURTESY OF NASA

FIGURE 1-2 Astronaut Neil A. Armstrong, Apollo 11 mission commander of the Lunar Module *Eagle*, became the first human being to step onto the lunar surface, July 21, 1969.

Engineers at the National Aeronautics and Space Administration (NASA) had to overcome many technical challenges to meet President Kennedy's important goal. To do so, they carefully planned, designed, and tested many projects. Each project brought them one step closer to their goal, until they were finally successful. After Apollo 11, astronauts flew six more missions to the moon, returning home safely every time.

Did You Know?

Engineers doing research for the space program also provided the world with many new products and innovations. Consumer products that use miniature electronics, smoke detectors for your home, polarized sunglasses, TV broadcast by satellite around the world, and battery-powered tools are all spin-offs of the space program.

SECTION 1: WHAT IS ENGINEERING AND TECHNOLOGY?

You have probably heard the terms **engineer, engineering,** and **technology** many times before. They are an important part of our modern society. Many people use the terms interchangeably, but they really have very different meanings.

Engineering professionals design solutions to problems to make people's lives better. The engineering profession includes many types of careers (see Section 3). In this book, you will read about real engineers who work in different types of jobs and use their valuable skills to benefit society.

Who Is an Engineer?

An **engineer** (used as a noun) is a person who designs products, structures, or systems to improve people's lives. You use many of these products every day, like hair dryers, curling irons, refrigerators, video games, skateboards, MP3 players, and cell phones. And you see engineered structures all around you, like buildings, bridges, and highways. Just look around your classroom. Tables, chairs, computers, DVD players, monitors, whiteboards, light circuits, and heating and air conditioning systems are all examples of the products, structures, and systems that engineers design (Figure 1-3).

engineer
An engineer is a person who designs products, structures, or systems to improve people's lives.

The eyeglasses or contact lenses you or some of your classmates may be wearing were also designed by an engineer. And look at the many different styles of shoes! All these products were probably designed by an engineer. Engineers make very precise drawings and models of products.

Engineers work together in teams. Learning to work well with others is important because teamwork makes a project successful, especially as engineering designs get more complex. Engineers use mathematical and scientific principles to design solutions to meet human needs and wants. Engineers design, plan, and often oversee the manufacture of products. But they don't stop at designing the products; engineers also create the machines that make the products. Engineers

FIGURE 1-3 Products like the Apple iPhone were designed by engineers.

Math in Engineering

Mathematics is the "language" of engineering and technology. The development of mathematics in earlier civilizations allowed people to build larger, more complex structures like cathedrals. Today, mathematics forms the backbone of engineering. The language of math can describe objects in great detail. Engineers can also use mathematical models to test designs without ever actually building the object.

design and oversee the building of roads, mass transit systems, buildings, and other structures. They design the electrical, plumbing, and heating and air conditioning systems for buildings. They also design vehicles for sea, land, air, and space. This includes ways to move not only people and products but also raw materials like water, oil, and natural gas.

Engineers also plan how to harness energy resources to create power to do work for us. Engineers are always studying better ways to convert sunlight, wind, water, coal, oil, and even geothermal steam into power we can use. Engineers seek to improve the quality of our lives and to protect the environment (Figure 1-4).

Engineers use computers and computer-aided-design (CAD) software to create designs and test solutions (Figure 1-5). They often design systems made up of many parts (Figure 1-6). These systems might include simple machines and mechanisms. Engineers study issues like cost, effectiveness, and efficiency.

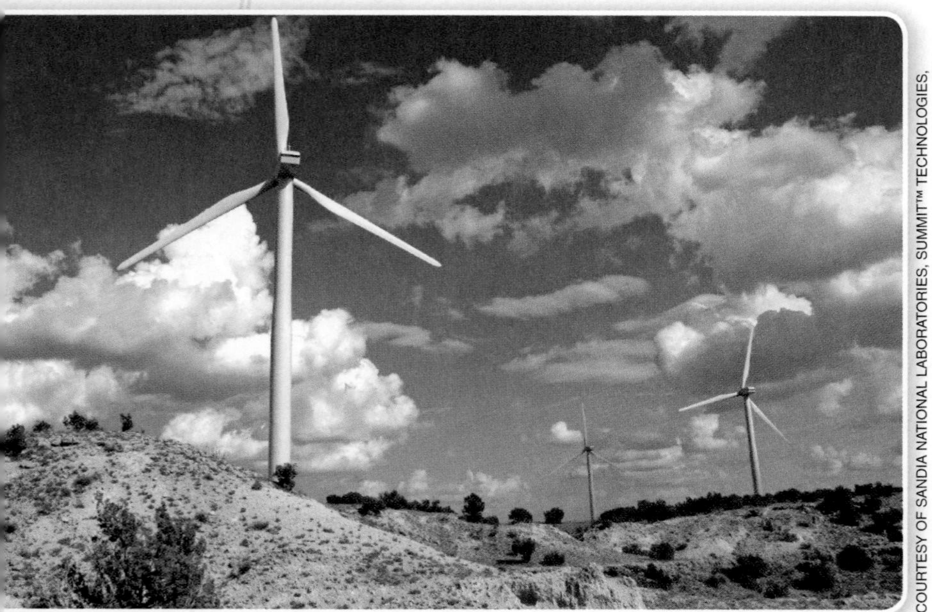

COURTESY OF SANDIA NATIONAL LABORATORIES, SUMMIT™ TECHNOLOGIES, WWW.MEMS.SANDIA.GOV

FIGURE 1-4 Engineers design ways to harness clean, renewable energy sources like these wind turbines near Albuquerque, New Mexico.

To help protect our environment, engineers look at the whole life cycle of the products they design. What happens to the product when its usefulness is over? Does it go to a landfill, or can it be recycled?

Today's engineers are designing our future. They design health care devices, consumer products, and robots. They design enormous spacecraft as well as

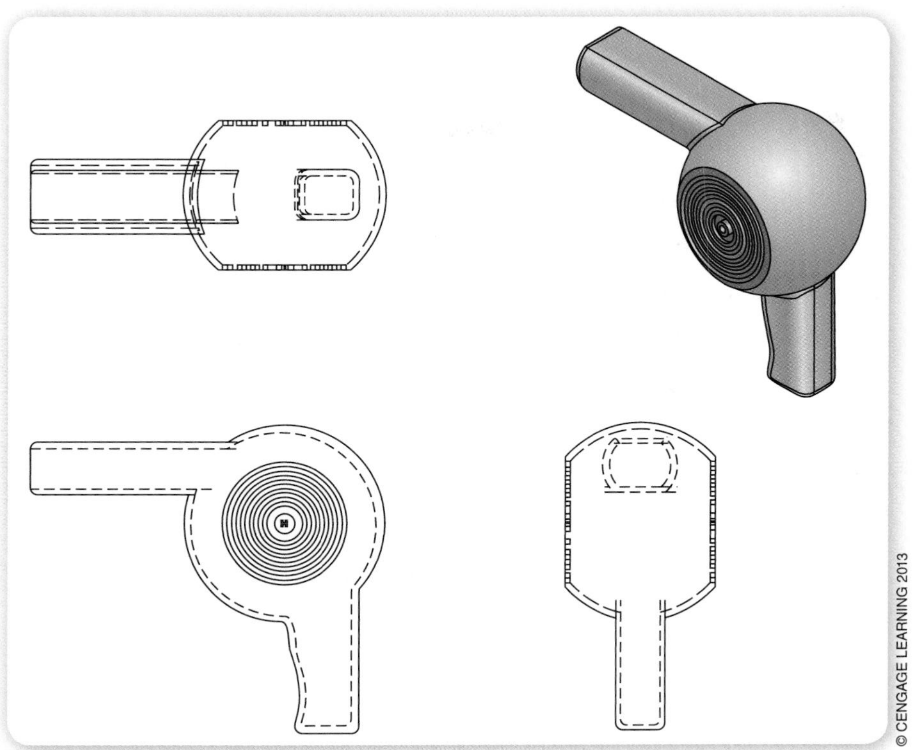

FIGURE 1-5 This picture is an example of a CAD drawing.

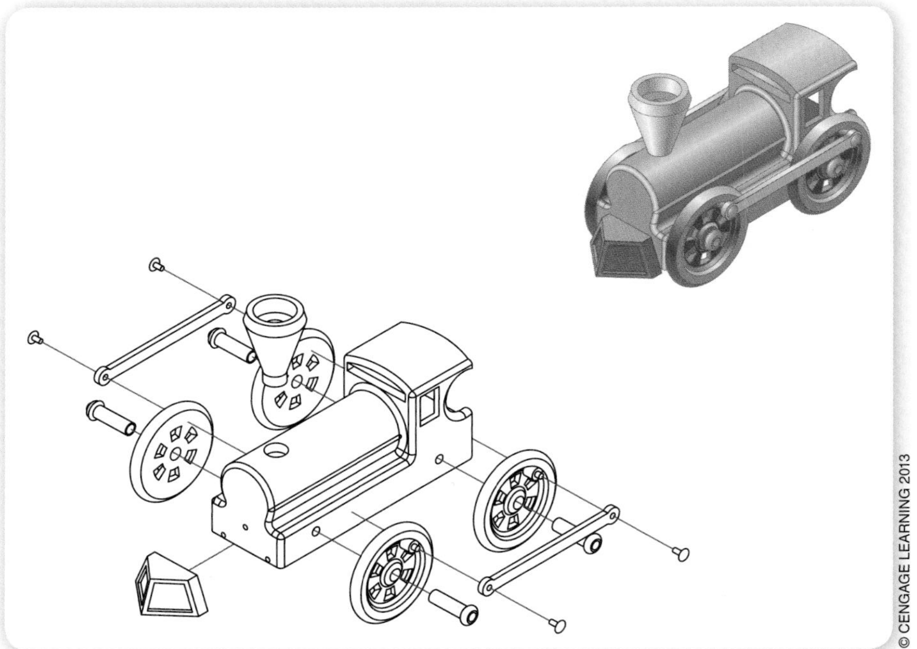

FIGURE 1-6 This CAD drawing shows how the final product is made up of multiple parts.

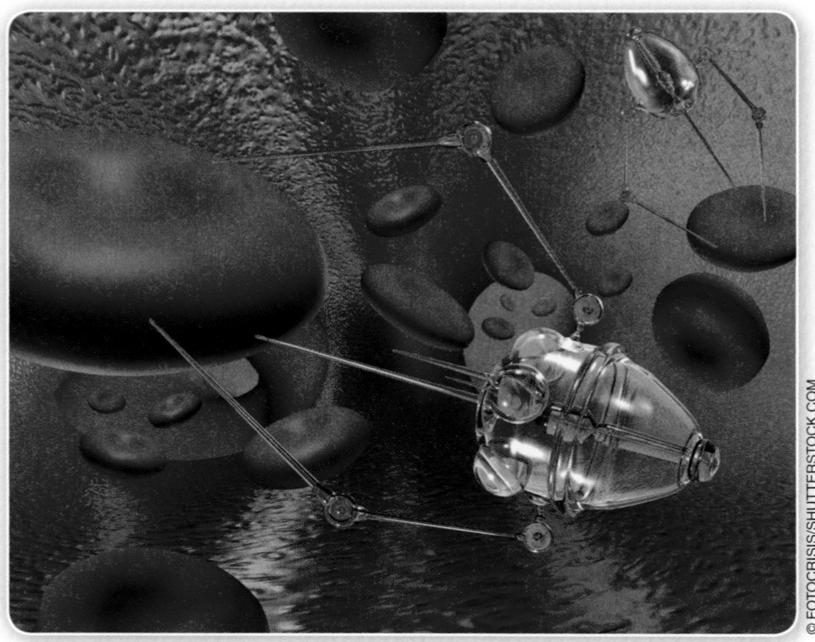

© FOTOCRISIS/SHUTTERSTOCK.COM

FIGURE 1-7 **Engineers are designing nanotechnology robots so small they will be able to travel in your bloodstream. Someday, a robot like this might be used to inject medicine into a single cell to fight cancer.**

1

engineering
Engineering is the process of designing solutions.

engineering technologist
An engineering technologist is a person who works in a field closely related to engineering. The technologist's work is usually more applied or practical, whereas the engineer's work is more theoretical.

nanotechnology devices that are smaller than the point of a pin (Figure 1-7). Engineering is an exciting career that shapes our way of life.

What Is Engineering?

Engineering (used as a verb) is the process of designing solutions. The engineering process includes making the drawings and models necessary for others to make the product. These drawings must be very precise and provide complete information. Often, the person who designs and/or oversees the production process for making these various products is an engineering technologist.

Who Is an Engineering Technologist?

Engineering technologists are more practical and less theoretical than engineers. They are the ones who work out the production details and design the systems for manufacturing a product. They have extensive knowledge of the tools, machines, and processes used to make things. In manufacturing, engineering technologists often serve in a role between the engineer and production workers. They may also be involved with product design.

What Is an Engineering Problem?

The term **problem** will be used throughout this book to mean a challenge that needs to be solved. A problem, as used in engineering, is not necessarily negative. For example, a problem might have been "How can I carry hundreds of CDs with me so I can listen to my music away from home?" One very good solution to this problem was the MP3 player.

Technology

technology
Technology is (1) the processes humans use to develop new products to meet their needs and wants and (2) the products or artifacts actually made.

Technology is different than engineering. Most people use the term *technology* to refer to computers or other electronic gadgets. But "technology" has a more specific meaning. Why don't you take a minute and look it up in the dictionary. You will probably find that the definition, simply stated, is the "practical application of knowledge" or the "art and science of producing products." It involves "know-how" and the ability "to do." The word *technology* is derived from Greek roots and refers to the art of producing products. Technology is a human process and is closely linked to invention and innovation. Today, technology has a dual meaning. It refers both to the study of innovation and invention, and also to the products or artifacts actually made.

Technology and Science **Science** is the study of the natural world and the laws that govern it, while *technology* makes use of human knowledge to develop the new products that people need (Figure 1-8). These products are sometimes called "artifacts." This means that people use their knowledge to manipulate, control, or change the world around them. Archeologists have found artifacts from every civilization. This proves technology is truly a human endeavor. Technology is part of what makes us human and separates us from animals.

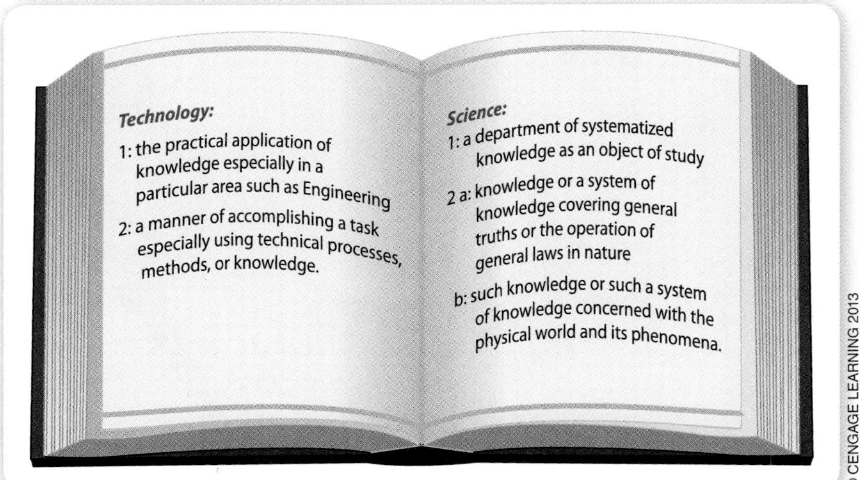

Technology:
1: the practical application of knowledge especially in a particular area such as Engineering
2: a manner of accomplishing a task especially using technical processes, methods, or knowledge.
Science:
1: a department of systematized knowledge as an object of study
2 a: knowledge or a system of knowledge covering general truths or the operation of general laws in nature
b: such knowledge or such a system of knowledge concerned with the physical world and its phenomena.

FIGURE 1-8 **Definitions of technology and science.**

Science in Engineering

Science and technology complement each other. Science explains why technological devices work. In the past, people developed technological devices before they could scientifically explain how they worked.

Today science and technology fuel each other. Scientists use technology to further their scientific investigations. Engineers and technologists use scientific principles to improve their designs.

When Did Technology Begin? Technology began when the first humans learned to manipulate the world around them. It is as old as the human race. When our ancestors discovered how to attach a rock to a stick to make a hatchet or a spear (Figure 1-9), a technological development was born. Learning how to cultivate the land by planting and harvesting crops was also a technological development. The rate of technological development is increasing at an **exponential** rate, due in part to the development of computers and the World Wide Web. This means that technology is developing at a faster and faster rate.

Is Technology Good or Bad? Technology itself is neither good nor bad. It is the *use* of the technology by humans that can have positive or negative

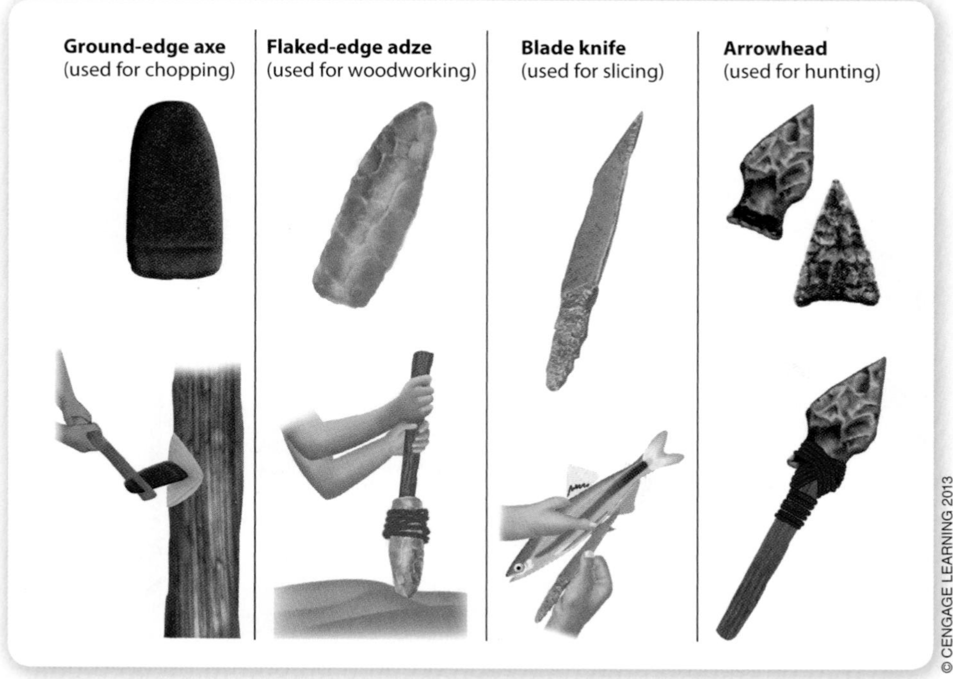

FIGURE 1-9 Prehistoric tools used by early humans.

Exponential Rate

The term *exponential rate of change* is often used simply to mean a very fast rate of change. In a mathematical sense, *exponential rate* has a very specific meaning. It means that something is increasing at a set pattern. The number sequence 2, 4, 8, 16, 32, 64 is an example of an exponential increase.

consequences. For example, when people invented the automobile they intended it to be a good thing. The automobile helped people move about the country easier than on bicycles or horseback. The inventors of the automobile did not anticipate the air pollution that would result from so many people using automobiles or the number of people killed in automobile accidents.

Technological developments can have four possible outcomes. First, there are the intended positive uses or consequences. Second, there may also be positive uses that were not intended by the original inventor. Third, there are the negative consequences that the developer expected. Finally, there are negative impacts that were not expected. These consequences can affect individuals, society, or the environment. Oftentimes, the development of technology involves ethical issues as well (Figure 1-10).

Invention

Invention is the process of developing something that has never been made before. An invention can be simple, like the ballpoint pen. Or it can be more complex, like a computer. Inventions can have large impacts on society.

invention
A new product, system, or process that has never existed before, created by study and experimentation.

The development of the transistor in 1947, for example, changed the world forever. Its invention opened the door to solid-state circuits and digital electronics. Every major electronic device we have today was made possible by the transistor (please see Chapter 16 for more detailed information about transistors). Another common example is the invention of the phonograph by Thomas Edison. Before Edison, there was no recorded sound.

Orville and Wilbur Wright developed the first powered airplane (Figure 1-11). Although other people were also working with gliders at the same time, the Wright brothers were the first to fly a powered aircraft. Their famous flight occurred on December 17, 1903. Just think about all of the commercial applications that have evolved in the 100 years following their historic flight!

The compact disc (CD) was invented in 1965 by James Russell, who was actually granted 22 different patents for the various parts of the CD playing and recording system. But it was not until the early 1980s, when Philips

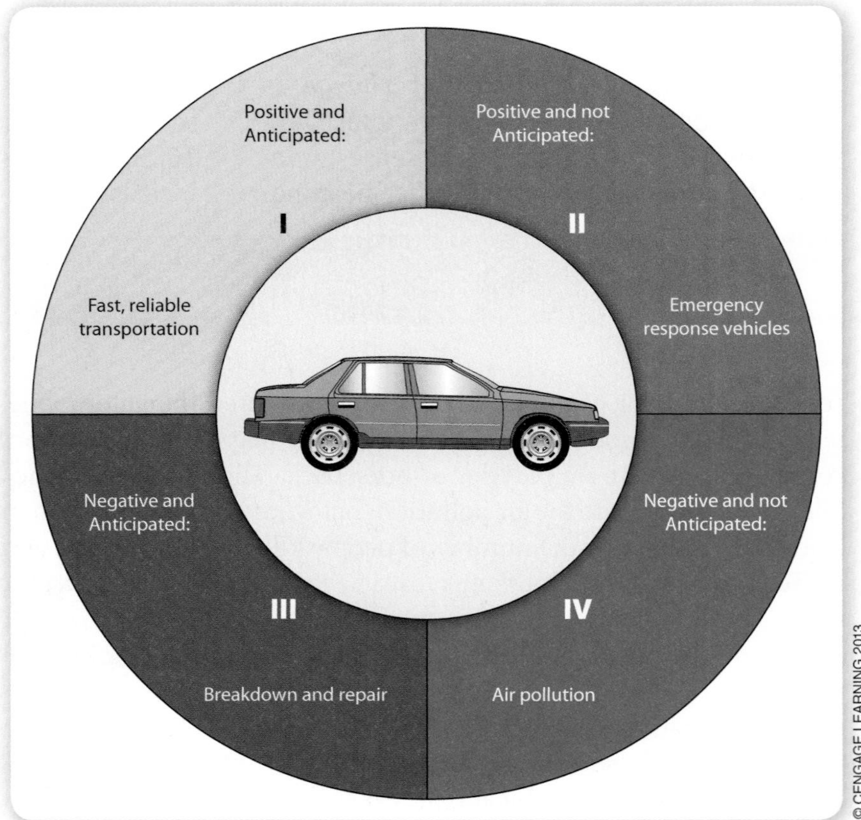

FIGURE 1-10 Technology can have positive and negative outcomes. What other positive and negative outcomes has the automobile produced?

FIGURE 1-11 The Wrights' first powered flying machine.

FIGURE 1-12 **Recording technologies have come a long way from Edison's early phonograph to today's CD.**

Corporation marketed the invention commercially, that the CD became commonplace (Figure 1-12). Inventions do not become popular until they are marketed by a company, usually in the form of consumer products.

There are many books, videos, and Web sites about inventions. Inventors can be found in every country or society. There were more inventions in the last century than the previous 1000 years. Some inventions are spectacular, like atomic energy or space travel, while others such as paper clips and vacuum cleaners are mundane. Even though paper clips and vacuum cleaners might not seem as exciting as space travel, they still had a huge impact on our daily lives.

Patents A **patent** is a unique number assigned to an invention by the U.S. Patent and Trademark Office that protects the inventor's idea. The purpose of a patent is to prevent other people from stealing the inventor's idea and making money from it. A patent application must include a drawing of the object and a description of how it works.

The first woman to receive a patent in the United States was Mary Kies on May 15, 1809. Kies invented a method of weaving straw using silk or thread to make hats. Today, thousands of women recieve patents each year. David Crosthwait, a famous African American inventor, held 39 U.S. patents and 80 international patents related to heating systems, refrigeration, and vacuum pumps. He is well known for designing the heating systems for Radio City Music Hall and Rockerfeller Center in New York City.

patent
A contract between the federal government and the inventor that gives the inventor exclusive rights to make, use, and sell the product for a period of 17 years.

innovation
An innovation is an improvement of an existing product, process, or system.

Innovation

An **innovation** is an improvement of an existing product, process, or system. An innovation is far more common than an invention. Many companies have become very successful by making innovations on existing products. A simple example is a bicycle. An 18-speed bicycle is simply an improvement

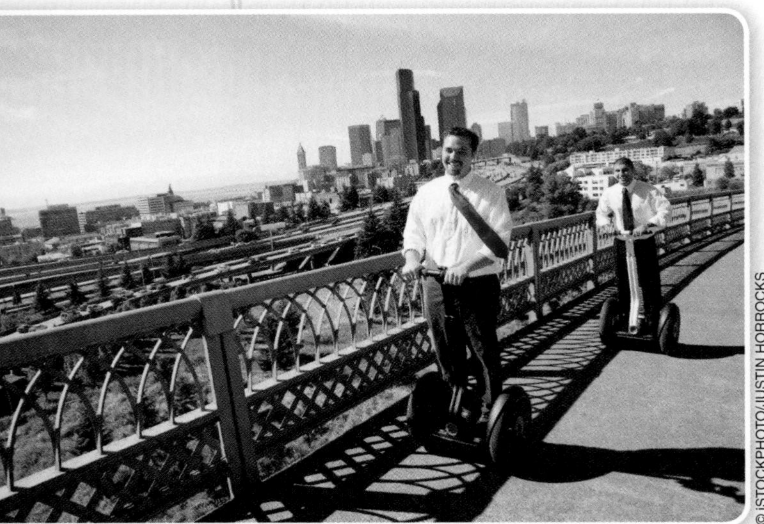
© ISTOCKPHOTO/JUSTIN HORROCKS

FIGURE 1-13 The Segway® has revolutionized personal transportation.

on a 3-speed bicycle. Thus, it is an innovation. However, a personal vehicle like the Segway is an invention because it is a completely new technology for human transportation (Figure 1-13). Innovation is the cornerstone of America's competitiveness in the global economy.

You might think the CD was just an improvement of a phonograph record. If that were true, the CD would be an *innovation* and not an *invention*. But the CD was an invention and not an innovation because it was a completely new form of recording sound. Even though both devices play back recorded music, they are two completely different technological processes. Thus, both are classified as inventions.

Why Study Engineering and Technology?

Engineering and technology have significantly altered the world, in both good and bad ways. Engineering allowed us to put a man on the moon. We have been able to explore the depths of the ocean and to see inside our bodies without ever making an incision. Engineering affects every aspect of your life (Figure 1-14), from your daily routines and modes of transportation to how you listen to music and communicate with your friends! Can you imagine what your life would be like without engineers?

Think about everything you do and use during the day. Do you wake to an alarm clock? Where does the electricity come from? How does it get to your alarm clock? Do you have hot running water in your home to take a shower? How does the water get to your home? How does the water move through your house? How is it heated? Do you use utensils to cook and eat food, or a microwave oven? Is perishable food stored in a refrigerator? Do you sit on a chair at a table? Is your home heated and cooled to keep it comfortable? Are you protected from the elements by your house?

Most people take all these things for granted. But the truth is they were all designed and produced by engineers and engineering technologists. Our lives would be quite miserable without the professionals who design and produce products we need! Studying engineering and technology will prepare you for a career designing ways to make life better for other people.

In addition, the lifestyle that we are accustomed to in the United States is the result of an economic system that has grown steadily through invention and innovation. Designing and producing products are in large part what has driven the economic engine of our modern world.

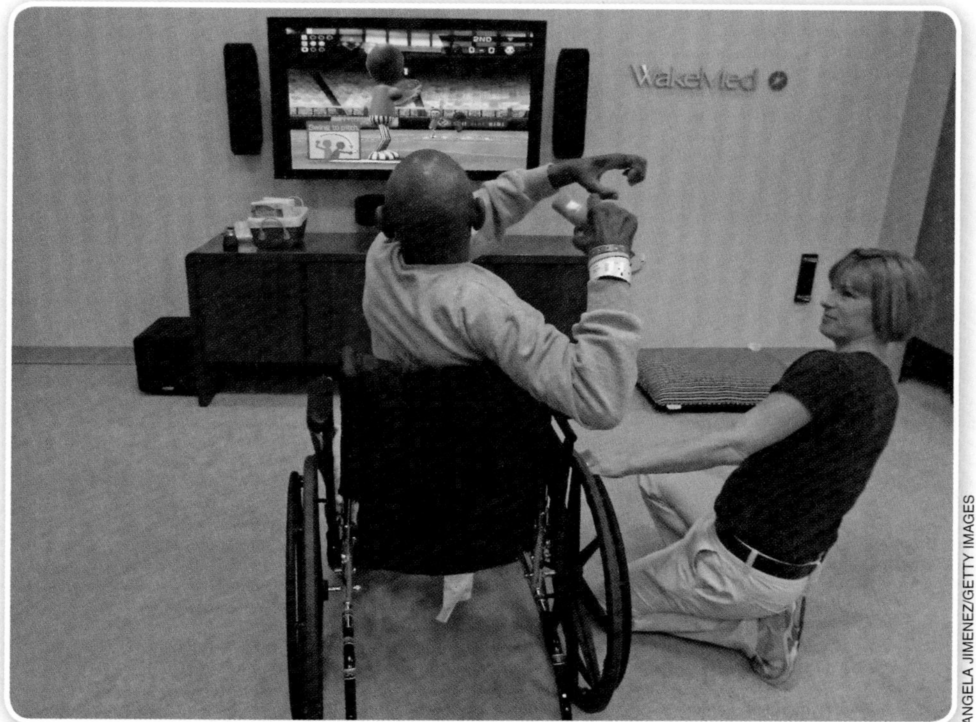

ANGELA JIMENEZ/GETTY IMAGES

FIGURE 1-14 **Inventions and innovations developed by engineers impact our daily lives. Nintendo's gaming technologies were developed for entertainment, but many nursing homes and rehabilitation centers are using Wii® games to provide physical therapy for their patients.**

SECTION 2: THE EVOLUTION OF TECHNOLOGY

Technology has defined societies and civilizations since the beginning of time. We use technology to describe different periods of time. When we refer to the *Stone Age,* for example, it is because early humans had only stones, bones, and sticks to make tools and weapons. In the *Bronze Age,* people had learned how to melt copper and pour it into molds, as well as to shape it in other ways. During the period described as the *Iron Age*, humans had discovered how to fashion iron ore into tools, weapons, and other products. As the name implies, the *Industrial Age* was the period of time when factories and manufacturing were growing rapidly. The *Information Age* is the age that we live in today. It is called the Information Age because information has become more valuable than manufactured products.

Society's Needs and Wants

People develop technology to satisfy their needs and wants. A need is something essential to maintaining life, while a want is something nice to have to make life more enjoyable. These needs and wants, beyond basic

© .SHOCK/SHUTTERSTOCK.COM

© DMITRY VINOGRADOV/SHUTTERSTOCK.COM

© MANA PHOTO/SHUTTERSTOCK.COM

FIGURE 1-15 Are these recreational technologies the result of a human need or a want?

survival, vary somewhat among different societies. A society's cultural values also help determine the type of technologies it develops and how these technologies are used (Figure 1-15).

The available natural raw materials in a region also influence how technology develops. Just think about the many different types of homes people live in around the world. People built homes in the past from grass thatch, mud, logs, or even ice, depending on what natural materials were available to them. The design and construction of a home are the results of people's needs and wants, as well as the natural materials available to them.

Technology and Society

Technology is part of every culture today, and it has been part of every civilization in human history. However, different cultures and civilizations have developed technology differently. Thousands of years ago, people developed incredible structures to worship their gods. In modern times, we use technology to improve people's lives. This happens in many different ways, from the developments in medicine to the availability of personal appliances around the home.

Egyptian Civilization Many ancient societies made significant contributions to the field of engineering. The pharaohs of Egypt commanded their engineers, craftsmen, and slaves to design and construct massive monuments to their gods. You are probably familiar with the pyramids of Egypt; they are known as one of the wonders of the ancient world. In this example, the needs of the Egyptians were religious in nature, to worship their gods and preserve their rulers for the afterlife (Figure 1-16).

Roman Civilization The Roman period of early civilization is famous for its civic buildings. Well-known examples include the many amphitheaters and aqueducts that have survived the past 2000 years (Figures 1-17 and 1-18). These engineering examples relate more to improving people's lives than to religion. As society developed into a more democratic form of government than in earlier civilizations, structures to be used by the citizens of the country became increasingly important. People living in Rome benefited from aqueducts for supplying water and sanitary systems for removing human waste. The system of aqueducts they built across their empire is considered a major engineering accomplishment.

FIGURE 1-16 Ancient Egyptians designed and built incredible structures to serve religious purposes. Their engineering efforts were driven by cultural values.

The Renaissance Period Leonardo da Vinci, who lived during the Renaissance period, was a prominent engineer, inventor and innovator, and architect in addition to being a famous painter. He designed many inventions. Many of these could not be made in his lifetime, such as a helicopter, a

FIGURE 1-17 Roman society developed public works projects like this aqueduct 2000 years ago. Some of these structures are still in use today.

FIGURE 1-18 This Roman amphitheater is still standing in Arles, France. In what ways does it remind you of today's large stadiums?

rotating bridge, and a flying machine, because people lacked the materials or know-how to make them. Other examples of his inventions that were built include a machine for testing the strength of wire and a machine for grinding concave lenses for telescopes.

Galileo, another famous Renaissance inventor, developed the telescope and a primitive thermometer that allowed temperature variations to be measured for the first time. Many of his inventions were not marketable. It was actually the development of a compass that brought financial success to Galileo. His compass was sold for three times the actual cost of production. The connection between engineering (design and invention) and the marketplace grew significantly during this period. Thus, economics was becoming a driving force behind engineering.

The Industrial Revolution The Industrial Revolution began in Europe during the seventeenth and eighteenth centuries and in the United States during the eighteenth and nineteenth centuries. Many types of machine tools were invented during this period, some of which we still use today. Because of the Industrial Revolution, people were able for the first time to create large-scale manufacturing plants (Figure 1-19). The first manufacturing plants were water powered. This meant that factories and mills had to be located on a river to harness the energy of moving water

FIGURE 1-19 The invention of the steam engine contributed significantly to the Industrial Revolution.

to power their machines. But the invention of the steam engine made power mobile. This allowed factories to be located almost anywhere. Steam engines were soon put on wagons, which led to the development of railroads.

Another major development of the Industrial Revolution was the standardization of parts (making parts interchangeable) through mass production. Prior to this time, each individual craftsman made his or her own parts. You could not just go to the hardware store for a replacement part because there were no standardized replacement parts! The interchangeable part is one of the major engineering developments that have contributed to our high standard of living today.

Today's Society Modern society has seen an explosion of new inventions and innovations in the past century. Invention and innovation are at the heart of our modern economy. These inventions have been commercialized and are such a big part of our lives that we take them for granted (Figure 1-20). These inventions and innovations are more than just conveniences and entertainment. They have fueled the world's economy and provided the way of life that we are accustomed to living.

Many believe that the development of the computer was the most important invention of the twentieth century. Some form of computer chip is present in almost every electronic device we use today, affecting almost every aspect of our lives. But, in fact, the most important development of the past century is probably the electrification of the world. Without readily available electricity everywhere, we would not have computers and the Internet, refrigeration and sanitation, or important life-saving medical devices.

FIGURE 1-20 **We sometimes take innovations for granted, not realizing the engineering behind the product.**

Career Spotlight

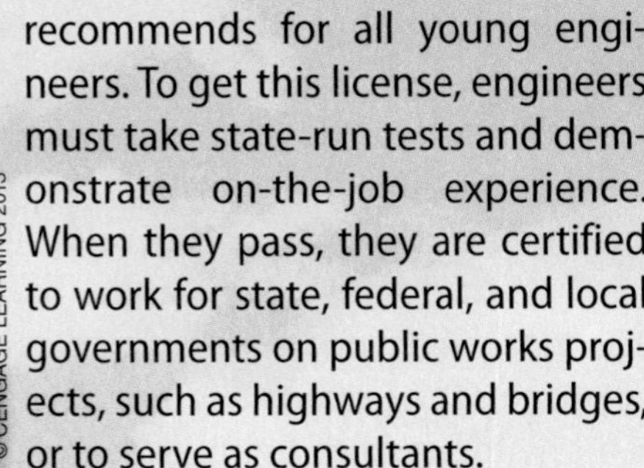
© CENGAGE LEARNING 2013

Name:
James W. Forbes

Title:
Technical Leader, Ford Motor Company

Job Description:
Forbes works in early product development, designing vehicle concepts. "We come up with the ideas for vehicles that would be appealing to customers," he says.

Along with Ford's marketing department, Forbes interviews customers about the kinds of cars and trucks they would like to drive. He then translates their words into engineering language. For example, if customers say they want a car for themselves and their family, Forbes must figure out how much typical family members weigh and how tall they are. This information will affect the engine size, the tire size, and the interior space.

Sometimes, Forbes asks customers questions that do not relate specifically to vehicles, such as the clothing designers they like. This gives him clues for creating a vehicle's interior.

Education:
Forbes received bachelor's and master's degrees in mechanical engineering from Worcester Polytechnic Institute. He also has a professional engineering license, which he recommends for all young engineers. To get this license, engineers must take state-run tests and demonstrate on-the-job experience. When they pass, they are certified to work for state, federal, and local governments on public works projects, such as highways and bridges, or to serve as consultants.

Advice to Students:
Forbes wants students to understand that engineering in school is different from engineering in the workplace. In school, it is a solo effort, where the student must find key pieces of knowledge. On the job, however, it is very much a team sport. At Ford, hundreds or thousands of people contribute to every project. Engineering, he emphasizes, will provide plenty of opportunities for interaction.

Forbes urges students to attend a college with a project-based curriculum. His own college took such an approach, and it helped prepare him for the workplace. Ford hired him right out of school, impressed by his master's thesis on advanced measurement techniques.

Forbes feels that language courses are especially important now that so many employers have overseas interests. "Engineering is a completely global enterprise today," he says. "Every day we're working with Europe, South America, and Asia."

SECTION 3: ENGINEERING CAREERS AND TECHNOLOGY EDUCATION

6

Engineering includes many different career paths that apply math and science through creative design to solve real-world problems. Educational programs that prepare students for careers in engineering and technology are available in most high schools, colleges, and universities.

Careers in Engineering and Technology

Engineers are the people who design solutions to technical problems. They work in both governmental agencies and private companies. Engineering is an exciting, diverse field with many different career paths. It is also a very rewarding career because you know you are working to make the world a better place for people to live.

Aerospace Engineering Aerospace engineers design machines that fly. These machines can be as small as an ultralight glider or as large as the International Space Station (Figure 1-21). Aerospace engineers design and develop aircraft, spacecraft, and missiles. They may specialize in a particular type of aircraft, such as helicopters, commercial passenger jets, military

aerospace engineer
An aerospace engineer designs machines that fly.

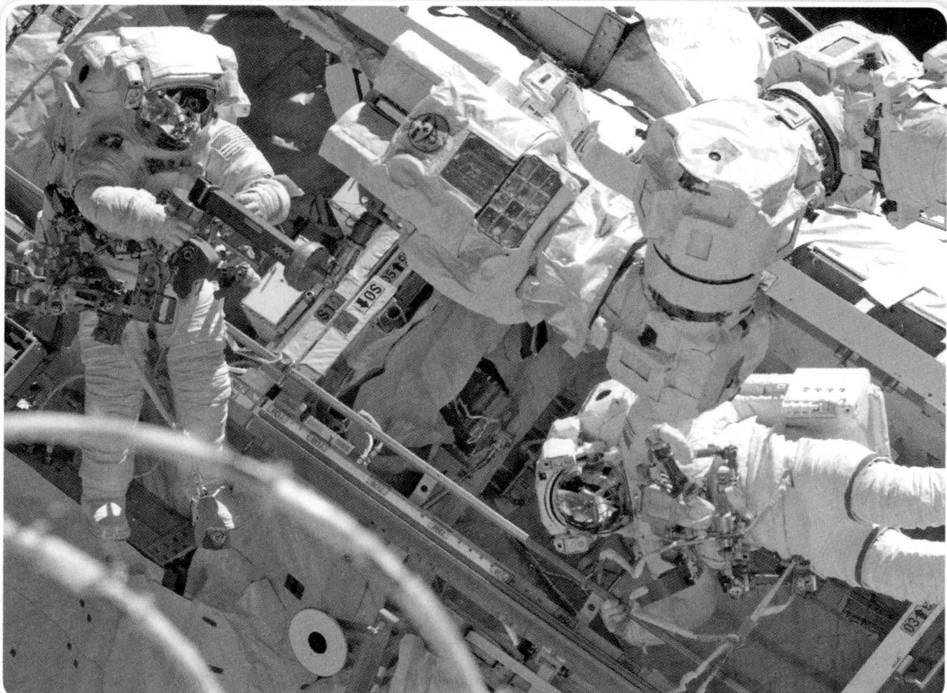

COURTESY OF NASA

FIGURE 1-21 The International Space Station is a very complex engineering project involving teams of aeronautical engineers from different countries.

fighter jets, or space transportation. There are two types of aerospace engineers: aeronautical and astronautical. Aeronautical engineers work with aircraft, while astronautical engineers work specifically with spacecraft.

civil engineer
A civil engineer is a person who designs and supervises the construction of public works projects (such as highways, bridges, sanitation facilities, and water-treatment plants).

Civil Engineering Our modern world has many spectacular buildings and bridges that illustrate some of the exciting work done by **civil engineers**. Civil engineers directly impact our quality of life. Civil engineers solve a variety of problems. They design highways and bridges (Figure 1-22). They plan communities. And they even determine how to store and deliver clean drinking water to your home through a vast network of pipes.

Civil engineers also work to preserve the environment by controlling flooding and designing less polluting power plants. The work of civil engineers strengthens the economy and improves the quality of life for everyone.

electrical engineer
An electrical engineer designs electronic systems and products.

Electrical Engineering **Electrical engineers** design and develop electronic systems and products. These systems can range from global positioning system (GPS) technology to electrical power generating stations. Some of the products they design become part of your everyday life, like televisions and MP3 players. Others are with you every day but are more hidden from your view, like the television broadcasting system or the electronic thermostat for your heating and air-conditioning system. Many electrical engineers also work with products and systems related to health care and computer technology.

FIGURE 1-22 Bridges like the Golden Gate in San Francisco, California, are designed by civil engineers.

Environmental Engineering As you might guess, **environmental engineers** solve problems related to the environment. For this reason, they also need a strong background in biology and chemistry. Protecting and maintaining our environment, controlling air and water pollution, waste disposal, and public health issues are all part of the work of environmental engineers. Municipal waste-treatment plants, for example, are critical to maintaining public health. Environmental engineers must truly think globally and act locally!

environmental engineer
An environmental engineer designs solutions to protect and maintain the environment.

Industrial Engineering Once products have been designed, they must be produced in a manufacturing plant. **Industrial engineers** design the most efficient means of producing products. Manufacturing is a very competitive global enterprise today. To increase productivity, industrial engineers look at how people and machines interact. They try to make sure that materials are processed in the most efficient manner possible. Industrial engineers might answer questions like "Would investing millions of dollars in new robotic equipment improve productivity enough to be worth the investment?"

Mechanical Engineering **Mechanical engineering** is one of the broadest fields of engineering. Mechanical engineers deal with a wide variety of problems involving energy, materials, tools, and machines. They design all kinds of products, ranging from simple toys to very large and complex machines. They may design products and machines that work with fluids, solids, or gasses. Mechanical engineers also design machines like internal combustion engines, elevators and escalators, and refrigerators and air conditioners.

mechanical engineer
A mechanical engineer designs products ranging from simple toys to very large and complex machines.

Engineering Technology Engineering technologists work in fields closely related to engineering, but their work is usually more practical than the work of engineers, which is more theoretical. Often engineering technologists work with engineers and scientists in determining the best method to accomplish a task because they have a more detailed understanding of machines and processes. They can work in a variety of fields ranging from construction to manufacturing. In manufacturing, engineering technologists may be involved with design, production, testing, maintenance, or quality control.

High School Courses

7

There are often courses available to students in high school that will help them prepare for careers in engineering and technology fields. Even if you do not know what area of engineering you will study after high school, or whether you will be attending a community college (two-year program) or university (four-year program), taking certain courses in high school will provide you with a solid background and preparation for your future.

In particular, you should consider combining courses in math and science with courses that apply technology and engineering principles. These high school courses will explore topics like engineering principles and design, drafting, digital electronics, computer-integrated manufacturing, aerospace technology, civil engineering, and architecture in much greater depth. These courses will have a laboratory component that will help you build hands-on knowledge and skills in science, technology, engineering, and mathematics (STEM). Taking courses like this in high school will help you decide whether a career in engineering and technology is right for you (Figure 1-23).

FIGURE 1-23 **Hands-on projects are an important part of many engineering and technology courses. The problems and projects you will work on in this course will help you understand how engineering ideas are applied in your everyday life.**

College Programs

Now is the time for you to begin the career planning process. Make an appointment to visit with your school counselor, and be sure to tell him or her that you are interested in learning more about careers in engineering and technology. When it comes to a career in engineering or engineering technology, there are hundreds, if not thousands, of colleges and universities to choose from.

Two-Year Colleges Two-year colleges typically offer Associate of Applied Science (AAS) degrees in fields related to engineering and technology. Some of these programs are articulated to university-level engineering programs. That means you can transfer easily to the corresponding university program when you finish your degree at the community college. Engineering technology degrees are often similar to engineering degrees, with very similar titles and work environments.

Four-Year Universities Four-year universities offer baccalaureate degree programs in both engineering and engineering technology. A bachelor's degree in engineering will be more theoretical in nature than an engineering technology degree (Figure 1-24).

- Aerospace Engineering
- Agricultural Engineering
- Architectural Engineering
- Bioengineering
- Chemical Engineering
- Civil Engineering
- Computer Engineering
- Electrical and Electronic Engineering
- Environmental Engineering
- Industrial Engineering
- Manufacturing Engineering
- Materials Science and Engineering
- Mechanical Engineering
- Mining Engineering
- Nuclear Engineering
- Petroleum Engineering
- Software Engineering

FIGURE 1-24 **Degrees in engineering.**

Typically, engineers are more concerned with designing solutions, while engineering technologists are more often involved with taking the engineer's design and making the final product. The two work hand in hand to produce solutions to help people and society.

In addition to visiting with your school counselor about career options in engineering and planning your high school courses accordingly, you should also talk with your parents and consult career planning resources. Now is the time to begin thinking about an exciting and rewarding career in engineering or engineering technology.

CAREER SEARCH

Research one of the engineering fields listed in this chapter. Prepare a written report answering the following questions, plus any other interesting facts you learned about this career field.

1. What engineering (or engineering technologist) field did you research?
2. Please list the sources of information you used for this activity.
3. Provide at least four activities this person would do on the job.
4. What physical conditions would you work under? For example, would you work in a manufacturing plant, sit behind a desk, or work outside at a job site?
5. Would you work primarily on a team or as an individual?
6. What specific skills or level of education would you need in each of the following areas: communication, mathematics, science, technical knowledge, and tool skills.
7. What level of education is needed (i.e., high school, community college, or university)?
8. What would the hourly wage and annual salary be for a typical person in this field?
9. What is the employment outlook for this field? How many people will be needed in this field over the next 5–10 years?
10. What type of company would hire this type of engineer?

SUMMARY

In this chapter you learned:

- An engineer is a person who designs products, structures, or systems.
- Engineering is the process of designing solutions to technical problems to improve people's lives.
- Science is the study of the natural world and the laws that govern it, while technology is how humans develop new products to meet their needs and wants.
- Technology may be used for good or bad purposes, and it may have both anticipated and unanticipated consequences.
- Invention is the process of developing a completely new solution.
- Innovation is the process of improving upon an existing solution.
- Engineers and the invention process are critical to growing the economy.
- The engineering profession is important to our country's economic competitiveness and our standard of living.
- There are many different types of engineering careers.

VOCABULARY

Write a definition for these terms in your own words. After you have finished, compare your answers with the definitions provided in the chapter.

Engineer
Engineering
Engineering technologist
Technology
Invention
Patent
Innovation
Aerospace engineer
Civil engineer
Electrical engineer
Environmental engineer
Mechanical engineer

STRETCH YOUR KNOWLEDGE

Please provide thoughtful, written responses to the following questions.

1. Prepare a class presentation about what you believe was the most important invention or innovation of the past 100 years. Explain your answer by describing some of the direct results from this invention or innovation. How did it affect people or the environment?
2. Compare and contrast science and technology.
3. How has technology impacted our society? Describe both positive and negative effects.
4. How has technology affected our environment? Describe both positive and negative effects.
5. What is the primary difference between invention and innovation?
6. What is the purpose of a patent? Why do you think a patent is important to our society?

Onward to Next Destination

CHAPTER 2
Technological Resources and Systems

Menu

Before You Begin

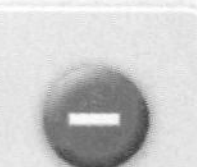

Think about these questions as you study the concepts in this chapter:

1. What resources are needed in engineering and technology?
2. What are technological systems?
3. What is a closed-loop system?
4. How do engineering and technology affect the environment?
5. How do engineering and technology shape society?
6. How does society influence engineering and technological development?

POPPERFOTO/GETTY IMAGES

FIGURE 2-1 Henry Ford designed a better way to make cars.

Engineering in Action

S T E M

A Better Way to Make Cars

Henry Ford was a brilliant problem solver who loved to tinker and experiment. Contrary to what some people believe, Ford did not invent cars or mass production. He did, however, develop a new and revolutionary way to mass produce cars (Figure 2-1). His improvement brought the parts to the workers on a moving assembly line. In other factories at that time, the workers moved and the parts stayed in one place. Ford's production technique involved an elaborate conveyor system. This process allowed workers to specialize in certain operations.

The full-scale production of the Model T automobile made Ford famous, but the Model T was not his first attempt at mass producing cars. In fact, there was a Model A, a Model B, a Model C, and so on. Persistence and problem solving are important in the engineering process.

SECTION 1: TECHNOLOGICAL RESOURCES

As you learned in the previous chapter, engineers design products, structures, or systems. They apply mathematical and scientific principles to solve problems. A resource is something that has value and that can be used to satisfy human needs and wants. The fields of engineering and technology use seven key resources:

(a) people,
(b) energy,
(c) capital,
(d) information,
(e) tools and machines,
(f) materials, and
(g) time.

technological resources

A resource is something that has value and that can be used to satisfy human needs and wants.

These **technological resources** are the backbone of engineering and technology. Which one do you think is the most important? Think back to the first chapter. There we defined technology as a human activity. People have been "doing" technology since the beginning of time. The engineering process today depends on the same seven resources that our ancestors used (Figure 2-2).

(a) People

(b) Energy

(c) Capital

(d) Information

(e) Tools and machines

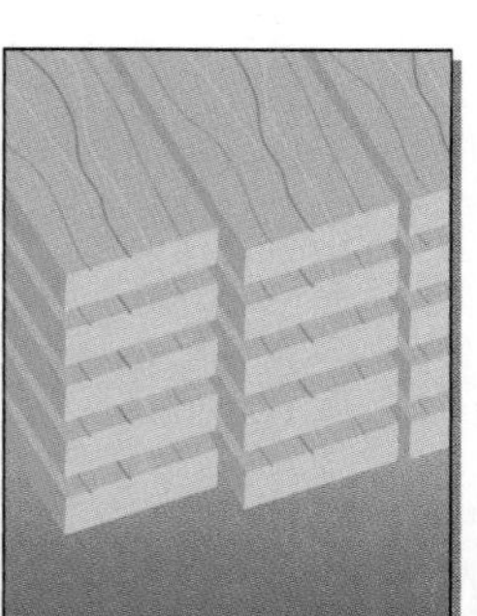

(f) Materials

(g) Time

FIGURE 2-2 *Seven technological resources.*

Math in Engineering

Mathematics is critical in engineering. Through math, each technological resource is developed, measured, evaluated, or allocated.

Technological Resource 1: People

People are involved in every aspect of technology. When you think about it, engineering is really a form of problem solving. Humans are natural problem solvers. Without us, there would be no engineering or technology. So people are the most important resource (Figure 2-3).

People are creative. In fact, the International Technology and Engineering Educators Association (ITEEA) defined *technology* as "human ingenuity in action." Ingenuity is the natural ability of people to solve problems. It is the force that drives engineering.

People also have many technical skills. They bring their skills to every aspect of producing a product. Sometimes a product may be something like a cellular phone or an MP3 player. Other times, a product may be an electronic signal or program code that helps display a Web page.

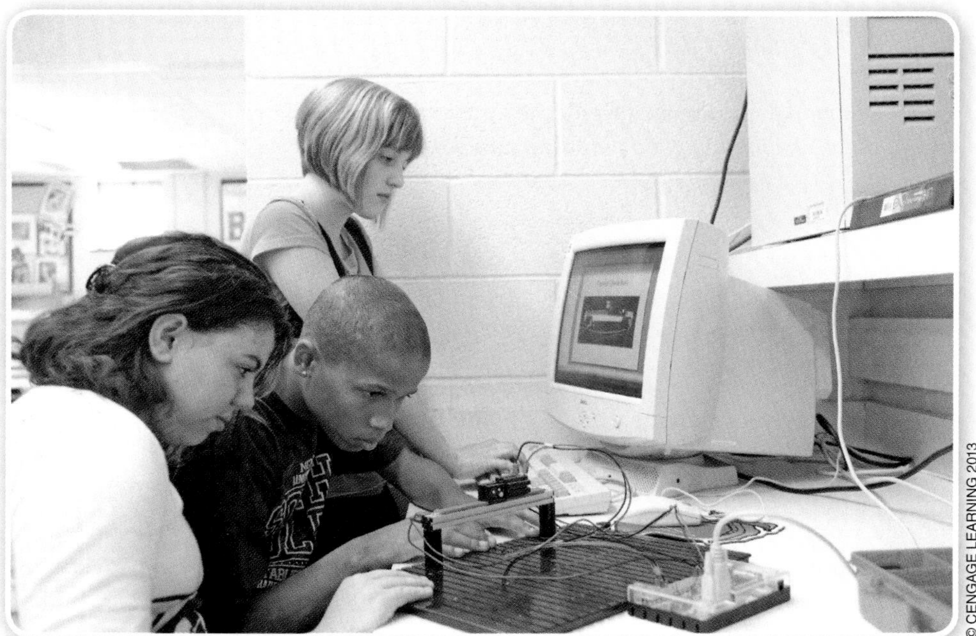

FIGURE 2-3 *People are the most important resource in engineering and technology.*

Did You Know?

Henry Ford did *not* invent the automobile! Ford acquired his wealth by inventing a better way to make cars (Figure 2-4). As a result, he became one of the richest men in the world. Ford was able to substantially reduce costs by increasing the efficiency of the manufacturing process. More people could afford to buy a car, thanks to Henry Ford.

GETTY IMAGES

FIGURE 2-4 *Although he did not invent the automobile, Henry Ford developed a more efficient way to manufacture automobiles.*

energy
The ability to do work.

work
The term engineers use to describe how much force it takes to move some object a set distance.

Technological Resource 2: Energy

Simply stated, **energy** is the ability to do work. In this case, work is not a job or someplace you go. Instead, **work** is the term engineers use to describe how much force is needed to move some object a specific distance. Energy makes it possible to move things. You do work when you jog or ride your skateboard or bicycle. Your body needs energy to do the work. A car uses energy when it takes us from one location to another. Energy provides much more than movement. It can also provide light, heat, and sound (Figure 2-5). In a large concert hall or factory, for example, you could have all three of these forms of energy—light, heat, and sound—present at the same time. We discuss energy in Chapter 9.

Technological Resource 3: Capital

Capital is a word that describes financial assets such as money, stocks, property, and buildings. Capital allows industry to purchase the technological resources needed to produce a product or create new technology. These resources include people with certain skills and expertise. Large corporations use *investment capital* to help entrepreneurs start new businesses. Investment capital has become increasingly important.

FIGURE 2-5 This parrot is doing work.

Many products today needed millions of dollars to be developed. The manufacturing process includes purchasing raw materials, producing the product, and paying all the people involved in making the product. Manufacturing also includes marketing or advertising the product and packaging and shipping it to market. Despite these heavy uses of capital, many modern inventions were designed in people's garages and basements. Have *you* ever thought about inventing something new?

Technological Resource 4: Information

Information has become increasingly important, and technical know-how has always been crucial in manufacturing. As noted in Chapter 1, knowledge has increased *exponentially,* and new materials have significantly affected the success rate for new inventions and innovations. Engineers have to be current with the latest information about available materials and their properties.

Information is critical in the engineering field. Some people are experts at researching patents and copyright information. A company would not want to invest money and time on a product already patented by another

Did You Know?

Work is a good thing. In engineering, work is how we measure the force required to move things. Engineers design machines to do work so we do not have to work!

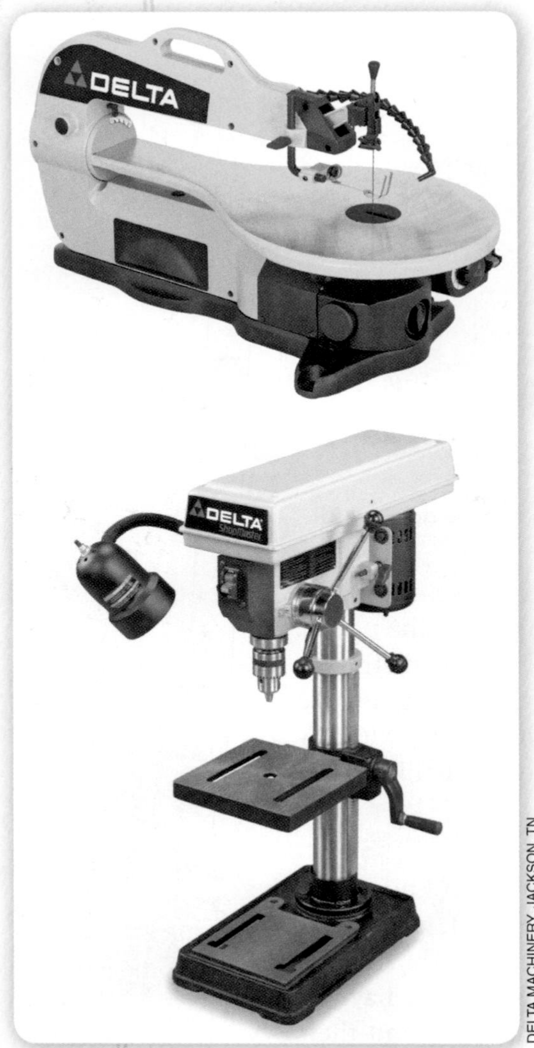

FIGURE 2-6 Tools and machines are important for shaping materials.

company. Engineers also need information about market demand and the preferences of customers. Others must have current information about the supply and availability of materials and other resources. The need for current information cuts across many aspects of the engineering field. Thus, information is one of the key resources.

Technological Resource 5: Tools and Machines

Tools and machines are a critical part of the story of the human race. Since our beginning, the development of tools and machines has defined each society. Tools are much more than things such as hammers and saws. Sewing machines, plows, tractors and combines, and food processors are also examples of tools—and so are tape measures, rulers, scissors, ballpoint pens, computers, and computer-aided-design (CAD) software. Tools and machines are often used to process materials and are a critical resource for engineers and technologists (Figure 2-6).

The late 1700s saw an explosion in the development of new machines and tools. The steam engine helped spur the Industrial Revolution. This invention was significant because, for the first time in history, people had a portable power source. Before the steam engine, machines had to be either powered by animals such as mules and oxen or located near rivers or streams to take advantage of the energy in falling water. With the invention of the steam engine, large factories could be built almost anywhere. New industries began to emerge. The development of industry created new cities and provided new kinds of jobs for people.

Technological Resource 6: Materials

Engineers must know a lot about the characteristics and properties of the materials they use in their designs. For example, why are cooking and eating utensils not made of lead? We can classify materials as natural or synthetic. Natural materials are those used directly from their natural state such as lumber. We create synthetic materials by changing the state of the raw material. Plastic, which is made from petroleum, is a common example.

Engineering Challenge

EXPONENTIAL RATE

Which one of the following two allowances would you rather have? The first allowance begins at one penny ($0.01) and then doubles each day. The second allowance begins with one dollar ($1.00) and increases each day by one dollar.

As the following table shows, the difference over time is big. The exponential rate of growth for the one-penny allowance passes the one-dollar allowance in 11 days—even though the dollar allowance was 100 times greater the first day!

	Day											
	1	**2**	**3**	**4**	**5**	**6**	**7**	**8**	**9**	**10**	**11**	**12**
Option 1 ($)	0.01	0.02	0.04	0.08	0.16	0.32	0.64	1.28	2.56	5.12	10.24	20.48
Option 2 ($)	1	2	3	4	5	6	7	8	9	10	11	12

If you were to complete the table for the remaining days in a 30-day month, what would your allowance be per day at the end of 1 month? Explain how this exponential rate applies to technology.

Engineers have many types of materials to choose from today. Good designs must consider many factors, including a material's characteristics (strength versus weight, for example) and texture, the type of processing required to shape and form the material, and the type of finish it can receive. Engineering requires scientific knowledge about a material's properties (Figure 2-7).

Did You Know?

Today, people are inventing new ways to build machines. Nanotechnology, for example, involves making machines that are extremely small—less than 1/1000 the diameter of a human hair. Scientists and engineers are now developing tiny robotic machines that can be placed into your bloodstream to repair damaged cells in your heart or other organs. But can you imagine trying to find one of these robots if it were accidentally dropped on the floor?

Science in Engineering

Engineers and scientists study the molecular structure of materials to understand their properties. We call this field of study materials science. One way to define a material is by describing its *properties*. For example, is the material hard or soft? Is it heavy or lightweight? Rigid or flexible? The arrangement of the molecules in a material partly determines its properties and characteristics. Studying chemistry and the molecular structure of materials is important for engineers.

Technological Resource 7: Time

Every activity we undertake requires time. If you are a gamer and want to purchase a new game, then you would probably spend time researching its features. You might read magazines or articles on the Web or perhaps spend time trying the new game out in person at the store or at a friend's house. Either way, you need time to research and plan before your purchase. The same is true with engineering and technology. Time is an important element in each step in the process from design to distribution. Why do you suppose a video game is so expensive to buy when the actual disc and package probably cost less than one dollar?

In engineering, time is a key resource. You have probably heard the expression, "Time is money." This idea is especially true in engineering and

FIGURE 2-7 We can classify materials as natural or synthetic. Lumber is a natural material, whereas the materials inside this car are synthetic.

Science in Engineering

An *alloy* is a combination of two metals. Alloys are the most useful metals to engineers because they have specific properties. For example, you have probably noticed that iron rusts. But if you add nickel to the iron (at least 10 percent), it becomes stainless steel, which does not rust. Iron is a very soft metal; stainless steel is much harder. Thus, adding nickel to the iron not only prevents rusting but also makes the metal harder. Engineers take advantage of the properties of one material to change the characteristics of another.

manufacturing. Time is required to design products, develop production processes, train employees, manufacture products, and ship them.

Engineers typically spend time researching, visualizing, and sketching potential solutions before they actually begin their designs on a computer. What are the requirements of the device or structure? Who is the customer? What are the customer's needs? What different types of materials are available? How would a material affect the design or functionality of the product? How would different materials affect the manufacturing process?

SECTION 2: TECHNOLOGICAL SYSTEMS

A **system** is a set or group of parts that all work together in a systematic, organized way to accomplish a task. Systems are critical parts of technology. Engineers design systems to be as efficient as possible. All systems include an input, a process, and an output. A **subsystem** is a system that operates as part of another system (Figure 2-8). When parts of a system are missing or not operating properly, then the entire system may not function as intended. Some technological systems are modeled after systems that are found in nature.

system
A set or group of parts that all work together in a systematic, organized way to accomplish a task.

subsystem
A system that operates as part of another system.

Nature has many systems, including your body. Your circulatory system keeps blood moving through your body. Your respiratory system extracts oxygen from the air. Your digestive system processes the nutrients in food. In each system there is an input signal, some processing function, an output, and often a feedback message to let your body know if the process was successful or not. Some of the systems in your body are so automatic that they function without your even being aware of them. Society also has systems such as the government, legal, and educational systems.

In technology, we often refer to communication systems, production systems, and transportation systems. Each one can be broken down even more. For example, we could divide a transportation system into land, sea, air, and space transportation systems. Like those in your body, these systems

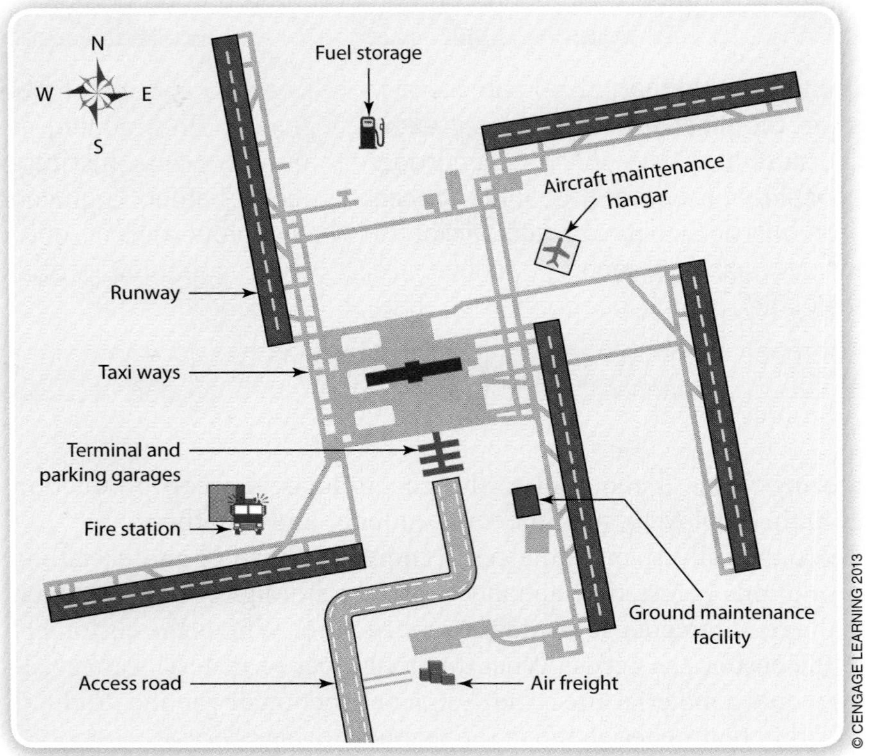

FIGURE 2-8 **An airport is an example of a very large system that includes many subsystems.**

open-loop system
An open-loop system is the simplest type of system and requires human action to be regulated.

closed-loop system
A closed-loop system includes an automatic feedback loop to regulate the system.

are designed to accomplish tasks as efficiently as possible. Each may include an input, a process, an output, and sometimes a feedback loop.

Open-Loop and Closed-Loop Systems

There are two types of systems: open loop and closed loop. An **open-loop system** is the simplest type of system. An open-loop system has no automatic regulatory process. Open-loop technology systems require human action to regulate the system. A light circuit in your home is an example. The light is on when you flip the switch. The light stays on until you turn the light off.

Input
Process
Output
Feedback

FIGURE 2-9 **A closed-loop system contains an input, process, output, and a feedback loop that adjusts the system's operation.**

A **closed-loop system** adds an automatic feedback loop to the system (see Figure 2-9). The purpose of the feedback loop is to regulate the system. In your home, a thermostat on the heating and cooling system provides a feedback loop and turns the heat or air conditioning on or off when the room reaches a certain temperature. Once the temperature has been set, the system requires no human intervention.

Do any of the rooms in your school have automatic lights that come on when you enter the room? These are examples of sensors added to a system. Motion regulates the lights: they come on when they sense movement, and they shut off when there is no movement for a certain period of time.

Have you placed items on the conveyor belt at a checkout line in a grocery store and noticed how the belt stops moving when your items reached the cashier? This is another example of a closed-loop system. A sensor detects when an item has reached the end of the conveyor belt and turns the belt off. Without this feedback loop, the conveyor belt would keep running and your groceries would spill onto the floor.

Communication Systems

A **communication system** is one type of technological system. It transfers information between people and other people or machines. This information can be in the form of words, symbols, or messages. Sometimes we can categorize communication systems as either electronic or graphical.

There is a difference, however, between *communicating* and a *communication system*. Two people talking face to face are communicating (if they are listening to each other), but they are not using a technological system. If those same two people are talking on their cellular telephones, then they are communicating by using a communication system (Figure 2-10).

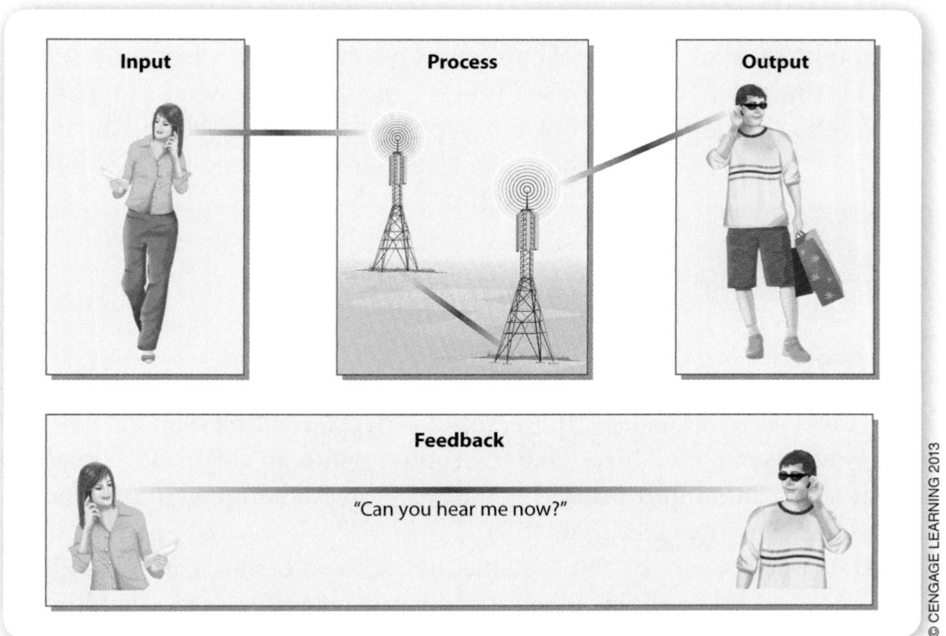

FIGURE 2-10 *Communication system involved in a cellular telephone call.*

Career Spotlight

Name:
Lora B. Freeman

Title:
Structural Engineer, Parsons Brinckerhoff

Job Description:
Freeman designs bridges and buildings. One of her main concerns is each structure's support system: How much will it weigh? How much weight can it take? When designing a bridge, for example, she must consider the pressure applied by cars, people, wind, and, in a worst-case scenario, earthquakes.

Luckily, Freeman does not have to make it all up from scratch every time. Design codes help guide her work.

Structural engineers are responsible for developing designs that not only safely serve the public but also are economical. When Freeman designs a bridge or building, she first develops a preliminary design. Then she evaluates the design to make sure it satisfies all the project's requirements. Finally, she helps develop a construction plan, and the design becomes a reality.

Freeman enjoys the challenge of a new project. "You're given a set of constraints for the problem, and you come up with interesting ways of applying the math you learned," she says. "It's all very fascinating in my mind."

Education:
Freeman got her bachelor's and master's degrees in civil engineering at West Virginia University. As a graduate student, she began her civil engineering career by working with the West Virginia Department of Transportation's Division of Highways. The department made use of her thesis, "Development of an Optimized Short-Span Steel Bridge Package."

Advice to Students:
Freeman tells students not to be intimidated by the engineering field. She herself was worried that she could not handle the work. "I initially questioned my ability when I started college," she says. "But I was pleasantly surprised that it wasn't over my head. I'm glad I took the steps to challenge myself and pursue the career I was really interested in."

For the two people talking on their cellular phones, the communication system involves several steps. In the input process, the first person's voice (a sound wave) is encoded through a microphone into an electronic signal for transmission. This input process actually involves changing sound energy into electrical energy (see Chapter 9).

In the process part of the system, the electronic signal is transmitted to and received by a cellular tower and then transmitted over a landline to another tower near the intended receiver. At the second cellular tower, the signal is broadcast and received by the other person's cell phone.

USED UNDER LICENSE FROM SHUTTERSTOCK.COM

FIGURE 2-11 **Sailors communicating with semaphore flags.**

The second person's cellular phone must convert the electrical signal through a speaker into sound waves that the person can hear. This is the output part of the system. Although this was a simple explanation of a very complicated process, it shows the basic concepts shared by all communication systems: encoding, transmitting, and decoding. Computers, printers, and all other electronic devices include the same processes.

Communication technology does not have to be electronic. Years ago, sailors communicated between ships at sea by using semaphore flags. The position of the flags represented letters (Figure 2-11). Sailors, soldiers, and others have also used blinking light sources to send Morse code. In both cases, however, they relied on *visual* signals, so they could communicate only as far as people could clearly see. Thankfully, today's technologies allow us to communicate around the world at the speed of electrons.

Production Systems

Production systems produce physical products efficiently and at the lowest cost possible. We use these systems to produce all kinds of products—from food that you or your pet eats to miniature electronic components that will become part of a consumer product like a video game. We can subdivide production systems into manufacturing and construction, depending on whether a product is built off-site or on-site. Manufacturing usually involves making multiple pieces in a plant. Construction typically involves building a single structure on-site. For example, we could manufacture a

© ISTOCKPHOTO/MOREPIXELS
© ISTOCKPHOTO/ANDREY VOLODIN

FIGURE 2-12 **Production systems can produce anything from potato chips to computer chips.**

house in a plant and then deliver it to a site to be set on a foundation. Or we could construct (build) the house on a foundation at the site.

All products are designed to meet a specific human need or want (Figure 2-12). A product that does not satisfy a need or want is not likely to be successful. Usually, a lot of research is done before a product is put into production. Developers must demonstrate that they have researched what consumers want and are willing to pay for a given product. One key technological resource discussed earlier in this chapter, investment capital, would be hard to get without this research.

Every product has many constraints placed on its design and must meet many **criteria**, special requirements a product must have. For example, for a remote control designed for an older person or someone who has impaired vision, one criterion might be larger and brighter number buttons. Perhaps this remote control would also be slightly larger and have a different texture on the surface to help people with arthritis. For every product, certain criteria must be met for the product to be successful in the marketplace.

constraints
A limit, such as appearance, budget, space, materials, or human capital, in the design process.

In addition to criteria, the design also has **constraints**, or limits, placed on it. A constraint might be the manufacturing process that is available, a budget, or the time needed to manufacture the product. Market research will determine, for example, how much people will be willing to pay for a new type of phone. The designers might have to cut back on different features and options to hold down production costs.

Transportation Systems

Engineers design **transportation systems** to move people or goods in an organized and efficient manner. Earlier in this chapter, you read about an airport as an example of a transportation system. Now consider a different type of transportation system. How are products moved (transported) around a large factory as they are being produced? Sometimes they move on a conveyor belt type of system (Figure 2-13). Other times a robotic arm places them on or removes them from a conveyor. In a large plant, a forklift might move the products to a different location.

We can also consider a department store's escalators or elevators as transportation systems. Obviously, there are many other transportation options. All are systems because they contain inputs, processes, outputs, and usually feedback loops that are designed to move products as efficiently as possible.

FIGURE 2-13 A conveyor belt is an example of a transportation system.

SECTION 3: IMPACTS OF TECHNOLOGY

All human engineering and technological activities affect people, society, and
the environment. As discussed in Chapter 1, sometimes such impacts are
planned and sometimes not. We call planned impacts *anticipated* and unplanned 4
impacts *unanticipated*. Anticipated impacts of technology can be either positive
or negative. Unanticipated impacts can also be positive or negative.

The Effect of Engineering on the Environment

Henry Ford developed a production system to lower the cost of producing automobiles. He did this in part so that more families could afford to buy automobiles. Neither Ford nor his engineers, however, anticipated the air pollution that would result from so many cars on the road many years later (Figure 2-14). This lack of foresight is a classic example of unanticipated and negative consequences.

FIGURE 2-14 Early automobile producers did not anticipate the amount of air pollution automobiles would produce.

We can find many other examples of unanticipated consequences, most on a smaller scale. Who would have predicted the amazing increase in the use of small

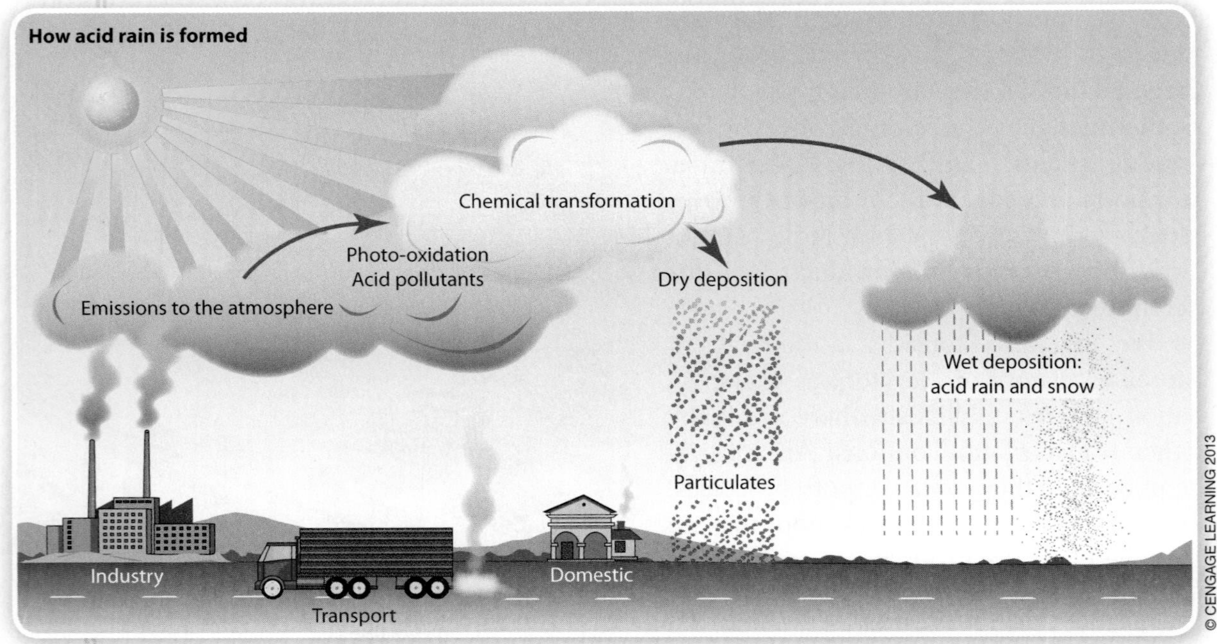

FIGURE 2-15 **Acid rain is an environmental problem caused by automobile and industrial emissions.**

personal electronic devices such as cellular telephones, MP3 players, BlackBerries, and the like? Batteries power each device. If these batteries are not properly disposed of, they become a serious threat to the environment.

An unanticipated consequence of air pollution from cars and industry is acid rain (Figure 2-15). The result of airborne toxins washed out of the air by rainfall, acid rain has particularly affected the northeastern parts of the United States and eastern Canada. These toxic chemicals have killed large areas of forested land. Acid rain falling into lakes and rivers can kill fish and other life forms that live in the water.

FIGURE 2-16 **Acid rain has many negative consequences on the environment.**

Large industrial operations in the northeastern United States put these toxic chemicals into the air. When these factories were designed and built, no one knew their emissions would cause so many problems. Their designers were focused on intended and positive outcomes: electricity generation and industrial production. These plants would provide cheaper products and employment for the people who lived in the area.

The Canadians did not benefit from the U.S. industrial plants, but they did suffer damage from the acid rain (Figure 2-16).

This example shows how the environmental impacts of technology do not respect boundaries between countries. The natural ecosystem of Canada suffered extensive damage as a result of technological activities in the United States.

The Effect of Engineering on Society

5

The development of technology affects society. The technologies a society develops partly determine how people live, work, and play. Today, because of our extremely rapid communications and international trade, developments in one country are quickly adopted by citizens in other countries unless there are governmental or religious restrictions. The leaders in some countries, however, fear that certain technologies offend their society's moral and cultural values, so they ban them from their societies.

Can you think of an engineering or technological development that had an unanticipated but positive impact? The Internet started as a network that was designed exclusively to allow communication among scientists and between scientists and the military. With the development of Hypertext Markup Language (HTML) in the early 1990s, the World Wide Web was born. The Internet found new uses in entertainment and education. Middle school students gaining access to millions of pieces of information from around the world is an example of an *unanticipated positive* impact of this development (Figure 2-17). Cybercrime and other harmful uses of the Web are unanticipated and negative impacts. As this example shows, technology is neither good nor bad; how people use it determines its value.

FIGURE 2-17 Using the Internet for school research projects is one example of a technology's unanticipated positive impact.

6

Societies determine which technologies are developed. Social customs and religious beliefs influence what products are developed and used by people in a given society. For example, the Old Order Amish form a society that does not use certain kinds of technology, including automobiles and electricity (Figure 2-18). To practice their religious beliefs, Amish people continue to use horse-drawn buggies for transportation, windmills to bring water to their homes and livestock, and horse-drawn machinery to plow and harvest their crops.

FIGURE 2-18 Social values help to determine which technologies a society will adopt.

Currently, stem cell research is being conducted around the world, but such research has been fiercely debated in the United States. To what extent should people be allowed to experiment with or manipulate human cells? We have successfully cloned animals. Should we be allowed to clone humans? Should we have been allowed to clone animals in the first place? These are difficult questions that have no easy answers. Who decides what engineers are allowed to develop?

Social values affect the extent to which engineers and scientists are allowed or encouraged to develop new technologies. In the United States, questions about the appropriate use of engineering and technology may be decided by Congress or in the courts. This is why it is so important that citizens be well educated about engineering and technology; only then can they make informed and rational decisions as voters and not fall prey to misinformation.

SUMMARY

In this chapter you learned:

- In the fields of engineering and technology there are seven key resources: people, energy, capital, information, tools and machines, materials, and time.
- A system is a set or group of parts that all work together in a systematic and organized way to accomplish a task.
- There are two types of systems: open loop and closed loop. To be regulated, an open-loop system requires human action. A closed-loop system is automated.
- A system requires an input, a process, and an output.
- Technological systems include communication systems, production systems, and transportation systems.
- All human engineering and technological activities affect people, society, and the environment.
- Sometimes the impact of technology is planned for (anticipated) and sometimes it is not (unanticipated).
- The consequences or impacts of technology may be either positive or negative.
- Technology is neither good nor bad. How people use technology determines its value.

VOCABULARY

Write a definition for each term in your own words. After you have finished, compare your answers with the definitions provided in the chapter.

Technological resources
Energy
Work
System
Subsystem
Open-loop system
Closed-loop system
Constraints

STRETCH YOUR KNOWLEDGE

Provide thoughtful written responses to the following questions.

1. Why is it important for students as citizens to have a solid understanding of technology?
2. Why are technological resources important in the field of engineering?
3. In what ways do engineering and technology affect the following?
   a. people
   b. society
   c. the environment
4. Explain the idea that engineering and technology can have anticipated, unanticipated, positive, or negative consequences.
5. Research a technological development and describe how it has affected people and the environment. Present your research findings to the class.

Onward to Next Destination

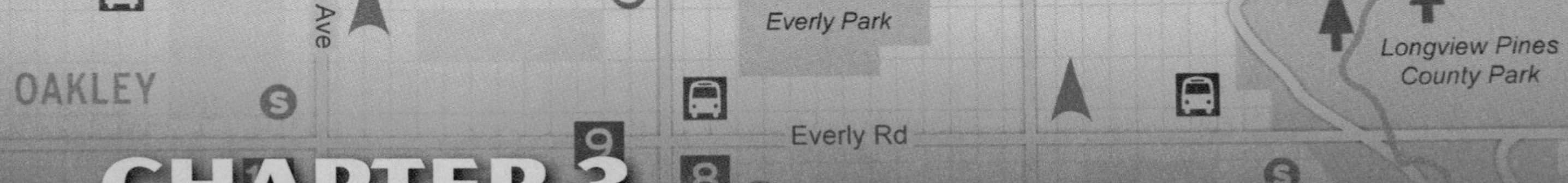

CHAPTER 3
The Engineering Design Process

Menu

Before You Begin

Think about these questions as you study the concepts in this chapter:

1. 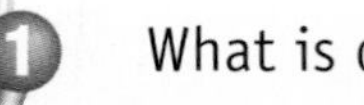 What is design?
2. 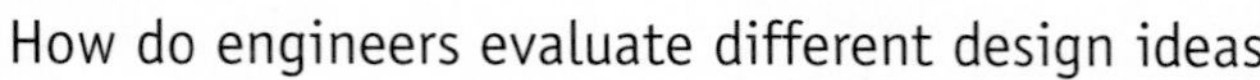 How do engineers evaluate different design ideas?
3. 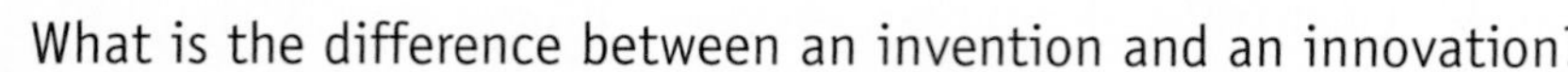 What is the difference between an invention and an innovation?
4. What is the engineering design process?
5. What is a patent?
6. What steps are involved in the engineering design process?
7. Why do engineers use an engineering notebook?
8. What are design constraints?

Engineering in Action

If you were asked to think about an engineered design, you would probably think about a new computer, a high-speed airplane, or maybe a cellular telephone. The engineering design process, though, is also used in common products you might find around your house. The upside-down ketchup bottle is a good example. For decades, the glass ketchup bottle of the H.J. Heinz Company was well known. If Heinz wanted to remain the leader in the ketchup industry, though, it had to focus on always improving its products. For this reason, Heinz operates a $100 million and 100,00-square-foot Global Innovation and Quality Center that is dedicated to product improvement.

Heinz's market research indicated that the company's customers were having trouble pouring ketchup from the traditional upright bottle. The research found that 25 percent of customers used a knife to help remove the ketchup. The research also showed that 15 percent of users stored

FIGURE 3-1 The Bigfoot upside-down ketchup bottle holder.

FIGURE 3-2 The Heinz Easy Squeeze® Ketchup container.

their bottles upside-down to make it easier to pour the ketchup. In fact, a company named Larien Products developed the Bigfoot fixtures shown in Figure 3-1 to hold ketchup bottles upside-down. So Heinz used the engineering design process to develop a new ketchup container that would allow the ketchup to be poured easily. The result of this engineering design process was the Heinz Easy Squeeze Ketchup (see Figure 3-2).

So, remember: the next time you put ketchup on your cheeseburger, the task was made easier because of engineering design. Just about every product our society uses was developed or improved through engineering design.

Did You Know?

British and Dutch traders originally imported ketchup from East Asia in the seventeenth century. The original product, called *ketsiap*, was made from pickled fish and resembled today's Worcestershire sauce. One hundred years later, farmers in Nova Scotia added tomatoes and sugar to this sauce, and modern ketchup was born.

SECTION 1: THE ENGINEERING DESIGN PROCESS

Introduction to Design

Many things influence the designs for today's products. Traditionally, design has included form, color, space, materials, and texture. However, things such as cost, safety, security, market trends, and service have become increasingly important in design. **Design** is turning a concept into a product that can be produced. In other words, it produces plans that help turn resources into desirable products or systems.

A design may improve an existing product, technological system, or method of doing something. This improvement is known as **innovation**. Innovation implies that people are willing to pay for the product or process. Innovation typically leads to economic growth. Most of the new products we see today are innovations. An **invention** is a new product, system, or process that has never existed before. Inventions typically happen after scientific experiments. An invention is often not planned, and over time some inventions become more useful.

engineering design process
The engineering design process applies mathematics, science, and engineering principles to help in decision making.

Design is really a series of steps. The **design process** is a systematic problem-solving strategy used to satisfy human wants or needs. The design process has evolved over time based on a society's needs to solve problems in a timely manner. Using the design process to address problems helps engineers achieve the best solution.

In the design process, engineers work in teams. The team is more effective than one or two engineers working by themselves (see Figure 3-3). Being a good team member takes work because many people are used to working alone and making their own decisions. Working in a team means finding solutions that all team members can support. This is why it is important that all members participate in team decision making. In this way, every team member has an interest in the design solution being successful.

Engineers have developed a set of steps termed the **engineering design process**. This process includes mathematics, science, and engineering principles to help make decisions. Figure 3-4 shows the 12 steps that the International Technology and Engineering Educators Association (ITEEA) includes in the engineering design process. Other organizations and writers have also developed engineering design process flowcharts. Most of these flowcharts have similar characteristics. ITEEA's 12 components are explained in three stages: concept, development, and evaluation. Each stage contains four design steps.

FIGURE 3-3 Teamwork is a good way to solve problems.

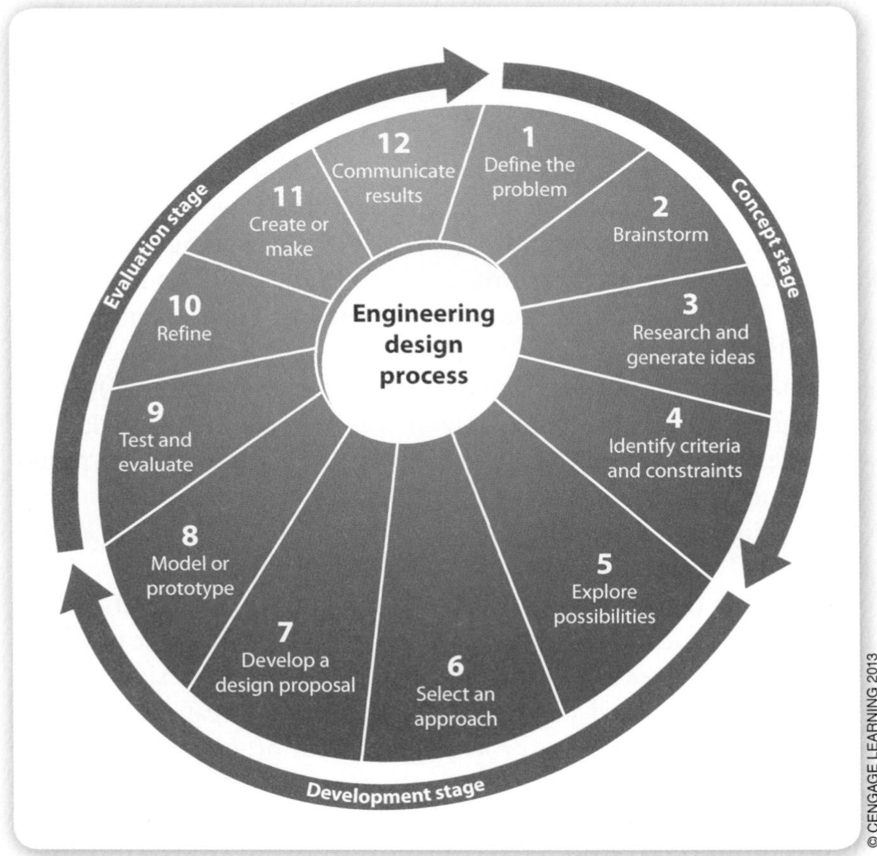

FIGURE 3-4 **The engineering design process.**

Concept Stage

The concept stage has four steps: (1) define the problem, (2) brainstorm, (3) research and generate ideas, and (4) identify criteria and specify constraints.

Before a problem can be defined, it must first be identified. Engineers themselves often do not identify the problem. Problems are typically noted first by society, which typically sees a need, not a problem. A *need* may refer to a group of people's desire, a shortage of a product or service, or possibly a luxury item. For the engineering design process, this need may then be termed a *problem*. Problem identification may come from research, from media exposure, or from client input to a company. The engineering design team must clearly define this problem in simple terms. The definition should not include a solution.

brainstorming

Brainstorming is a method of shared problem solving in which all members of a group engage in unrestrained discussion and spontaneously generate ideas.

Brainstorming is the second step of the engineering design process. In this step, problem solving is shared when all members of the design team spontaneously suggest and discuss ideas. The team makes a list of any and all possible solutions to the problem. Members must be open to new ideas and realize that there are no wrong ideas (see Figure 3-5).

© CENGAGE LEARNING 2013

FIGURE 3-5 **Brainstorming helps generate ideas.**

After the brainstorming session, the design team interviews people who are affected by the problem. After ideas to solve the problem have been listed, the design team conducts research to find out if any solutions to the problem already exist. Research is some of the best work an engineer can do during the design process. Research is essential because most new designs

ENGINEERING CHALLENGE 1

Your engineering design team was given this mountain bike problem. Your design team should brainstorm to come up with a list of possible solutions. Your engineering design team should then conduct research to find out if any solutions already exist. Your team's research could use the Internet, bicycle product catalogs, or a visit to the local bicycle shop, just to name a few. Your team should document its work in a notebook.

ENGINEERING CHALLENGE 1 Mountain bike supply storage.

are innovations, not inventions. Finding out what has already been done and is available is critical for the engineering design team.

design brief
A design brief is a written plan that identifies a problem to be solved, its criteria, and its constraints.

In the final step of the concept stage, the team must identify criteria for the product design. Remember from Chapter 2 that *criteria* are special requirements. During this step, the team also identifies the *constraints*, or limitations, that will influence the product design. At this point, the design team develops a **design brief**. This written plan states the problem, the criteria the design must meet, and the constraints on the solution. Figure 3-6 shows a sample design brief for the mountain bike.

Engineering Challenge

ENGINEERING CHALLENGE 2

Your engineering design team should now examine each solution found in Engineering Challenge 1. Do any of these products meet the constraints placed on the mountain bike as noted in the design brief? Your team should document its findings.

Design Brief

Problem:

There is no location to store supplies, such as water, food, or a first aid kit, on the mountain bikes currently being sold. This requires the rider to carry his or her supplies in a backpack. Wearing a backpack can make the rider unstable and can also put strain on his or her back muscles.

Criteria:

- Must carry food and water for one day of riding
- Must carry a first aid kit
- Must be a part of the bike, not an add-on

Constraints:

- Class has two weeks time to complete the development
- Only school fabrication faculties are available
- Material costs cannot exceed $20.00

FIGURE 3-6 The mountain bicycle problem's design brief.

Development Stage

The development stage has four steps: (1) explore possibilities, (2) select an approach, (3) develop a design proposal, and (4) make a prototype.

In the first step of development, we reexamine the list of possible solutions to the problem. This step could include more brainstorming by the design team and more research. The design team then evaluates each possible solution against the established design criteria and the imposed constraints.

A design evaluation matrix like the one shown in Figure 3-7 may help in selecting a solution, the second step of development. The design team collects and processes this information and data to determine how well each design meets the design requirements. A **matrix** is an arrangement of mathematical elements to help solve problems. The design team also may use some type of voting to help it select the design to adopt. The decision on the final design should be reached through group consensus, or general agreement. This requires that design team members work well with other people.

Design matrix			
Criteria	Design A	Design B	Design C
1	Good	Excellent	Good
2	Good	Poor	Excellent
3	Poor	Poor	Poor
4	Excellent	Good	Poor

FIGURE 3-7 A sample design matrix.

Math in Engineering

The design matrix in Figure 3-7 uses the terms *excellent, good,* and *poor* in evaluating designs. A true mathematical design matrix uses a mathematical rating system to evaluate the solutions based on numerical values. This type of matrix removes personal bias from the decision-making process.

Developing a design proposal is the third step of the development stage. In this step, detailed drawings are prepared. If the engineering design process continues, it will require drawings that provide a lot of detail about the product. The drawings in Figure 3-8 provide an example of the level of detail required. You can get information on preparing such detailed drawings in Chapters 5 and 6. In addition to preparing these drawings, the team must make other critical decisions at this point. Two of these decisions are (1) the type of material to produce the item from and (2) the type of fabrication processes that will be used to manufacture the product. To make these decisions, engineers require knowledge about both materials (Chapter 8) and the manufacturing process (Chapter 17).

The final step of the development stage is making a **prototype**, a full-size working model of the design. We explain prototype fabrication in Chapter 8.

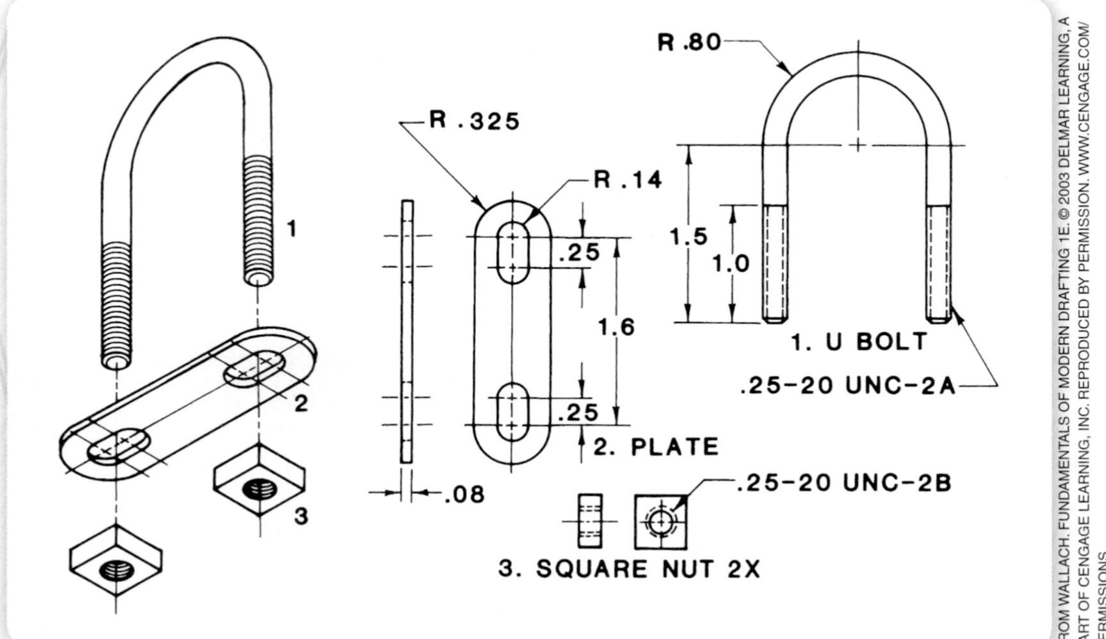

FIGURE 3-8 Detailed drawings are required.

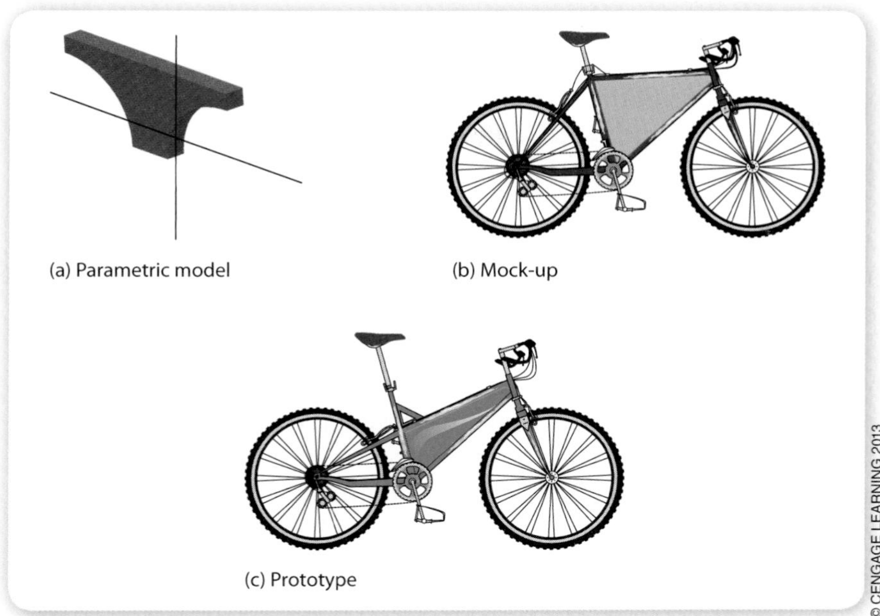

FIGURE 3-9 **The parametric model, mock-up, and prototype of the mountain bike.**

This working model of the design provides the engineering design team with a real-world product to test and evaluate.

The prototype must function like the intended final product. Sometimes designers use a **mock-up**, a model of the design that does not actually work, to see how the product might look. Figure 3-9 shows the parametric model of the mountain bike design, a mock-up (cardboard taped to a bicycle), and the fabricated prototype on the mountain bicycle design. The mock-up is for appearance only and may or may not be fabricated. Prototypes, though, are always made to test the design. Without a prototype, the engineering design team would not be able to see if its solution actually operates.

Evaluation Stage

The evaluation stage has four steps: (1) test and evaluate, (2) refine, (3) create or make a solution, and (4) communicate results. 5

Using the prototype made during development, the engineering design team now conducts experiments on its design. These experiments will allow the team to test and observe the design in a working environment. **Experimentation** is the act of conducting a controlled test on a prototype. **Testing** is the process of analyzing or assessing the performance of the design solution based on the established design criteria.

During testing, many design items are evaluated, including the design's functionality, the durability and workability of the materials, the availability of the fabrication methods required, and the product control systems. In **evaluation**,

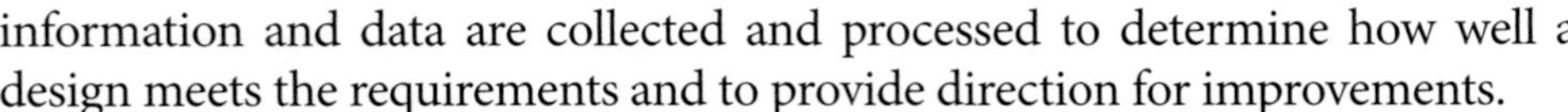

information and data are collected and processed to determine how well a design meets the requirements and to provide direction for improvements.

Based on the results of the prototype testing, the product may be refined or redesigned. This can mean only minor changes to or a complete redesign of the product. Companies typically also include customer feedback at this point. This process of **optimization** makes the design as effective and functional as possible. During this refinement step, all changes are documented and detailed product drawings are updated.

optimization
Optimization is an act, process, or methodology that is used to make a design as effective or as functional as possible within the given criteria and constraints.

Once the product has been refined, the engineering design team turns the product over to the production team, which will plan the steps in the manufacturing process. This should not be the first time the production team is involved with the design. Production also should have been discussed during product development.

In addition to making the product, the company will decide on how to package it. Information obtained earlier during problem definition may help. The marketing team will develop the product's sales and distribution plan.

SECTION 2: DESIGN CONSIDERATIONS

Constraints

The engineering design process must consider the constraints or limitations placed on the design. These may include resources (see Chapter 2) such as:

- the people available to work on the project and their skills,
- the capital or funding available,
- the reference information available,
- the types of fabrication process available (machines and tools),
- the types of materials available, and
- the amount of time available to work on the product.

Constraints can also include factors such as product appearance, safety requirements, and governmental regulations. Any factor that limits the product's design, production, or sale should be listed as a constraint.

Ethics in Engineering

Ethical behavior means acting on moral principles and values. Whether or not an unethical behavior is illegal, most people would agree it is not the right thing to do. Ethical behavior is more than simply following the rules. It also means acting in responsible ways that will not harm people, animals, or the environment. Most professions, including engineering, have sets of standards that define ethical behavior.

Career Spotlight

Name:
Merwin T. Yellowhair

Title:
President, Arrowhead Engineering Inc.

Job Description:
Merwin Yellowhair's home, the Diné Nation in northern Arizona, had few engineers to act as role models, but that only steeled his resolve to change the situation for the better. He worked through an engineering program, determined to show that civil engineering is a good career option for Native Americans.

Yellowhair founded Arrowhead Engineering Inc. in 2007. The company specializes in water resources, roadways, and site and construction plans for commercial and residential development.

Yellowhair is very knowledgeable about water resources. He uses mathematical equations to calculate the flow, speed, pressure, and capacity of fluids in various structures or pipes.

When working on site plans, Yellowhair takes an empty piece of land and decides how it might best be used. He picks a location for a building, a parking lot, and drainage facilities. Then he draws up a plan so his clients can decide if they want to go through with the project. When construction begins, Yellowhair's construction plan offers guidelines at every step of the way. He himself inspects the construction to make sure it is being done correctly.

Education:
Yellowhair received a bachelor's degree and a master's degree in civil engineering from Northern Arizona University.

"The fact that I had a structural emphasis for my bachelor's degree and a water-resources emphasis for my master's degree helped me become versatile in discussing all types of engineering principles with other engineers," he says. "Now that I own my own company, I can understand a lot more things than if I'd stayed in one area of civil engineering."

During his undergraduate career, Yellowhair enjoyed a project in which he helped create a 28-lot town-home subdivision for an engineering company. "That allowed me to get a feel for real-world design," he says, "because we actually had an engineer who supervised our work and used it for his client."

Advice to Students:
"The first thing I learned is that it's always good to ask a lot of questions," Yellowhair says. "You also have to be determined and focused on your specific goals. But don't forget to have fun, or else it won't be very interesting!"

Science in Engineering

To a biologist, a life cycle is a period of time from the generation of an organism to its death. Engineers, however, use the term *life cycle* to describe the period of time from a product's design through its use until it is discarded. The four stages of the engineering product's life cycle are similar to the biological life cycle: (1) introduction, or when the seed is planted; (2) growth, or when it begins to sprout; (3) maturity, or the period when leaves and roots reach their maximal growth; and (4) decline, or when the plant dies.

Would you knowingly design a product that might cause harm to young children? Of course not. The engineer, however, also must anticipate how consumers might use their products for things other than what they were designed to do. If you know that small children usually put toys in their mouths, then it would be unwise or unethical to design a small toy with parts that could easily break off in a child's mouth.

Engineers also have a responsibility to plan for a product's life cycle. What happens when a product's useful life is over? Can it be recycled? Can the used product be incorporated into a new or a different product? Will the product's material decompose naturally? Will the product remain in a landfill for thousands of years? Can a small rodent or animal get caught or trapped in the discarded product? These are only a few of the many ethical questions that engineers must ask themselves as part of determining the best overall product design.

7

PAUL J. RICHARDS/GETTY IMAGES

FIGURE 3-10 The U.S. Patent and Trademark Office.

Patents

After the design has been conceptualized, the design team typically files a patent with the U.S. Patent and Trademark Office (see Figure 3-10). A patent is a contract between the federal government and the inventor that gives the inventor exclusive rights to make, use, and sell the product for 17 years (see Figure 3-11). A patent can be granted to an individual, a group of people, or a company.

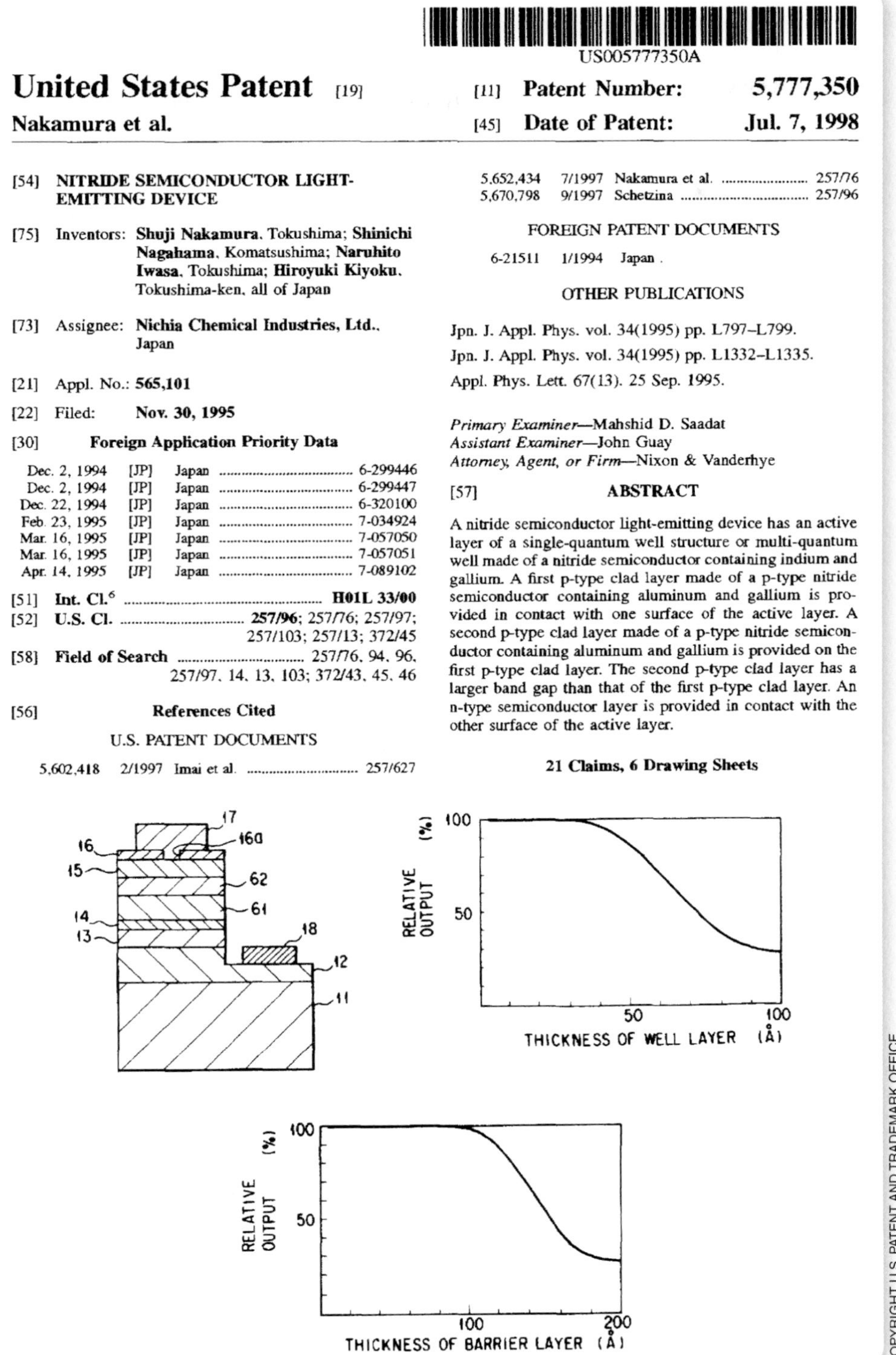

US005777350A

United States Patent [19]

Nakamura et al.

[11] **Patent Number: 5,777,350**

[45] **Date of Patent: Jul. 7, 1998**

[54] **NITRIDE SEMICONDUCTOR LIGHT-EMITTING DEVICE**

[75] Inventors: **Shuji Nakamura**, Tokushima; **Shinichi Nagahama**, Komatsushima; **Naruhito Iwasa**, Tokushima; **Hiroyuki Kiyoku**, Tokushima-ken, all of Japan

[73] Assignee: **Nichia Chemical Industries, Ltd.**, Japan

[21] Appl. No.: **565,101**

[22] Filed: **Nov. 30, 1995**

[30] **Foreign Application Priority Data**

Dec. 2, 1994	[JP]	Japan	6-299446
Dec. 2, 1994	[JP]	Japan	6-299447
Dec. 22, 1994	[JP]	Japan	6-320100
Feb. 23, 1995	[JP]	Japan	7-034924
Mar. 16, 1995	[JP]	Japan	7-057050
Mar. 16, 1995	[JP]	Japan	7-057051
Apr. 14, 1995	[JP]	Japan	7-089102

[51] **Int. Cl.**[6] **H01L 33/00**

[52] **U.S. Cl.** **257/96**; 257/76; 257/97; 257/103; 257/13; 372/45

[58] **Field of Search** 257/76, 94, 96, 257/97, 14, 13, 103; 372/43, 45, 46

[56] **References Cited**

U.S. PATENT DOCUMENTS

5,602,418	2/1997	Imai et al.	257/627
5,652,434	7/1997	Nakamura et al.	257/76
5,670,798	9/1997	Schetzina	257/96

FOREIGN PATENT DOCUMENTS

6-21511 1/1994 Japan .

OTHER PUBLICATIONS

Jpn. J. Appl. Phys. vol. 34(1995) pp. L797–L799.

Jpn. J. Appl. Phys. vol. 34(1995) pp. L1332–L1335.

Appl. Phys. Lett. 67(13), 25 Sep. 1995.

Primary Examiner—Mahshid D. Saadat
Assistant Examiner—John Guay
Attorney, Agent, or Firm—Nixon & Vanderhye

[57] **ABSTRACT**

A nitride semiconductor light-emitting device has an active layer of a single-quantum well structure or multi-quantum well made of a nitride semiconductor containing indium and gallium. A first p-type clad layer made of a p-type nitride semiconductor containing aluminum and gallium is provided in contact with one surface of the active layer. A second p-type clad layer made of a p-type nitride semiconductor containing aluminum and gallium is provided on the first p-type clad layer. The second p-type clad layer has a larger band gap than that of the first p-type clad layer. An n-type semiconductor layer is provided in contact with the other surface of the active layer.

21 Claims, 6 Drawing Sheets

FIGURE 3-11 A patent is good for 17 years. This is a facsimile of the patent awarded to S. Nakamura, whose work with blue lasers increased the storage capacity of optical disks.

Math in Engineering

Market research involves the mathematical concept of sampling. Sampling is a statistical process used to select individuals to contact to gain information. The statistician doing the research can choose between two different sampling processes: random sampling and stratified sampling. Random sampling is like choosing one M&M from an entire package of M&Ms. You may get a red one, a green one, or a yellow one. All candies have an equal chance of being selected. Random sampling would be used if the product could be used by everyone.

Stratified sampling is used for most products and involves choosing only from an identified group. Maybe you have designed a backpack for children 6 to 10 years old. The statistician would sample only that group and not ask adults if they would purchase the backpack.

The patent describes the product's design, includes detailed drawings of the product, and specifies the materials used to fabricate the product and any other details that are required to make the item operational. The patent office receives more than 400,000 applications for patents each year.

Sometimes people confuse patents with copyrights and trademarks. A copyright protects the authors of literary works, music, and artistic work. Others cannot duplicate their products. A trademark is a symbol or word that distinguishes one item or company. The trademark of McDonald's is its Golden Arches. When you see these arches, you know it means McDonald's.

market research
Market research is a survey of potential product users to find out their likes and dislikes about a product.

documentation
Documentation is the organized collection of records and documents that describe a project's purpose, processes, and related activities for future reference.

Marketability

Marketability refers to whether consumers will purchase the newly designed product. During the engineering design process, the design team should have conducted market research. **Market research** is a survey of potential users to find out their likes and dislikes about the product. This group should be a representative sample of the intended customers. Information from this market research should help in marketing the product for sale.

SECTION 3: THE ENGINEERING NOTEBOOK

Purpose

Documenting the engineering design process is essential. **Documentation** is the organized collection of records and drawings that describes the purpose, processes, and related activities of the process, to be used for future reference. Each engineering design project must have an engineering notebook.

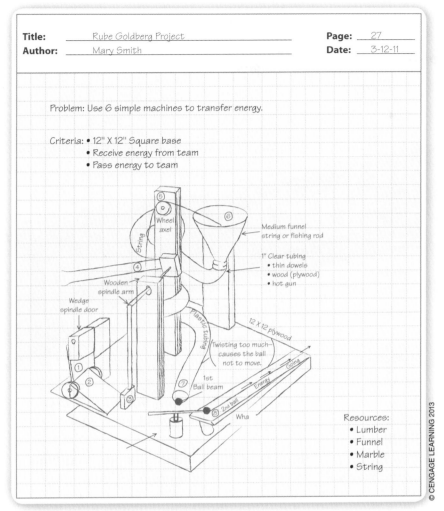

FIGURE 3-12 *A page from a well-prepared engineering notebook.*

An **engineering notebook** is the documentation of the steps, calculations, and evaluation of the engineering design process for a particular item. Figure 3-12 provides sample pages from an engineering notebook. Engineering notebooks are bound. Pages cannot be added, nor can pages be removed. Engineering notebook entries contain all sketches and calculations from the design. All notebook entries are dated. This engineering notebook is actually a legal document and can be used in patent disputes. The notebook could also be called into a court as evidence in various types of legal or safety cases.

engineering notebook
An engineering notebook is the documentation of all of the steps and calculations for and an evaluation of the engineering design process for a particular item.

Organization

The pages of an engineering notebook are part of a bound notebook. The pages should be dated and numbered sequentially. All design brainstorming ideas should be listed in the engineering notebook. In addition, all sketches and drawings of the design must be included in the engineering notebook.

During prototype fabrication and testing, digital pictures of the item should be added to the engineering notebook. Designers should also add observations, test results, and product evaluations that document the design process. Store the engineering notebook in a safe location once the design process has been completed.

SUMMARY

In this chapter you learned:

- The design process incorporates problem-solving steps to develop possible solutions to the problem.
- The engineering design process applies mathematics, science, and engineering principles to the design process.
- The engineering design process consists of 12 steps.
- The engineering notebook provides legal documentation of the design process.
- Ethical considerations are an important part of the engineering design process.
- A patent grants the inventor exclusive rights to the product for 17 years.

VOCABULARY

Write a definition for each term in your own words. After you have finished, compare your answers with the definitions provided in this chapter.

Engineering design process
Brainstorming
Design brief
Optimization
Market research
Documentation
Engineering notebook

STRETCH YOUR KNOWLEDGE

Please provide thoughtful, written responses to the following questions:

- Describe the types of interpersonal skills that a good design team member should have. Discuss why these interpersonal skills are important.
- Why is identifying the design criteria important in the engineering design process before the product is actually designed?
- Why is the evaluation of a design based on data and not on personal feelings?
- Describe the importance of a well-documented engineering notebook to both the design team members and their company.
- A person or company in the United States holds a patent for 17 years. Discuss whether this period should be longer or shorter.

Onward to Next Destination ▶

CHAPTER 4

Freehand Technical Sketching

Menu

Before You Begin

Think about these questions as you study the concepts in this chapter:

1. How does an engineer or a designer communicate design ideas?

2. What is the difference between art sketching and freehand technical sketching?

3. Why is freehand technical sketching important in the engineering design process?

4. What are anecdotal notes, and how are they used?

5. What are the different types of freehand technical sketches?

6. What important geometric terms are used when sketching design ideas?

7. How do you create freehand technical sketches?

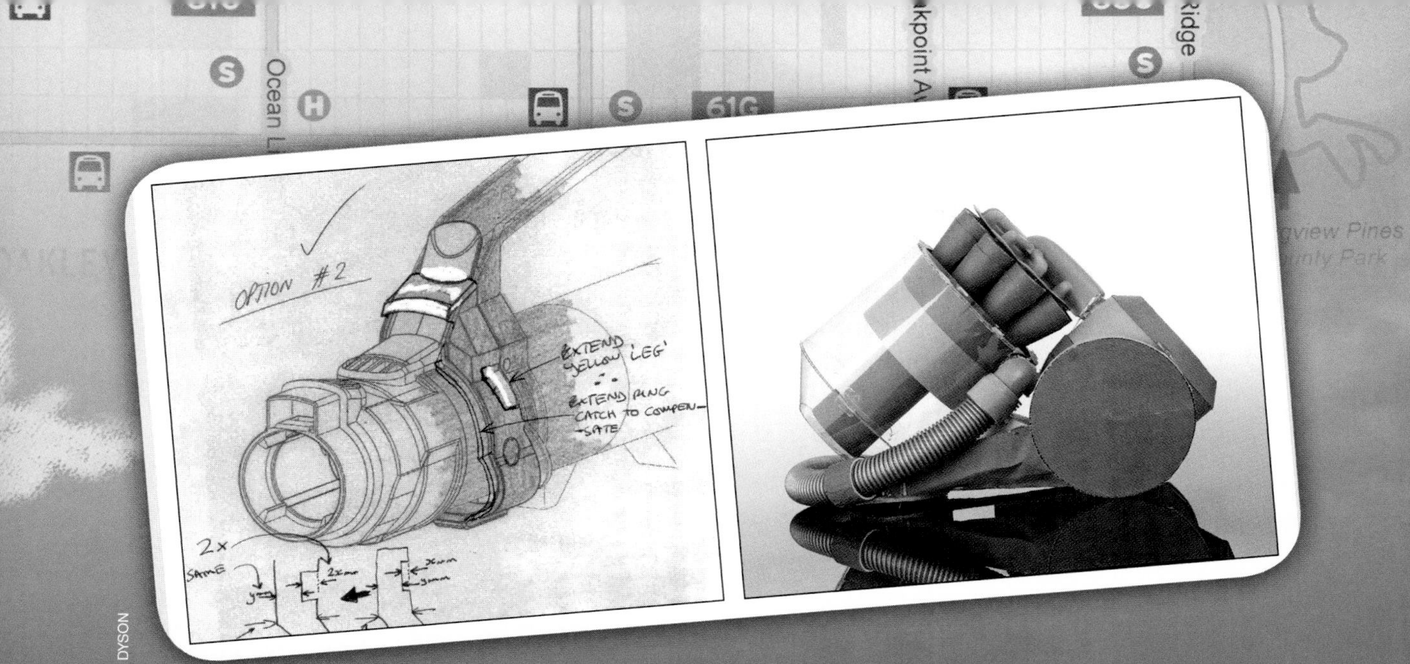

DYSON

Engineering in Action

James Dyson is the creator of the famous Dyson cyclone-style vacuum cleaner. He began creating his concepts by making rough thumbnail sketches. For Dyson, sketching is a key communication tool in idea development. They are a bridge between the concept in the design engineer's mind and the next vital step—creating basic 3-D models (www.jamesdysonfoundation.com).

Dyson wanted to solve a specific problem: the loss of suction in traditional bag-type vacuum cleaners. Whenever an engineer sets out to solve a problem, he or she begins by clearly defining the problem. Research follows so that the designer can learn as much as possible about the problem.

When engineers understand a problem, they develop ideas. Engineers often record their ideas as rough, freehand, thumbnail sketches as they discuss their ideas.

SECTION 1: SKETCHING, A GOOD WAY TO COMMUNICATE

Have you ever heard the expression, "A picture is worth a thousand words"? Pictures, or sketches, are very important to engineers and designers. Sketching can be the quickest and easiest way to record design ideas.

Communicating Ideas

Communicating is one of the most important things humans do (Figure 4-1). We communicate for many reasons. One reason is to tell others about our

© OLGA VASILKOVA/SHUTTERSTOCK.COM

© SOROK/SHUTTERSTOCK.COM

© GNUSKIN PETR/SHUTTERSTOCK .COM

© SUPRI SUHARJOTO/ SHUTTERSTOCK.COM

© TATJANA STRELKOVA/ SHUTTERSTOCK.COM

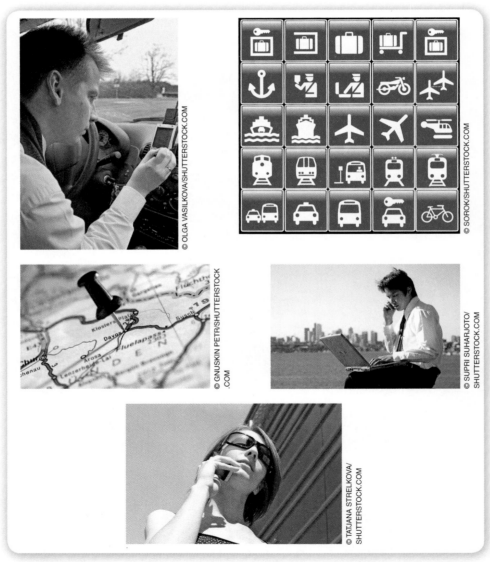

FIGURE 4-1 People communicate in different ways and for different reasons. These images show technologies that can help people get where they want to go.

FIGURE 4-2 **You can communicate in different ways: by speaking, by using hand signals, by writing, or using graphic images.**

ideas. Those ideas could be a map of how to get some place, directions on how to download songs from the Internet to an iPod, or how to build something. Communicating can happen through speaking (oral), gestures (hand signals), writing (text), and pictures (graphic) (see Figure 4-2).

People who design and make products or buildings that we use every day show and share their ideas by using graphic images (pictures) and written words. Before final mechanical drawings and specifications (technical notes) are developed on drafting boards or in computer-aided drafting (CAD) systems, designers and engineers use a type of sketching called *freehand technical sketching* to communicate ideas (see Figure 4-3b).

Freehand technical sketching is the process of drawing technical images without drafting tools such as a T square, a triangle, or a compass. This is different than the sort of sketching artists do. Artists use sketching to plan a design idea just as engineers do, but the artist's sketch will often be less exacting and more "free-flowing" (see Figure 4-3a). Things we build or construct such as automobiles, wheelchairs, skateboards, and houses all begin as ideas sketched on paper (see Figure 4-4). This type of sketching is often used to brainstorm ideas, record measurements and procedures, and begin the design process for new products.

freehand technical sketching

Freehand technical sketching is the process of drawing technical images without the use of drafting tools such as a T square, a triangle, or a compass.

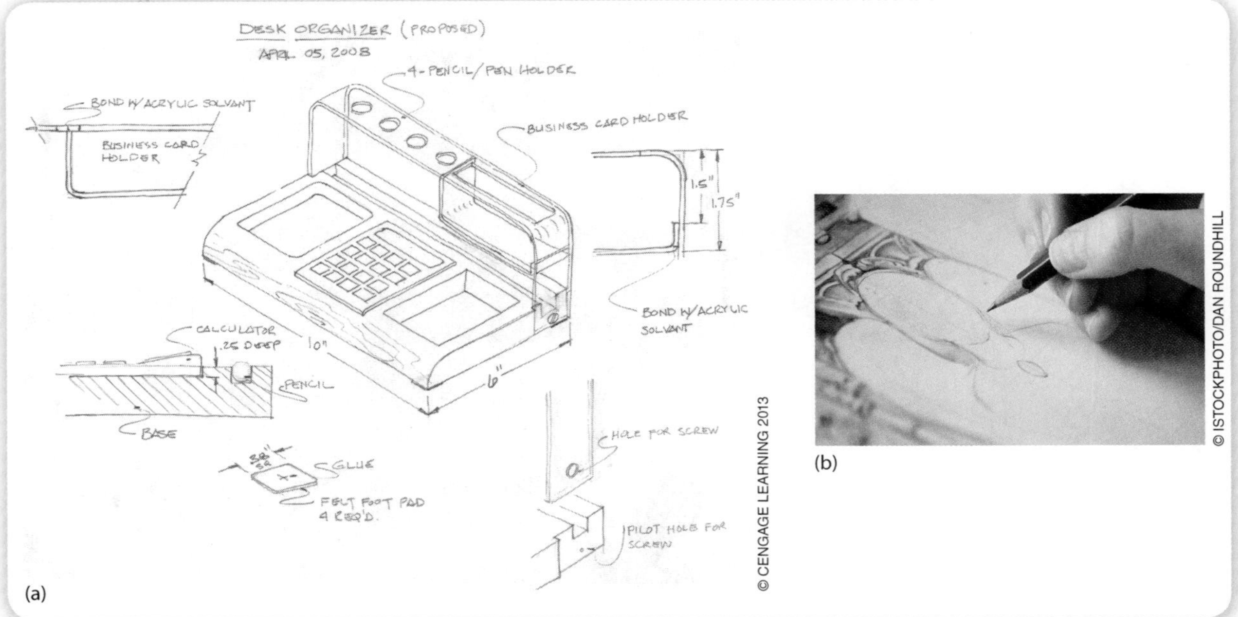

FIGURE 4-3 **(a) Engineer's freehand technical sketches of a proposed new product; (b) artist's sketch of figure study.**

Advantages to Sketching

Freehand technical sketching has some advantages over drawing with drafting tools or using a CAD system. First, you can sketch ideas or concepts anywhere and anytime an idea comes to you. You may think of ideas for a design while riding the bus to school, for example. You do not need mechanical drafting tools to get started. A pencil and a piece of paper to sketch on are the only tools you need. The same is true for people who design the products we use every day. Many great ideas were first recorded in a restaurant on a napkin, during airplane fights, or at kitchen tables (see Figure 4-5). So, you can sketch anytime, anywhere, and on most any kind of paper.

Who Sketches?

Sketching can be done by anyone, anywhere, at any time. People in engineering and design careers frequently use freehand technical sketching in a thought process called *ideation* or *conceptualization*. They refine their design ideas so others can understand them. People who develop mechanical drawings using traditional drafting tools or CAD programs use preliminary sketches to organize their thoughts before making formal drawings. This helps them reduce the number of mistakes in their engineering mechanical drawings (see Figure 4-6). Machine operators and people in building construction also use sketching. They can clarify operations and confirm design

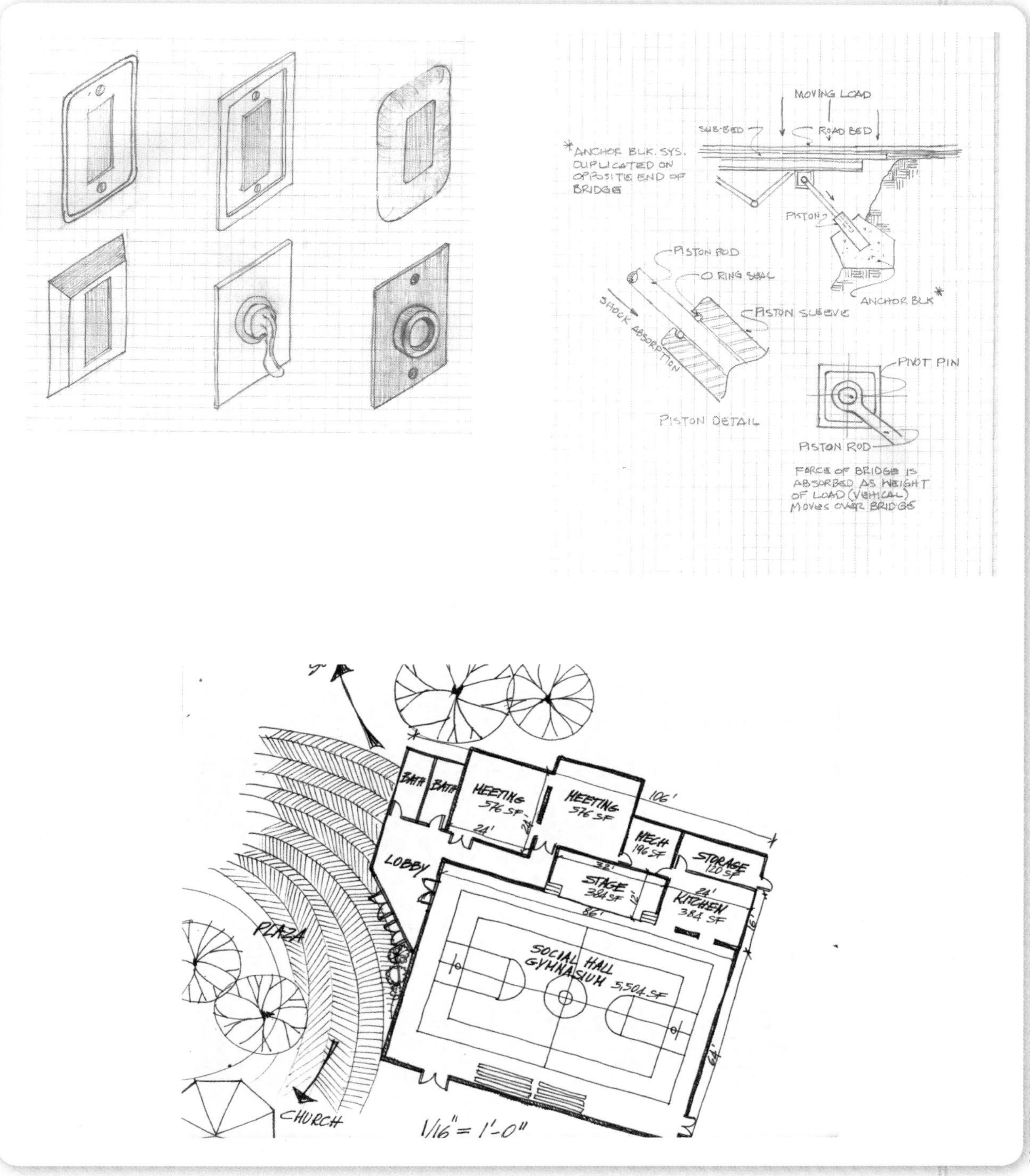

COURTESY OF ANNA VERALDI, INT'L. ASSOC. AIA - VERALDI RENOUARD ARCHITECTS

FIGURE 4-4 Engineers and designers use freehand technical sketching to record design ideas for products such as these light switch cover plates, these bridge parts, and this floor plan.

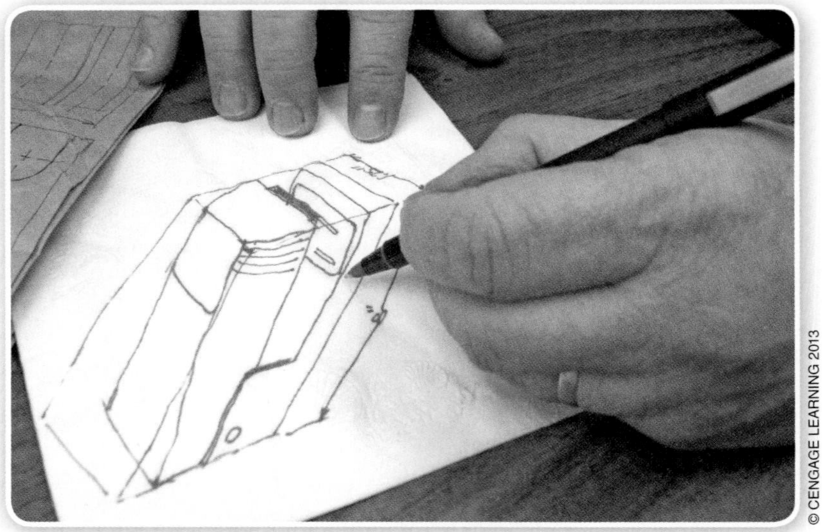

FIGURE 4-5 No matter where you are, you can quickly use freehand technical sketching to record any idea that comes to mind.

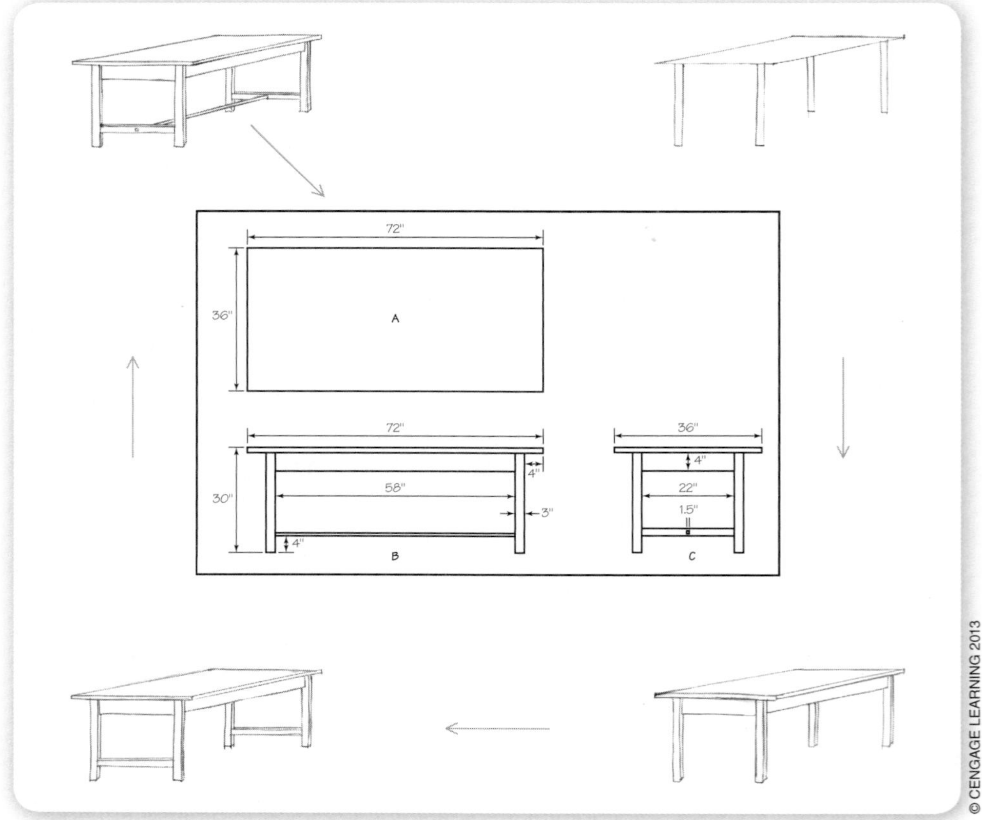

FIGURE 4-6 Engineers and designers use sketching in a thought process called *ideation* or *conceptualization.*

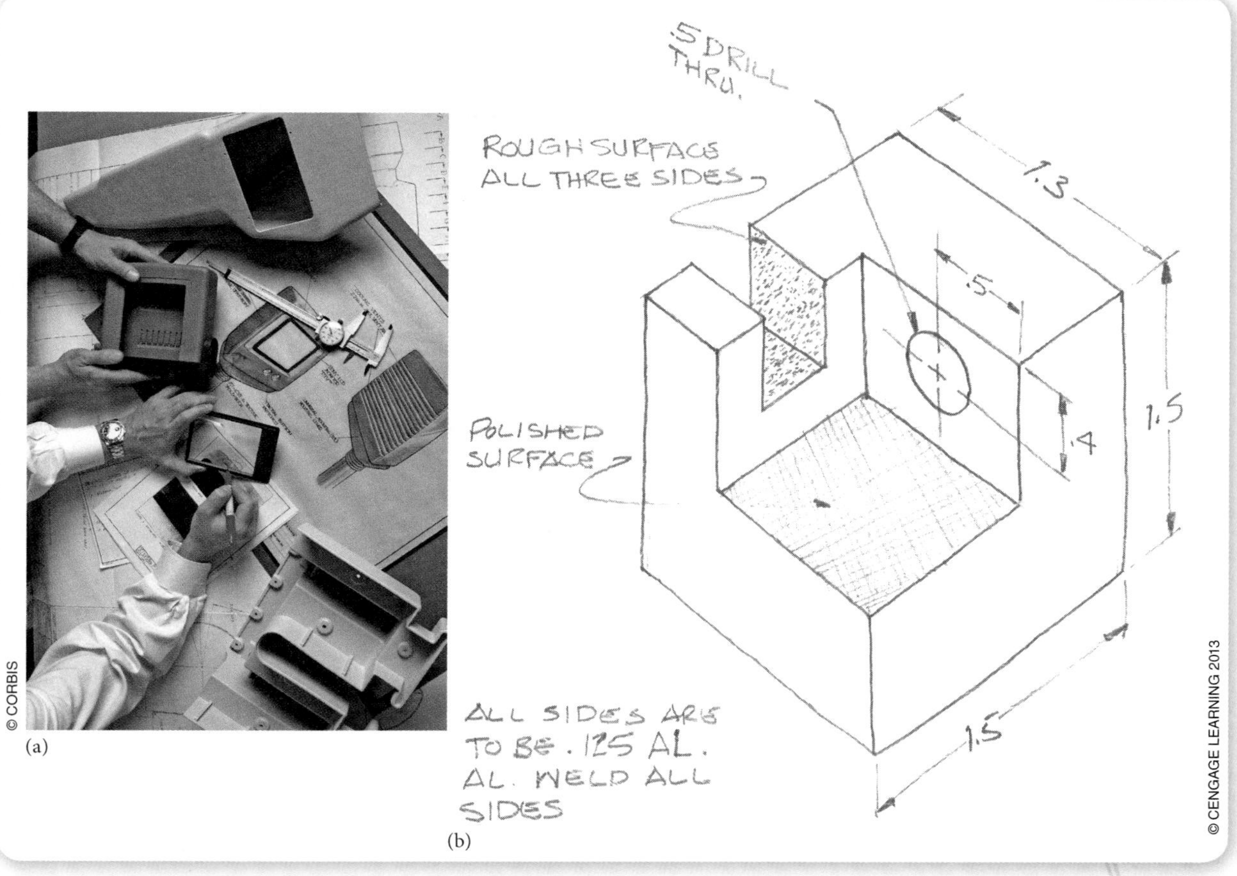

FIGURE 4-7 *(a) In reverse engineering, engineers disassemble a product, take measurements, sketch each part, and add dimensions and notes. (b) This detailed sketch shows annotations and dimensions. Can you pick out the annotations? Can you pick out the size dimensions or the location dimensions?*

procedures at the machine or on the construction site. Freehand technical sketching is also used during **reverse engineering**, which is the process of taking something apart and analyzing how it works in detail.

In this way, a designer or an engineer can understand the function of a device or a separate part in a device. As the engineer takes the object apart, each part is sketched with added notes, called **annotations**, and dimensions, giving sizes and locations of parts and features (see Figure 4-7). The reverse engineering process is discussed in more detail in Chapter 6.

annotations
Annotations are notes placed on an engineering sketch to clarify the viewer's understanding of the object or objects drawn.

Types of Freehand Technical Sketching

Freehand technical sketching can be divided into three types based on their use: thumbnail, detailed, and presentation.

Thumbnail Sketch A **thumbnail sketch** is usually small, simple, and with just enough detail to convey a concept. It is often used in brainstorming to record ideas quickly (see Figure 4-8a).

thumbnail sketch

A thumbnail sketch is usually small and simple and has just enough detail to convey a concept. It is often used in brainstorming to record ideas quickly.

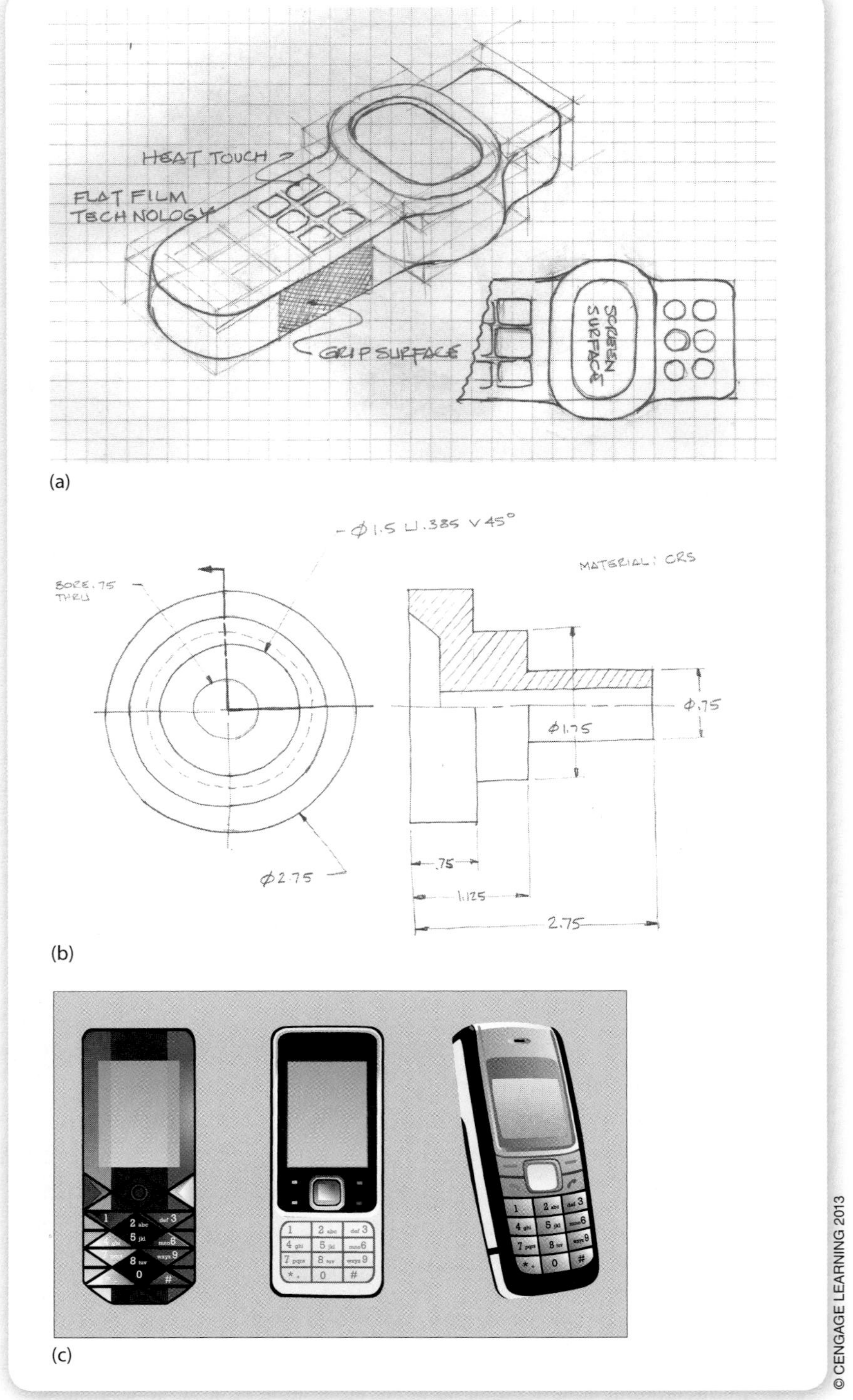

FIGURE 4-8 Three types of freehand technical sketches used by engineers and designers: (a) thumbnail sketch (rough layout conveying ideas), (b) detailed sketch (details of object with dimensions and notes, called *annotations*), and (c) presentation sketch (3-D detailed sketch often with color).

Detailed Sketch This type of sketch is used when a designer or an engineer wants to further develop a thumbnail sketch. A **detailed sketch** may be a three-dimensional pictorial or two-dimensional sketch that includes dimensions, annotations, and symbols. The sketch may also include shading or shadows for effect (see Figure 4-8b).

Presentation Sketch A **presentation sketch** is very detailed and made to look realistic. The use of color, surface texture, shading, and shadow is a process known as **rendering**. The sketch is usually a three-dimensional (3-D) pictorial view, such as an isometric or perspective drawing. You will learn how to develop these types of drawings in Chapter 5. These sketches are used by engineers and designers to make formal presentations (see Figure 4-8c).

SECTION 2: VISUALIZING—AN IMPORTANT SKILL FOR SKETCHING

Visualizing is the ability to form a mental picture of an object. Freehand technical sketching is the process of recording that mental picture on paper. Before you are able to do this, you must learn and be able to apply certain visualization concepts.

Proportion and Scale

The uses of *proportion* and *scale* help make objects look accurate and visually correct. **Proportion** is the relationship between one object and another or between one size and another size. It is often expressed in a mathematical relationship called a *ratio* (see Figure 4-9).

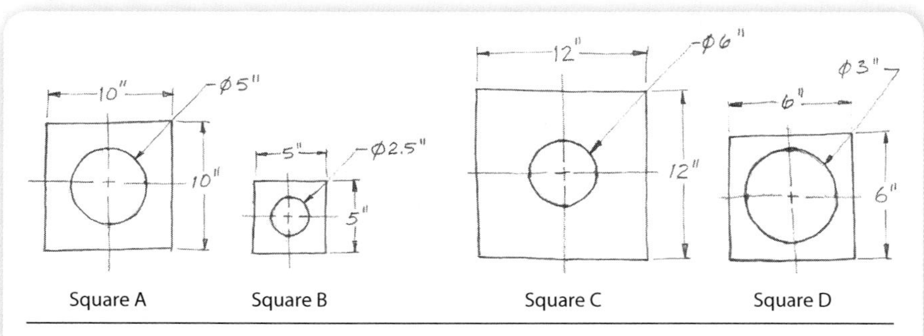

This sketch is said to be proportional

Square A is drawn twice the size of Square B just as Square B dimensions indicate it is half the size of Square A. The hole in Square A indicates a diameter of 5" and it is drawn at half the height of the 10" square. The hole in Square B indicates a diameter of 2.5" and it is drawn at half the height of the 5" square. We can then say that Square A is proportional to Square B and the holes are proportional.

This sketch is said not to be proportional

Square C is **not** drawn twice the size of Square D even though Square D dimensions indicate it is half the size of Square C. The hole in Square C indicates a diameter of 6" but it is **not** drawn at half the height of the 12" square. The hole in Square D indicates a diameter of 3" but it is **not** drawn at half the height of the 6" square. We can then say the squares are **not** drawn proportional to each other and each hole is **not** proportional to the corresponding square.

FIGURE 4-9 *Concept of proportion.*

detailed sketch
A detailed sketch is a type of freehand technical sketch that provides detailed information about the object, such as annotations (notes), dimensions, and shading.

presentation sketch
A presentation sketch is very detailed and made to look realistic, usually as a three-dimensional pictorial view such as an isometric or perspective drawing.

rendering
Rendering is the use of color, surface texture, shading, and shadow to give a sketched object a realistic look.

visualizing
Visualizing is the ability to form a mental picture of an object. Freehand technical sketching is the process of recording that mental picture on paper.

proportion
A proportion is the relationship between one object and another, or between one size and another size.

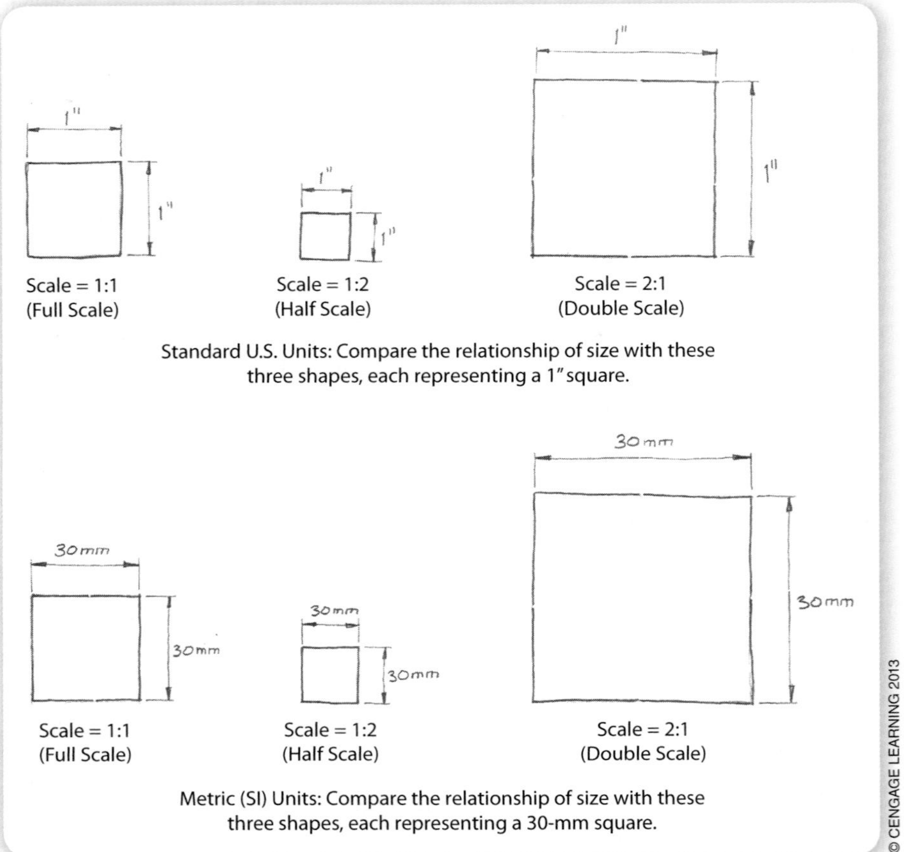

FIGURE 4-10 **Scale: size relationship based on full scale.**

scale
Scale is the mathematical relationship between the drawing of an object and its full size; it is expressed in a ratio such as 1:2 (read "1 to 2"), which represents one-half scale, or 1:1 (read "1 to 1"), which is full scale.

International System (SI)
The International System (SI), or metric system, is the system of measurement used in much of the world. (The abbreviation is from the French term, Systéme International d'units.)

This is also known as **scale.** If an object is drawn to full size, it has a ratio of 1:1 (read "one-to-one" or "full scale"). If the object is drawn at one-half the actual size, it has a ratio of 1:2 (read "one-to-two" or "one-half scale"). When doing freehand technical sketching, an engineer may need to draw a shape at full scale while on the same paper drawing the same shape at one-half scale. When doing this, the engineer makes sure the two sketches are proportional (see Figure 4-10). Note that Figure 4-10 shows two examples, one in standard U.S. units and one in metric or **International System (SI)** units.

The Language of Engineering Drawings

Creating a two-dimensional engineering sketch of your idea is like writing a sentence. Letters from the alphabet are put together to create words, and words are put together to create sentences. Sentences are put together to create paragraphs and tell a story. Engineering sketches are created by

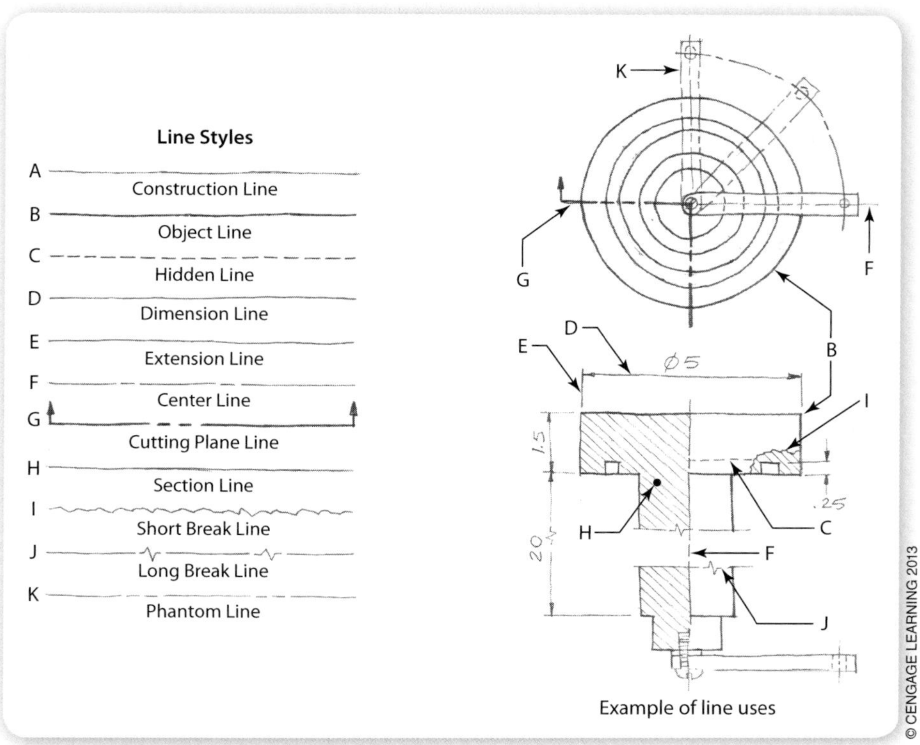

FIGURE 4-11 The alphabet of lines.

putting together lines and symbols from the **alphabet of lines**. These lines form shapes and provide a visual description of your design.

Eleven lines are used when sketching technical drawings. Each has a specific meaning and use. When you combine them, you describe an object. You must know the line types to describe the object just as you must know the letters in the alphabet. Study them and become familiar with their use (see Figure 4-11).

alphabet of lines
The alphabet of lines shows the various line types that, when used together, create an engineering drawing of an object.

Geometric Language

There is more to freehand technical sketching than drawing lines. You need to understand and use the language of geometric relationships between lines and shapes. You may have already studied these terms in a mathematics course. **Geometry** is a branch of mathematics concerned with the properties of space. This includes points, lines, curves, planes, and surfaces in space, as well as figures bounded by them. A person from ancient Greece named Euclid was the first person to formally develop geometric concepts and laws called *axioms* in 300 B.C. The following terms are used to show how lines and shapes relate to each other. They are also shown in Figure 4-12. Review them to see how many you already know.

geometry
Geometry is a branch of mathematics concerned with the properties of space. This includes points, lines, curves, planes, and surfaces in space, as well as figures bounded by them.

Alphabet of Lines

- ***Construction line:*** a very light line used to lay out a preliminary shape
- ***Object or visible line:*** a thick dark line that outlines the object
- ***Hidden line:*** a thin but dark and dashed line that indicates the edge of a hidden surface
- ***Dimension line:*** a thin dark line with arrowheads on each end that indicates the direction and size of an object or feature on an object
- ***Extension line:*** a thin dark line extending from an object that shows the limits of a dimension line
- ***Center line:*** a thin dark line that indicates the center of a circle or arc and the center of a cylinder
- ***Cutting plane line:*** a thick dark line (thicker than the object line) that indicates where an object is cut in half to show the inside of the object
- ***Section line:*** thin light lines usually drawn at 45 degrees (45°) to indicate the surface that has been cut open and exposed
- ***Short break line:*** a wiggly line the same darkness and thickness as an object line that is used to indicate that a smaller object is only partially drawn
- ***Long break line:*** a thin dark line with spikes used to indicate a large or long object is only partially drawn
- ***Phantom line:*** light thin lines with one long and two short dashes alternately spaced that indicate different positions of moving parts

6

Geometric Line Terms

1. *Horizontal line*: line that is drawn flat and level from left to right (for example, the edge of a floor)
2. *Vertical line*: line that is drawn straight up and down (for example, the side edge of a wall)
3. *Inclined line*: line that is drawn straight at an angle so that it is neither horizontal nor vertical (for example, a handrail on a staircase)
4. *Parallel lines*: lines that are always an equal distance apart and going in the same direction (for example, the opposite edges of a square table)

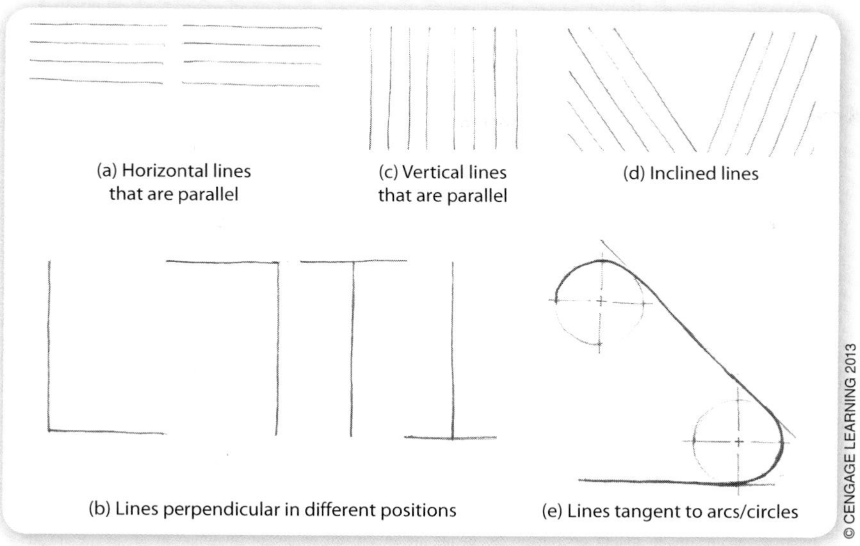

FIGURE 4-12 **Freehand sketched lines in geometric positions.**

Did You Know?

The first woman to make a significant contribution to the development of mathematics was Hypatia of Alexandria (A.D. 370–415). Her father, Theon of Alexandria, introduced her to mathematics and philosophy. Hypatia became the head of the Platonist school at Alexandria. She is recognized as the first woman to be a professional geometer and mathematician.

Career Spotlight

Name:
Nicole E. Brown-Williams

Title:
Project Engineer, Malcolm Pirnie, Inc.

Job Description:
Making sure cities operate smoothly and efficiently is a core component of engineering. In 2002, New York City suffered a severe drought. Nicole Brown-Williams now works to redesign the city's water supply to handle such future problems. The project has put her in the middle of an extraordinary engineering challenge.

Brown-Williams's work has focused on evaluating existing conditions and designing improvements for the city's groundwater system. Her work includes scheduling, estimating costs, research, design, fieldwork, technical writing, mapping, regulation compliance, interdisciplinary coordination, and coordination with clients, contractors, and subcontractors.

Brown-Williams also implements ongoing community outreach for the project. She often gives presentations at public meetings and cooperates with community members and elected officials.

Education:
Brown-Williams earned a bachelor's degree in civil engineering at Georgia Institute of Technology. Through her path of study, she gained a vast understanding of water-treatment processes and water-treatment system design. That knowledge base prepared her to meet New York City's civil engineering challenges.

Advice to Students:
As president emeritus of the New York City alumni chapter of the National Society of Black Engineers, Nicole Brown-Williams actively encourages minorities to join and excel in the engineering field.

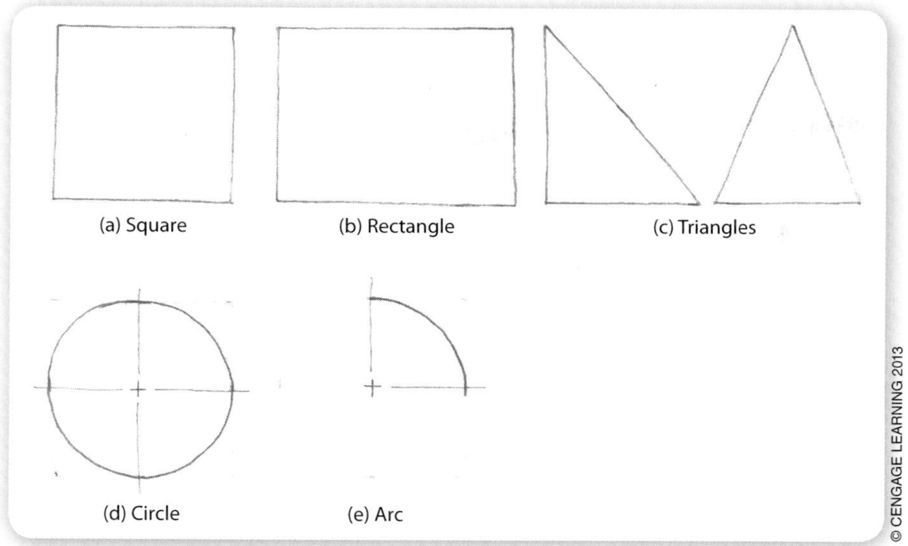

FIGURE 4-13 **Sketches of geometric shapes.**

5. *Perpendicular lines*: lines that intersect (meet) and form a 90° angle (also known as a *right angle*) (for example, the intersection of a wall and floor)
6. *Tangent line*: line that intersects an arc or circle but does not cross the arc or circle

Basic geometric shapes are also important to know in freehand technical sketching. How many of them do you remember from your math class (see Figure 4-13)?

Geometric Shapes

1. *Square*: A closed shape in which all four sides are of equal length, opposite sides are parallel, and all angles are 90° (right angles).
2. *Rectangle*: A closed shape with four side in which the opposite sides are of equal length, and all angles are 90° (right angles).
3. *Triangle*: A closed shape formed with three sides; when the three angles formed by the sides are added together they equal 180°.
4. *Circle*: A closed shape formed by a continuous line that maintains an equal distance (radius) 360° around a common center point.
5. *Arc*: Any segment of the circumference of a circle; the points on the arc are all an equal distance (radius) from a common center point from greater than 0° but less than 360° around the center point.

Math in Engineering

Freehand technical sketching is the process of recording graphic information. Products from engineers and designers begin as concepts and ideas recorded as sketches. Fundamental shapes—*geometric shapes*—are used to represent these concepts and ideas. It is important for engineers and designers to know and understand these geometric shapes and how they relate and interact with each other.

Geometry is all around us. Look around you. Everything made by humans and everything in nature has a geometric relationship. Everything can be reduced to points, lines, and shapes such as squares, rectangles, circles, arcs, spheres, cubes, and cones. The relationships among these can be expressed in mathematical terms, such as the Cartesian coordinate system or the Pythagorean theorem.

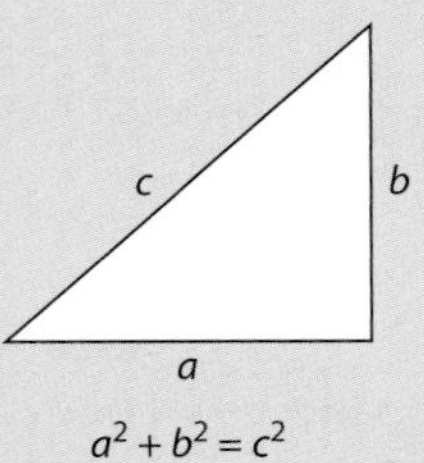

$a^2 + b^2 = c^2$

7

SECTION 3: SKETCHING TECHNIQUES

You may be saying to yourself, "I can't draw. My drawings never look right!" Freehand technical sketching is an important skill. It is a simple one to learn with the right amount of practice. If you follow the easy techniques, discussed next you will soon be able to create freehand technical sketches.

First, remember that neatness is necessary when communicating your ideas. Can you hear music clearly on the radio if you hear static in the background? It is difficult, too, to understand a one-on-one conversation when you are standing in a crowded room and several conversations are happening around you. Static and background noise both limit successful communication. Messy sketches are similar to static: they interfere with the communication of your ideas. By taking the time to create neat sketches, you will be a successful communicator (see Figure 4-14).

Second, a major advantage of freehand technical sketching is that it can be done quickly. Practicing this skill will allow you to increase the speed at which you draw. This will give you the ability to clearly and accurately communicate your thoughts.

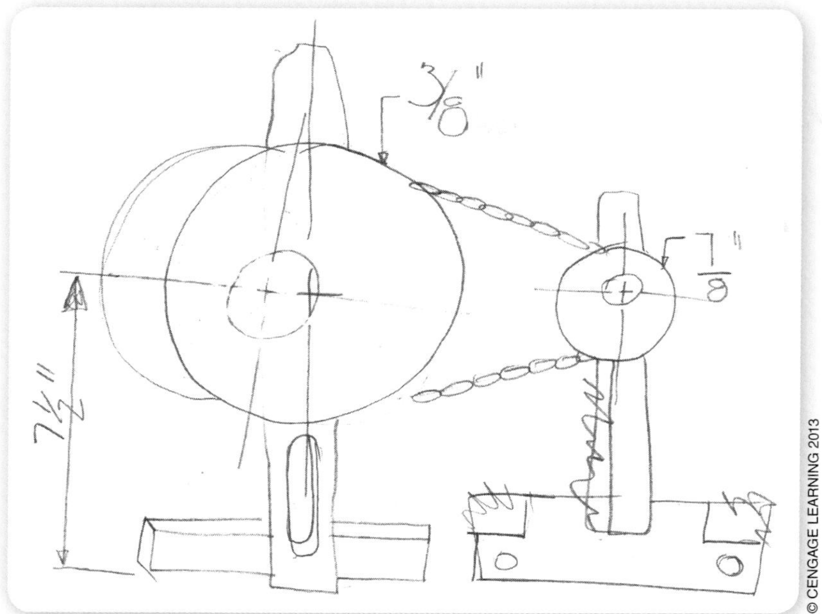

FIGURE 4-14 **Poor sketching can cause poor communication.**

Tools You Need

Freehand technical sketching can be done anywhere. All you need is a no. 2 or softer lead pencil, a soft eraser, paper (grid or plain), and a hard smooth surface. This is one reason why designers and engineers often carry clipboards or engineer's notebooks with them (see Figure 4-15).

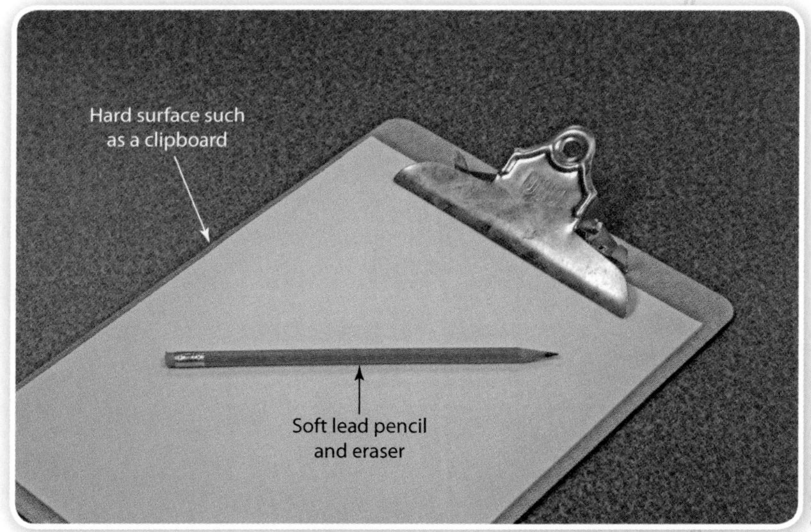

FIGURE 4-15 **The only tools you need to create freehand sketches are a soft lead pencil, a soft eraser, a hard surface, and paper.**

An engineer's notebook is usually a bound book with graph paper and lined paper, numbered pages, and a place to write important information such as the name, date, and title of the sketch. This information should be written with an ink pen; the freehand technical sketches should be drawn with pencil. This book becomes a permanent record of the engineer's concepts and ideas (see Figure 4-16).

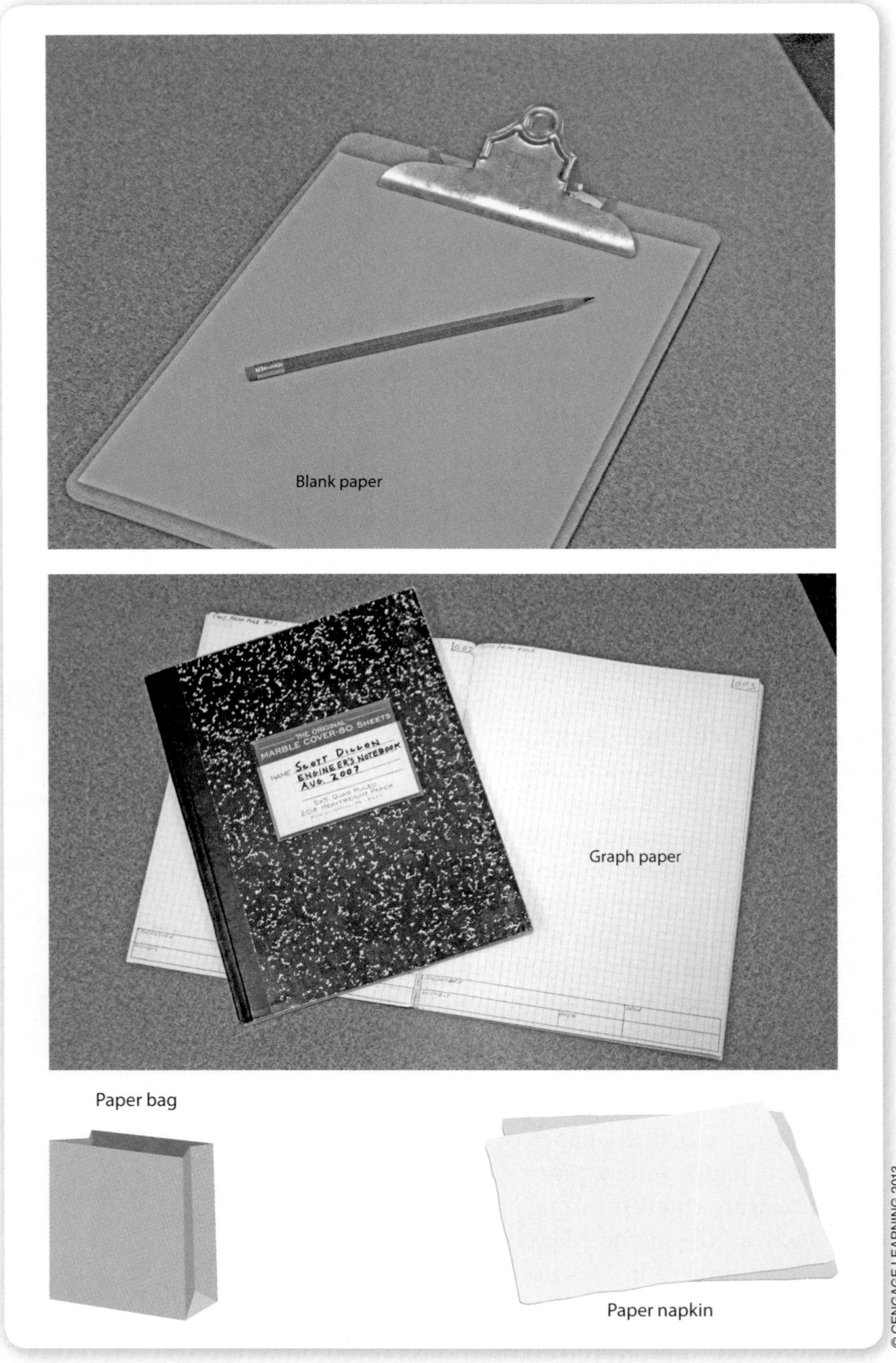

FIGURE 4-16 Blank paper or graph paper is the usual medium to sketch on, but even paper bags or paper napkins can be used when a great idea comes to mind.

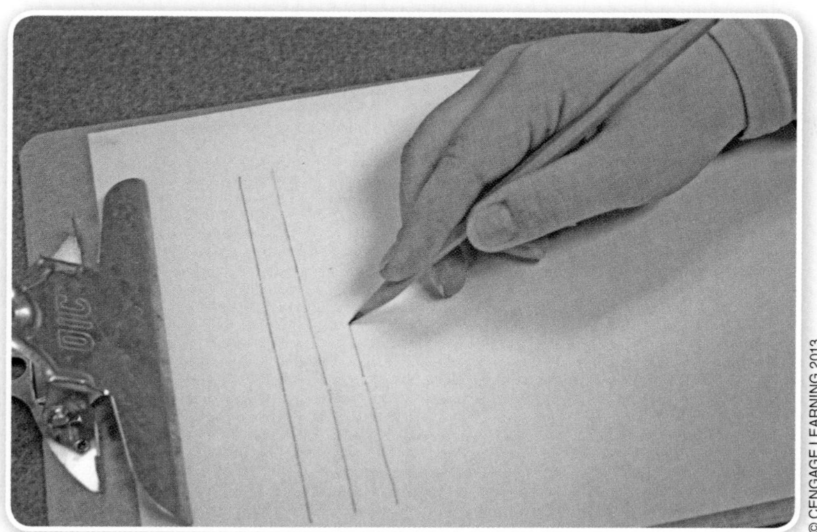

FIGURE 4-17 Hold the pencil so it is an extension of your index finger. Grip only enough to keep the pencil from slipping out of your hand.

Using the clipboard or notebook allows you to rotate the paper to any position you need. For example, it may be easier for you to draw horizontal lines than vertical lines. Just turn your paper or clipboard 90°, and you will be able to draw the vertical lines horizontally. Of course, ideas can be sketched even on paper bags or napkins when a thought comes to mind (see Figure 4-5).

Sketching Lines

Good sketching begins with good pencil control. Hold the pencil so it becomes an extension of your index finger as shown in Figure 4-17. Hold the pencil loosely, gripping it only enough to keep the pencil from slipping out of your hand. This will help keep your hand and arm relaxed and give you better control of your drawing.

Learning to sketch objects starts with learning to sketch lines. Sketch very lightly. This will cut down on the need to erase and will save time. Once you have completed your sketch, you can trace the lines you want to keep so they will be darker and stand out (see Figure 4-18).

Sketching quality lines is easy if you follow these simple steps (Figure 4-19):

1. Begin with the pencil point slightly dull, not sharp.
2. Sketch lines in short segments and connect them to create longer lines.
3. Sketch lines from point to point. Place small, light points at the beginning of your line and where you want to end the line.

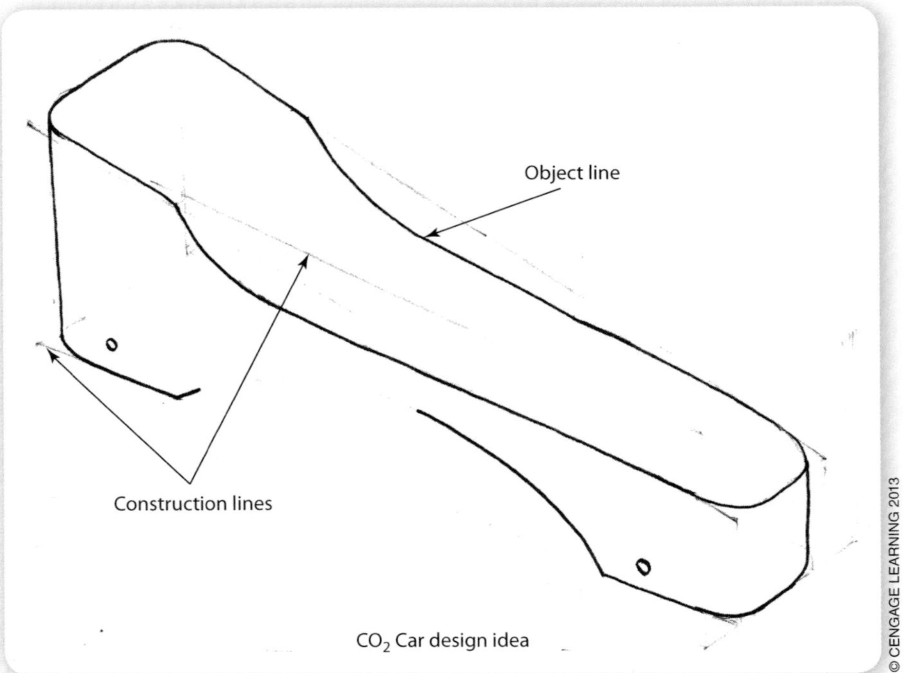

FIGURE 4-18 A freehand sketch laid out with light construction lines and then darkened with heavier object lines.

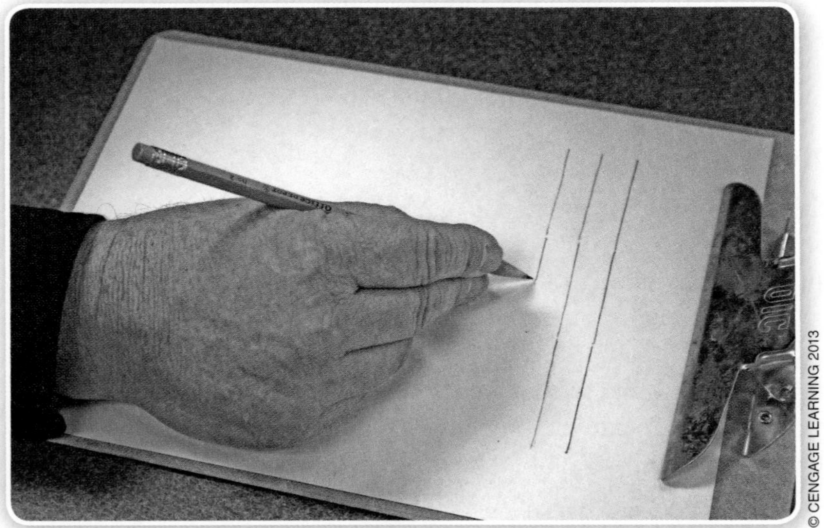

FIGURE 4-19 Begin sketching a line by placing a start point and an end point on your paper. Keeping your eyes on the end point, move your pencil from the start point to the end point.

4. Place your pencil point on the beginning point and keep your eyes on the ending point—do not look back!
5. Keeping your eyes on the end point, move your pencil point to it. As you move your hand, keep it off the paper.
6. When you move your hand to sketch the line, keep it parallel to the line you are drawing. Remember, keep your eyes on the point you are drawing to and do not watch your pencil move.

Sketching Arcs and Circles

Sketching quality arcs and circles is easy if you follow these steps (see Figure 4-20):

1. Begin with the pencil point slightly dull, not sharp.
2. Sketch two lines the same length perpendicular to each other (see Figure 4-20, photo 1), then sketch two lines 45 degrees through the intersection of the perpendicular lines.
3. Mark off the length of the circle's radius from the point of intersection on all lines (the radius will be one-half the diameter of the eventual circle). For this example, that will be one-half inch, making the diameter of the circle 1 inch (see Figure 4-20, photo 4).
4. Placing your pencil point on one radius mark, sketch an arc to the next radius mark. Remember that any point on the arc will be a distance from the center that is equal to the radius (see Figure 4-20, photo 5).
5. Repeat step 4 until you have completed all eight arcs to sketch the circle. Remember, you can rotate your paper to make it easier for you to sketch (see Figure 4-20, photo 6).

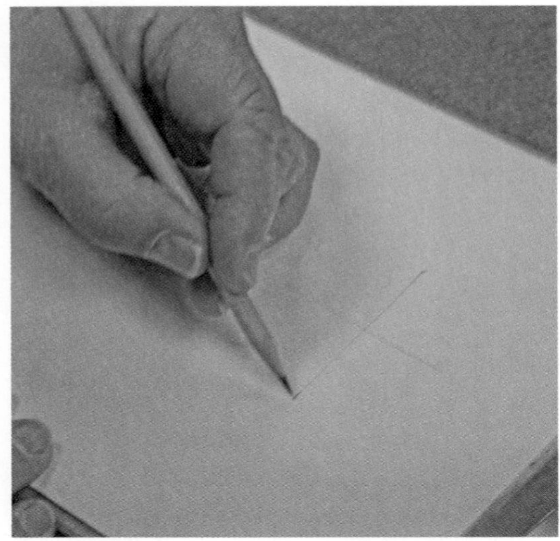

1. Sketch two lines perpendicular to each other

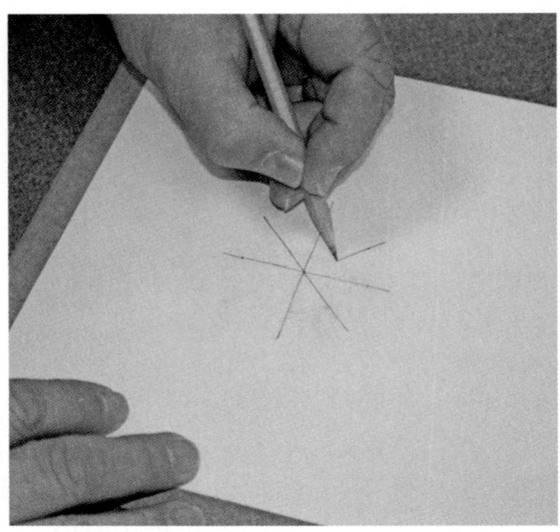

2. Sketch two lines at 45° to the perpendicular lines

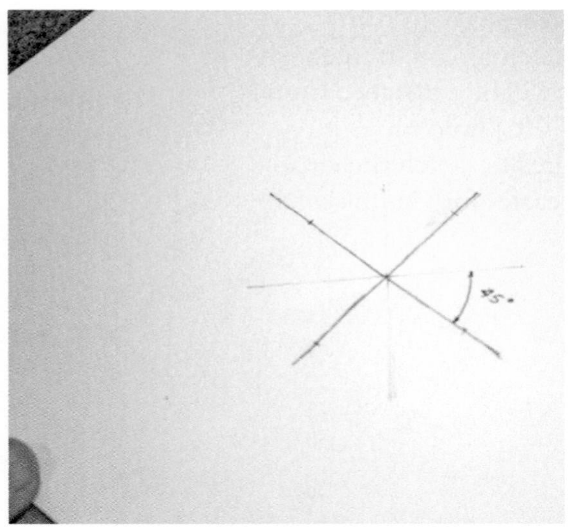

3. 45° lines to the perpendicular lines

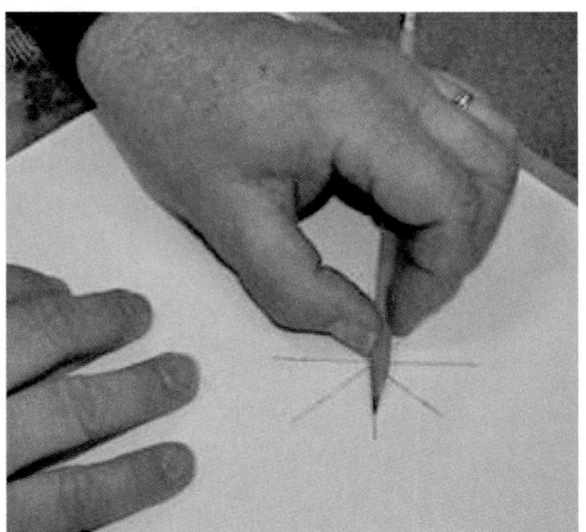

4. Mark of radius on each line from the center point using the point of the pencil and your thumb as a ruler

FIGURE 4-20 Steps to sketch circles and arcs.

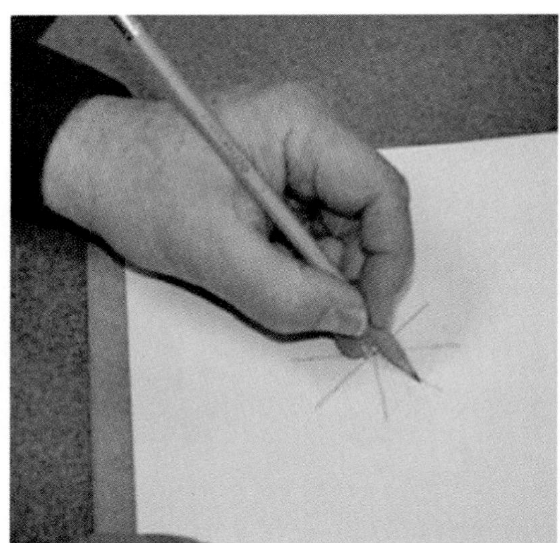

5. Using your wrist as a pivot point, sketch an arc from one radius mark to the next

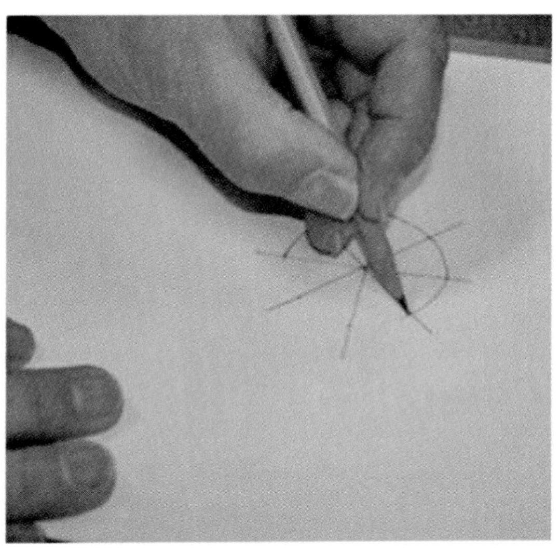

6. Continue sketching each arc

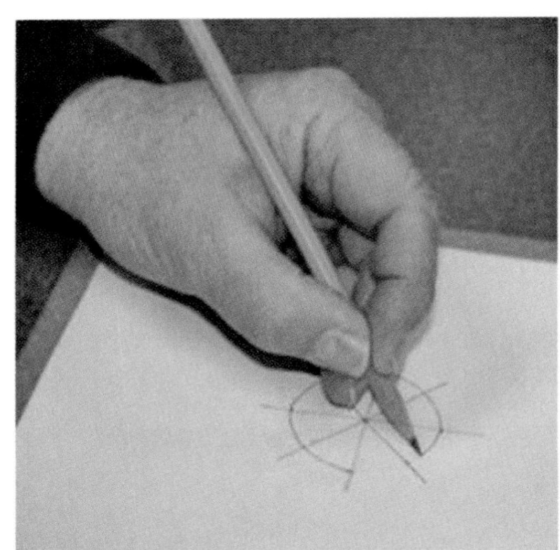

7. Rotate your paper to complete all four arcs to complete the circle

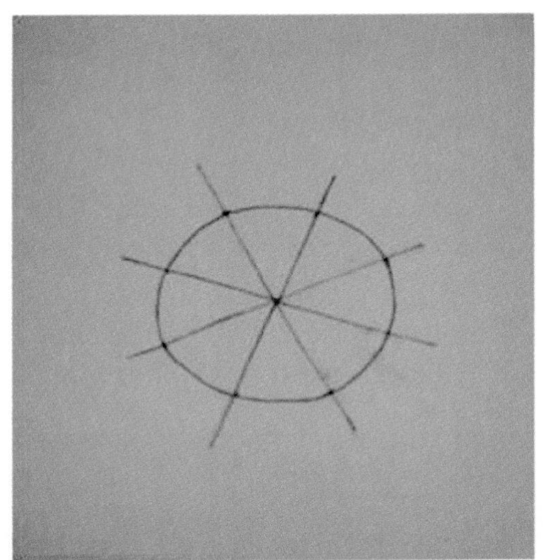

8. Complete circle

FIGURE 4-20 *(continued)*

SUMMARY

In this chapter you learned:

- Freehand technical sketching is the process of drawing images without the use of drafting tools or a CAD system.
- Freehand technical sketching is often used to brainstorm ideas, document measurements, record procedures, and begin the design process for a new product.
- Freehand technical sketching can be used to quickly and accurately communicate ideas and concepts.
- Freehand technical sketching is an important skill used by engineers, designers, drafters, CAD operators, machine operators, and people in the construction trades.
- There are three types of freehand technical sketching: thumbnail, detailed, and presentation.
- Engineers and designers must be able to visualize objects before they begin to sketch.
- Engineers and designers must understand the concepts of proportion and scale to be successful at sketching.
- Engineers must be able to communicate graphically using the alphabet of lines.

VOCABULARY

Write a definition for each term in your own words. After you have finished, compare your answers with the definitions provided in this chapter.

Freehand technical sketching
Annotations
Thumbnail sketch
Detailed sketch
Presentation sketch
Rendering
Visualizing
Proportion
Scale
International System (SI)
Alphabet of lines
Geometry

STRETCH YOUR KNOWLEDGE

Provide thoughtful, written responses to the following questions and assignments.

1. Why is it important for engineers and designers to understand geometry and geometric terms when sketching product ideas?
2. Discuss the advantages of reverse engineering.
3. What possible problems may occur if an engineer's freehand sketch is cluttered and unorganized?

Use sketch paper to complete the following exercises as assigned by your instructor:

4. Using correct freehand technical sketching techniques, practice sketching horizontal lines. See Figure 4-21 for a practice example.

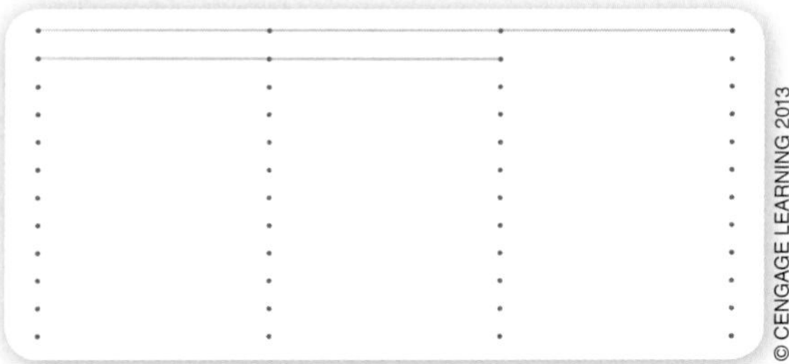

FIGURE 4-21 Sketch horizontal lines from point to point using correct freehand technical sketching techniques.

5. Using correct freehand technical sketching techniques, practice sketching inclined lines. See Figure 4-22 for a practice example.

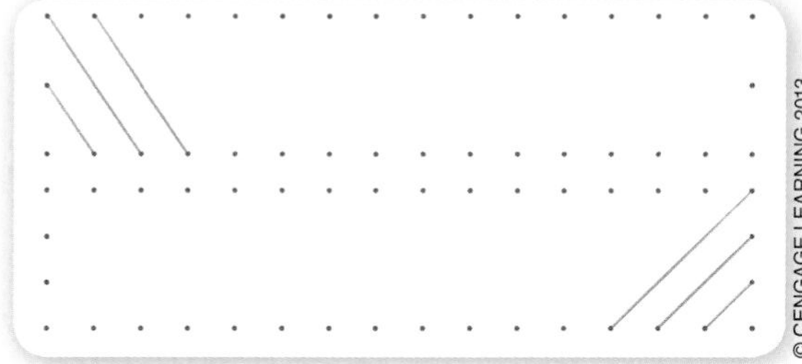

FIGURE 4-22 Sketch inclined lines from point to point using correct freehand technical sketching techniques.

6. Using correct freehand technical sketching techniques, practice sketching vertical lines. See Figure 4-23 for a practice example.

FIGURE 4-23 Sketch vertical lines from point to point using correct freehand technical sketching techniques.

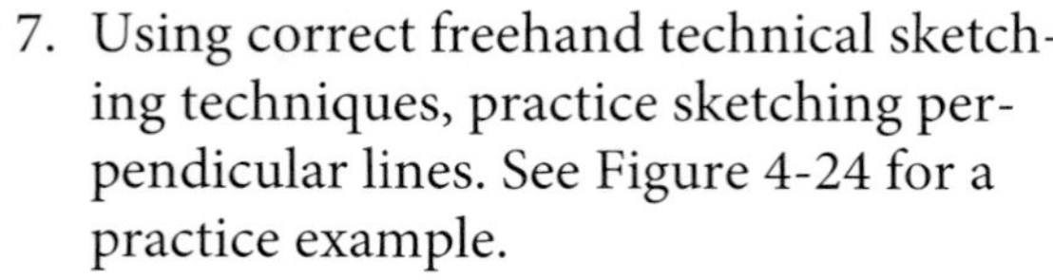

7. Using correct freehand technical sketching techniques, practice sketching perpendicular lines. See Figure 4-24 for a practice example.

FIGURE 4-24 Sketch perpendicular lines from point to point.

8. Using correct freehand technical sketching techniques, practice sketching circles and arcs. See Figure 4-25 for a practice example.

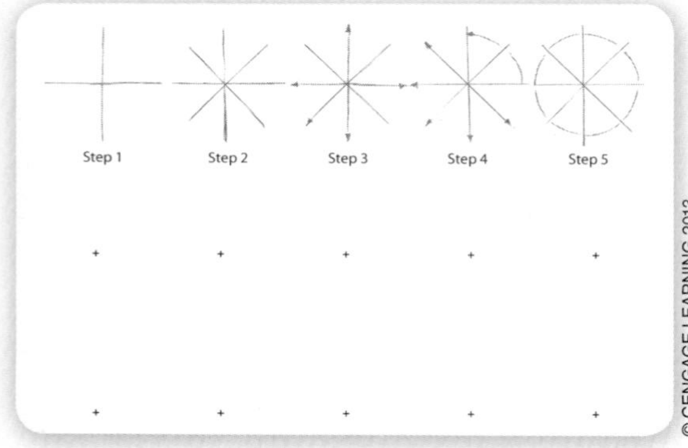

FIGURE 4-25 Steps for sketching circles.

9. Using correct freehand technical sketching techniques, practice sketching squares. See Figure 4-26 for a practice example.

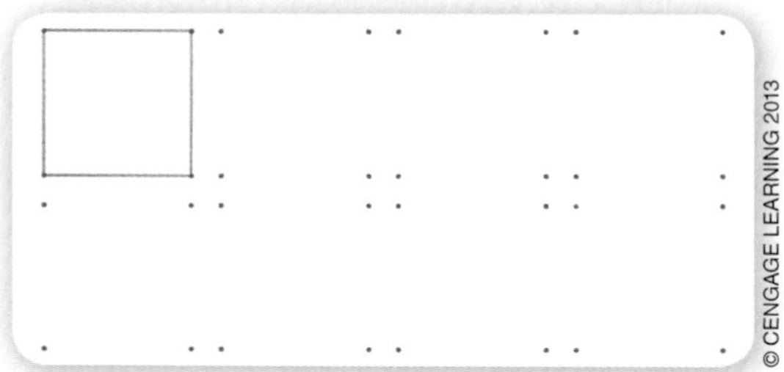

FIGURE 4-26 Sketch squares in the first two rows using the dots as points for each square. Sketch four squares in the third row using the line as a base for each square.

10. Using correct freehand technical sketching techniques, practice sketching rectangles and triangles. See Figure 4-27 for a practice example.

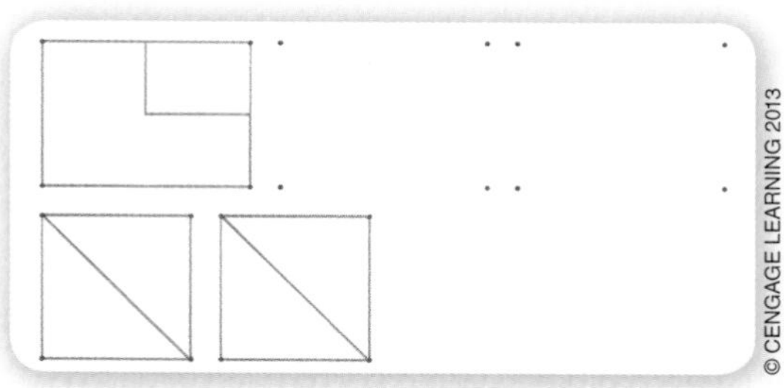

FIGURE 4-27 Copy the rectangular shape with a notch two times in the first row. Copy the two triangles in the second row.

Engineering Challenge

Engineers who design the products we use every day begin with ideas or concepts in their minds. Engineers often must work through complicated mechanical system concepts to make sure their designs will work correctly. Freehand technical sketching allows an engineer to visually record ideas and work out preliminary designs. This requires the engineer to have basic sketching skills and a good understanding of geometric shapes.

A gear box—a box that contains a series of connecting gears—is one example of a mechanical system that an engineer might have to design. This box must be of a certain maximum size and must contain a certain number of connecting gears. Assume you are an engineer and must solve the following design challenge. Create a freehand technical sketch of your solution using the skills and knowledge you have gained while studying this chapter.

Sketch your solution in a two-dimensional drawing using the following criteria:

1. The gear box must be 4 inches wide by 3.5 inches tall (4″ × 3.5″).
2. The gear box must contain four gears with the following dimensions.
   a. gear 1: 2″ diameter
   b. gear 2: 0.5″ diameter
   c. gear 3: 1″ diameter
   d. gear 4: 0.75″ diameter
3. Use circles with center points to represent the gears.
4. Each gear can touch only one other gear.
5. There must be at least 0.5″ clearance (space) between the edge of any gear and the sides of the box.
6. The center points of the gears must not be in a single straight line.
7. Inside each circle representing a gear, label the gears as GEAR 1, GEAR 2, GEAR 3, and GEAR 4.

Remember to use your best freehand technical sketching.

Bonus

Gears turn in a *clockwise* (↻) or *counterclockwise* (↺) direction. When a gear turns in one direction, the gear it touches turns in the opposite direction. So, if a gear turns clockwise, the gear it touches turns counterclockwise. If you set gear 1 to turn clockwise, in which direction will gear 4 turn? Place arrows on each circle (gear) to indicate which direction it will turn if gear 1 turns clockwise.

Onward to Next Destination

CHAPTER 5
Pictorial Sketching

Menu

Before You Begin

Think about these questions as you study the concepts in this chapter:

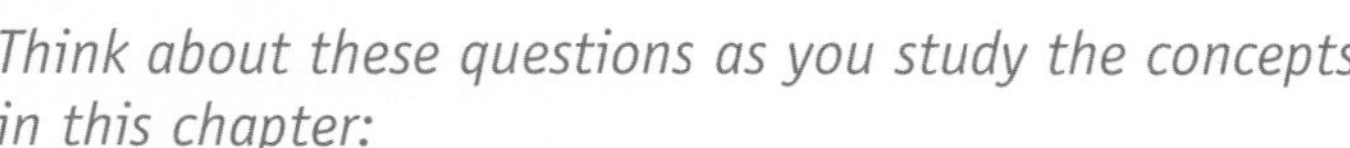

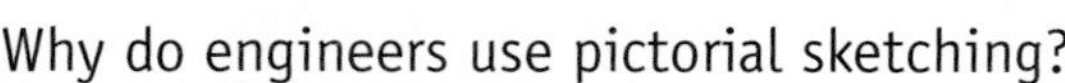

1. Why do engineers use pictorial sketching?
2. How do we perceive objects in our world?
3. What is pictorial sketching?
4. What is axonometric drawing?
5. What is isometric drawing?
6. What is perspective drawing?
7. What are presentation drawings?

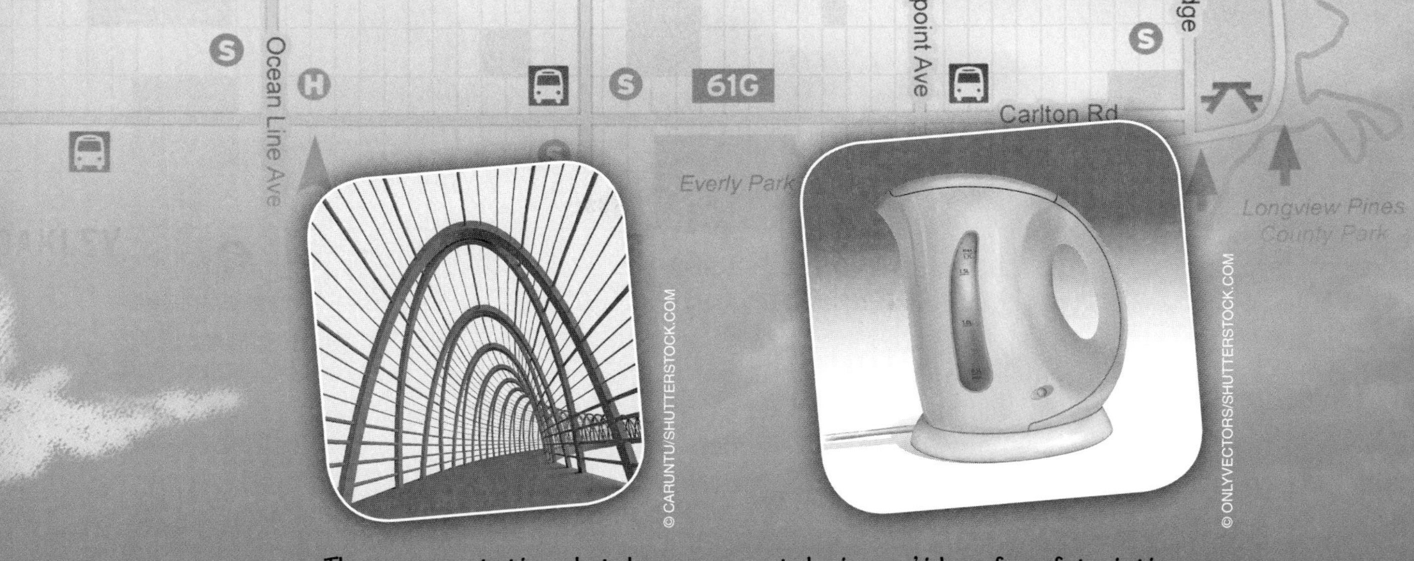

These presentation sketches represent designers' ideas for a futuristic bridge and an electric tea kettle.

Engineering in Action

Engineers often make presentations, both formal and informal, to explain or defend their design concepts or projects. They might present plans to a client who is paying them to design a new bridge. Or they might present designs to a company committee that is developing a new appliance. Presentations like these usually include presentation sketches. These important documents, sometimes called *renderings*, may be prepared by the engineer or developed by a technical illustrator.

1 As we saw in Chapter 4, presentation sketches are very important in the design process. Imagine how difficult it would be for the engineers to explain ideas using only words. Clients or committee members would have a hard time visualizing the engineer's ideas. Presentation sketches help the clients or committee members "see" the engineer's ideas and plans.

SECTION 1: WHY DO ENGINEERS AND DESIGNERS USE PICTORIAL SKETCHING?

In Chapter 4, you learned and practiced the techniques and processes of developing freehand sketches. You also learned that many people, including engineers and designers, use freehand sketching to communicate ideas quickly and accurately. In this chapter, you will put this knowledge and skill to work by learning to sketch objects in *three dimensions*. You will also learn how engineers and designers can represent objects realistically by using *pictorial sketching* and rendering techniques (see Figure 5-1).

How We See the World

When you look at the world around you, do objects appear to be flat? Can you tell that objects have depth or that some objects are close to you and others are farther away? Of course you can. As humans, we see the world around us in **three dimensions**, which is often just called **3-D** (see Figure 5-2). This is the natural way our brains interpret the objects we see. The three dimensions we see are called *width*, *height*, and *depth* (see Figure 5-3).

three-dimensional (3-D) sketching
In three-dimensional sketching, sketched objects show width, height, and depth.

Three-dimensional sketching is the quickest way to convey ideas to others. It also is the least likely method to create misunderstanding. Remember from Chapter 4: clear communication is most important to the design process.

What Is Pictorial Sketching?

pictorial sketching
Pictorial sketching is the process of graphically representing objects in three-dimensional form.

Pictorial sketching is the process of graphically representing objects in 3-D form. The opposite of this is graphically representing objects in *two-dimensional* (2-D) form; these are often referred to as *multiview drawings.* Two-dimensional drawings show only one side of an object and are used by engineers to accurately communicate how an object should be made

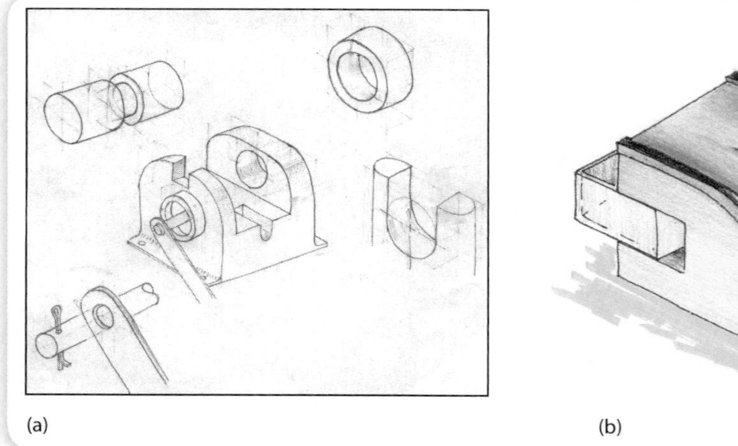

FIGURE 5-1 Engineer's pictorial sketches representing parts to a machine.

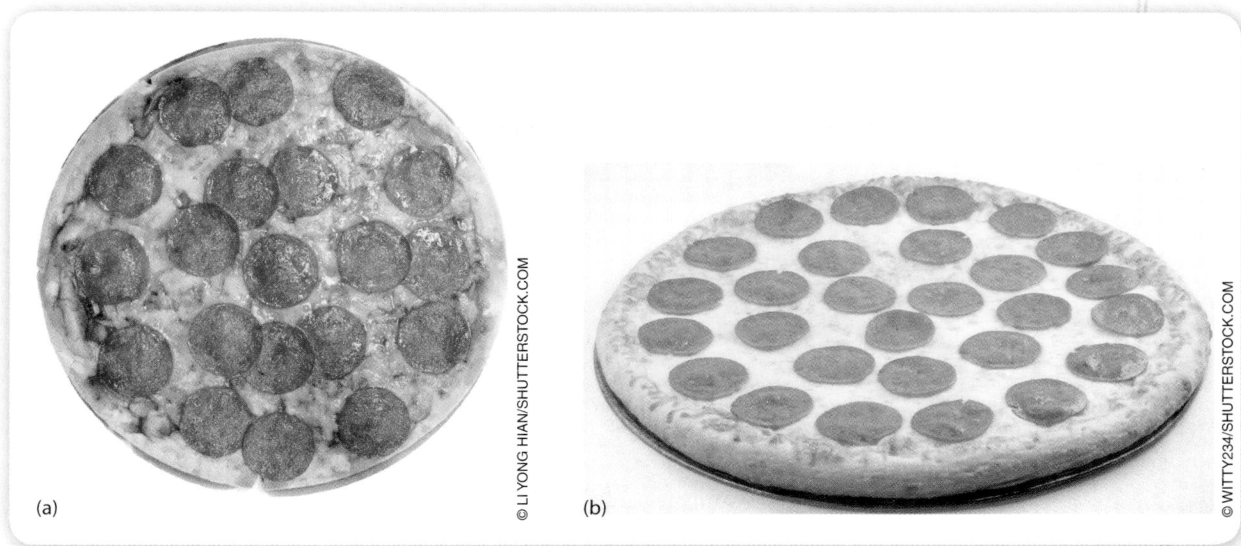

FIGURE 5-2 (a) A 2-D view of a pizza. Notice you cannot see depth in the pepperoni slices. (b) A 3-D view of a pizza. Notice the depth of the pepperoni slices.

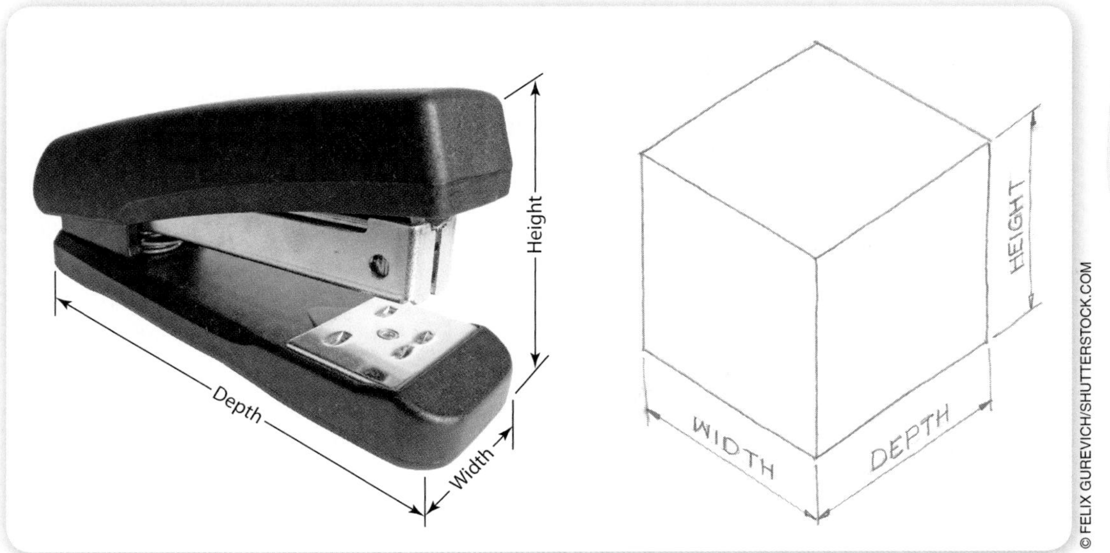

FIGURE 5-3 The objects around us have width, height, and depth. When we sketch in three dimensions, we also show width, height, and depth.

(see Figure 5-4). There is also a mathematical difference between 2-D and 3-D. The concept of *area*, measured in square units (square inch, square foot, square centimeter), is a 2-D mathematical relationship. The concept of *volume*, measured in cubic units (cubic inch, cubic foot, cubic centimeter), is a 3-D mathematical relationship. Take a look at the 3-D object in Figure 5-4a. It has the appearance of a real object sitting on your desk. You can see the width,

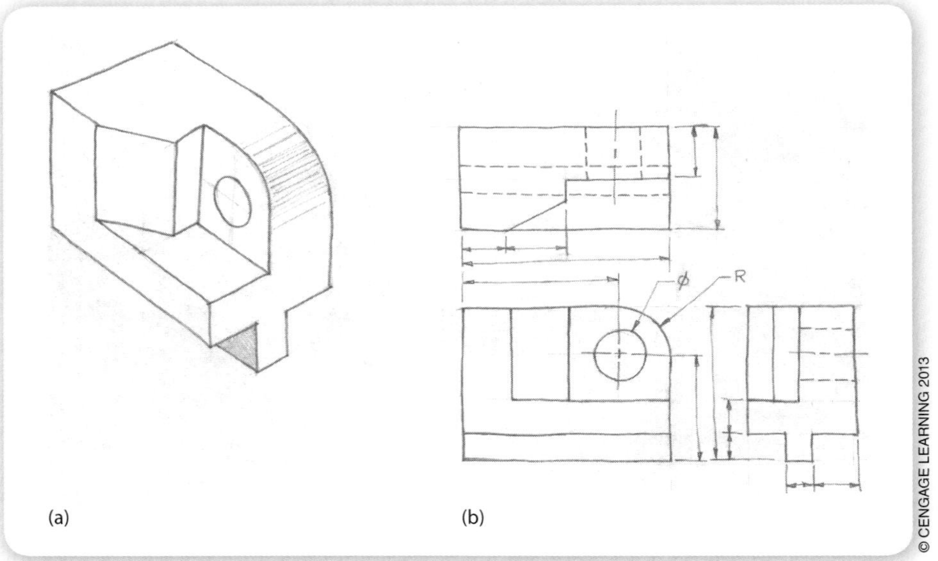

FIGURE 5-4 (a) A 3-D pictorial sketch of an object; (b) a 2-D multiview sketch of an object.

height, and depth of the object in the single sketch. Now, take a look at the 2-D sketch in Figure 5-4b. The way this sketch is drawn, you can see only three faces (sides) per view. The most common views used are the front, top, and right side views. This type of 2-D drawing is called orthographic projection.

plane
A plane is a geometric flat surface.

The Six Sides of 3-D Objects No matter the type of pictorial sketch, every object is defined by six sides or views (also referred to as **planes**). This allows the viewer to understand the orientation and the relations of each side to the other sides. The six planes are front, right side, left side, top, back (or rear), and bottom. The front view is the anchor view; all other views are established by identifying the front view first. But any pictorial sketch, no matter what the type, will only show three sides. Figure 5-5a demonstrates the relationship of each view (side) to the front view. The relationship of views also helps demonstrate the development of 2-D drawings, which show only two dimensions (see Figure 5-5b).

Did You Know?

Many computer and video games have their beginning in axonometric, oblique, and perspective drawing concepts. Graphic designers need to have a solid understanding of these pictorial concepts.

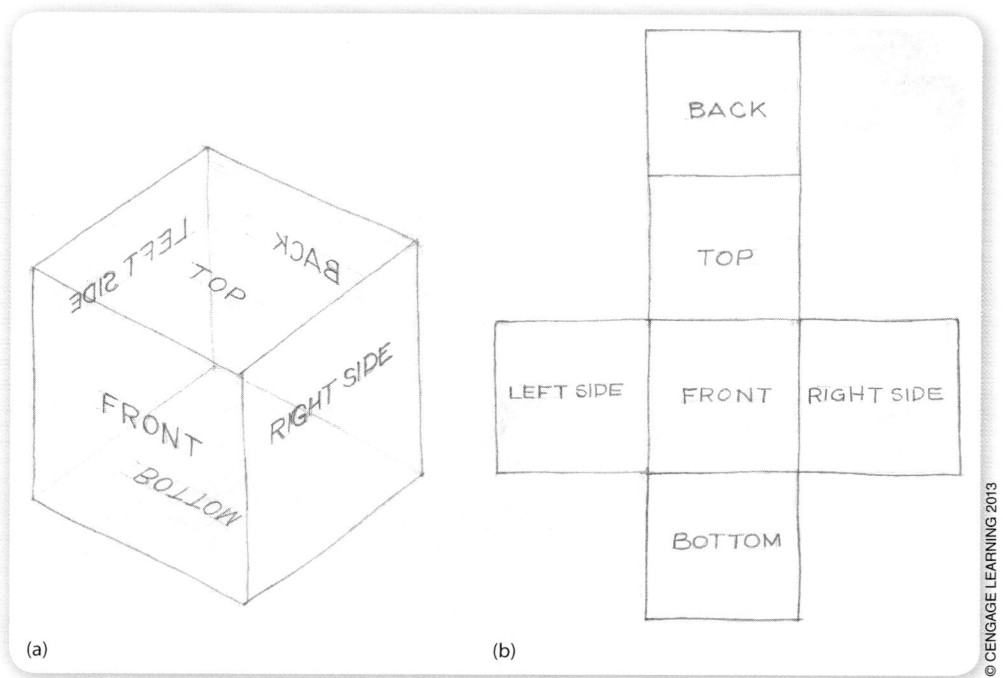

FIGURE 5-5 (a) No matter how complex its shape, any object has six sides. (b) The 2-D representation of any object begins with an unfolded box, with each view in the order shown here.

Types of Pictorial Sketches

Pictorial representation is divided into three groups. Two of these groups are demonstrated in this chapter. The three types of pictorial sketches are *axonometric* (meaning to measure along the axis), *oblique*, and *perspective*. Of the axonometric family of drawings, **isometric sketching** is the one used here (see Figure 5-6).

isometric sketching

Isometric sketching is a method of representing three-dimensional objects with three axes—*x*, *y*, and *z*. The *x*- and *y*-axes are always drawn 120° apart.

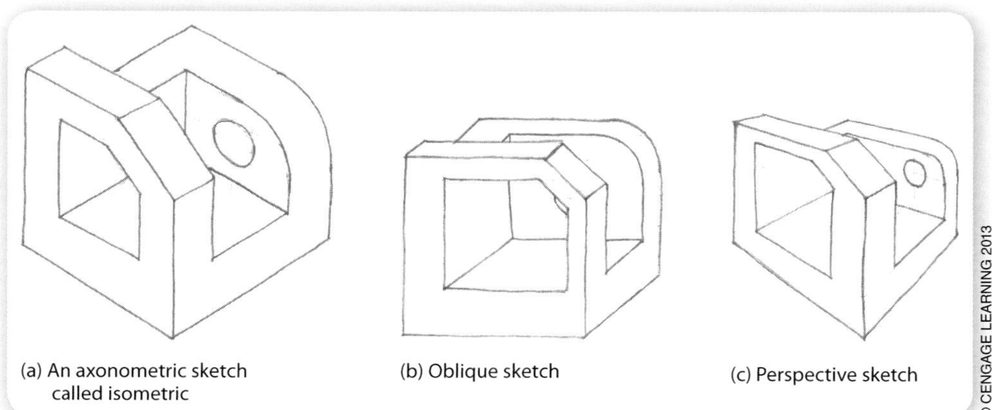

FIGURE 5-6 Types of pictorial sketches.

Isometric Sketching

Isometric drawing is a method of representing 3-D objects with three axes—*x*, *y*, and *z*. The *x*- and *y*-axes are always drawn 120° apart. These three axes relate to the Cartesian coordinate system you may have already studied in your math courses. The *x*-axis represents the width of the object. The *y*-axis represents the height of the object and the *z*-axis represents the depth of the object. The Cartesian coordinate system is discussed in more detail in Chapter 7. Figure 5-7 illustrates this method. Isometric shapes begin with lines projecting from the top of a vertical line, up and to the left and right. Each line is separated by 120° equally. The base lines of the front and side views of the object form 30° angles to the horizontal. Measurements are made along each of these lines to determine width, height, and depth.

Sketching an Isometric Cube An easy way to learn to sketch in isometric is to learn to sketch a cube in isometric. Remember, a cube is a three-dimensional shape that has squares on all its faces. Figure 5-8 illustrates the steps you should take to sketch the three-unit cube (3 × 3 × 3). In Chapter 4, you learned about the geometric term *parallel*. Now is the time to put this term to work. When you sketch, make sure all opposite edges are parallel.

The sketched isometric cube is formed by three isometric square planes.

Sketching an Isometric Rectangular Solid Sketching a rectangular solid in isometric is very similar to sketching a cube. In fact, a cube is a special type of rectangular solid (remember, all its faces are square). A rectangular prism has six rectangular sides. However, two sides opposite each other can be square sides. Study Figures 5-8 and 5-9 to see the difference between the cube and the rectangular solid. Notice that the steps are the same in developing both shapes.

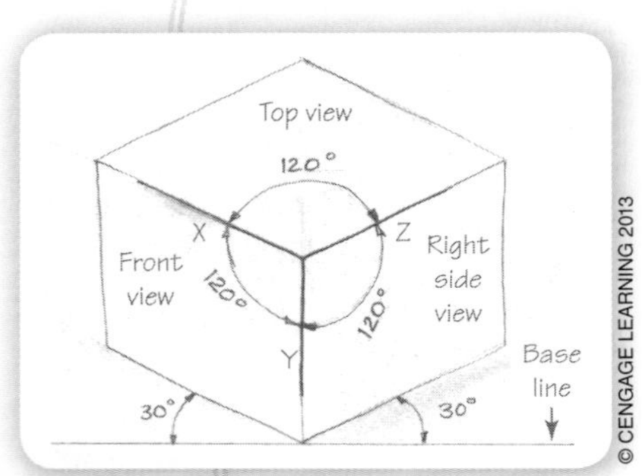

FIGURE 5-7 All isometric shapes begin with three axis lines, which are often referred to as the *x*, *y*, and *z* axes, separated by 120°. Notice the base of the object is sketched at 30° to the left and to the right along the horizontal.

Sketching a Rectangular Solid with Parts Removed or Added All objects begin as rectangular solids, even circular shapes. Simple or complex shapes begin here. The process of sketching a complex shape begins by sketching a simple shape and either "subtracting" or "adding" material to the shape. Think of the process as cutting out pieces of the object with a tool such as a saw. Begin with the larger cuts and then proceed to the smaller cuts. Follow the steps in Figure 5-10 as the process moves from simple shape to complex shape.

Sketching a Rectangular Solid with Inclined (Slanted) Surfaces We know the world is not made up of rectangular solid shapes—that is, not all sides are perpendicular (at right

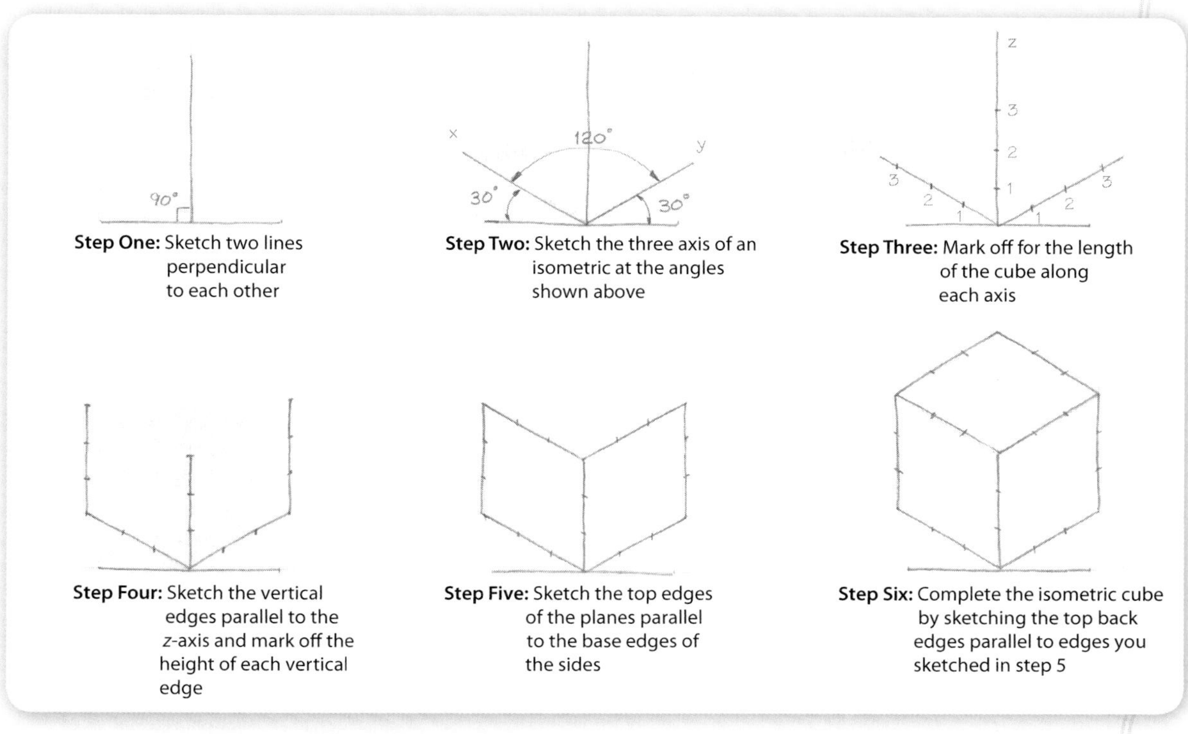

FIGURE 5-8 Steps in developing an isometric cube.

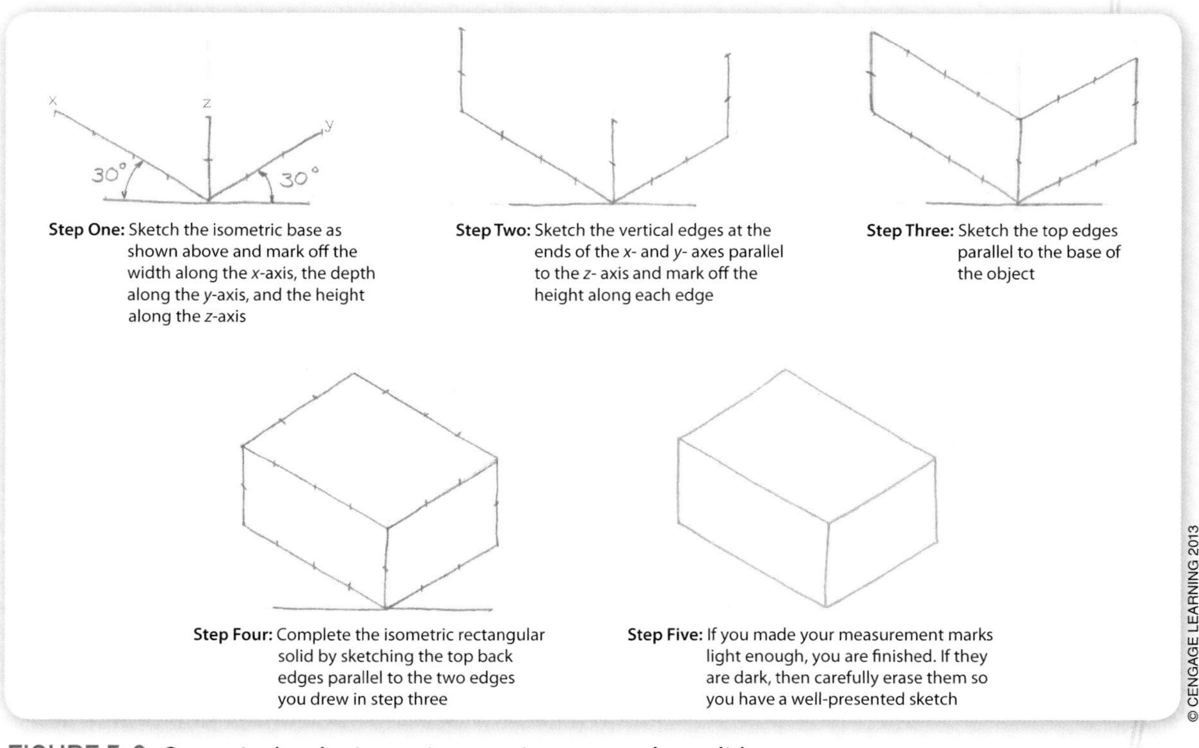

FIGURE 5-9 Steps in developing an isometric rectangular solid.

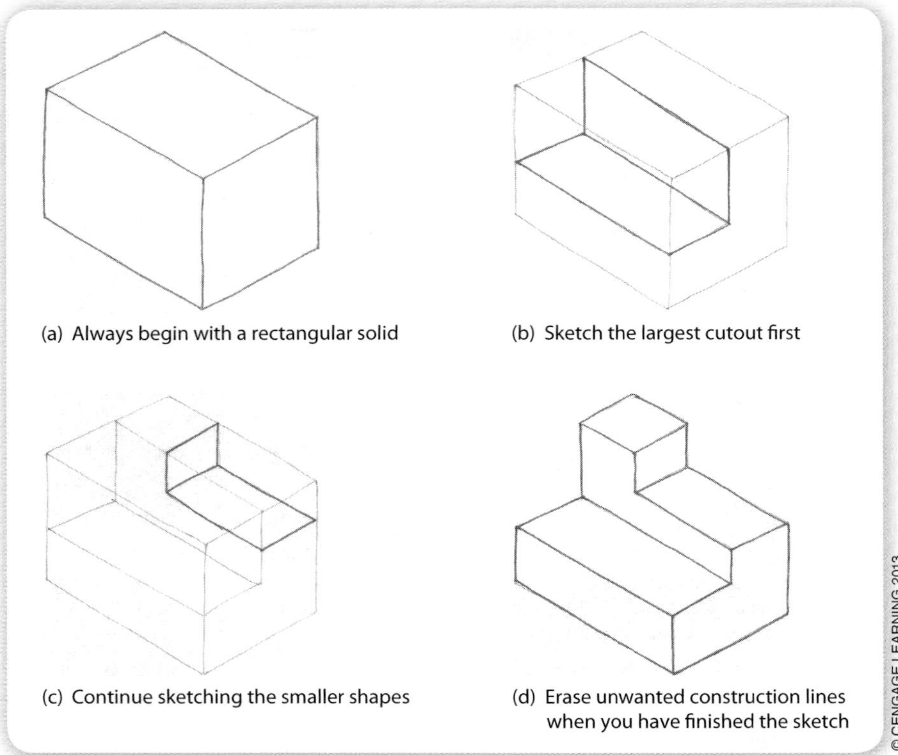

(a) Always begin with a rectangular solid
(b) Sketch the largest cutout first
(c) Continue sketching the smaller shapes
(d) Erase unwanted construction lines when you have finished the sketch
© CENGAGE LEARNING 2013

FIGURE 5-10 *Developing simple rectangular shapes.*

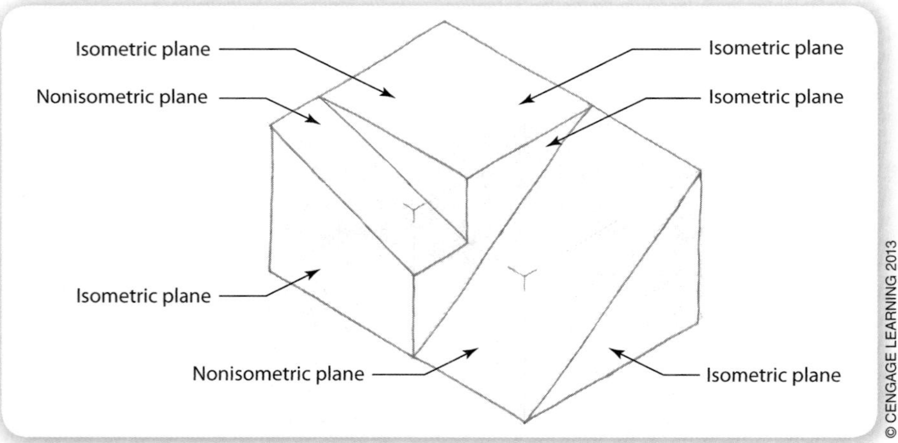

FIGURE 5-11 *Isometric and nonisometric planes.*

nonisometric planes
Nonisometric planes are inclined surfaces that do not lie within an isometric plane.

angles). Look around you. You will see objects that have curved and inclined surfaces. In isometric sketching, surfaces that are inclined are called **nonisometric planes** (see Figure 5-11).

Inclined planes can be sketched by locating and connecting points at the corners of the inclined plane. Follow the steps shown in Figure 5-12 and observe how the inclined plane is developed.

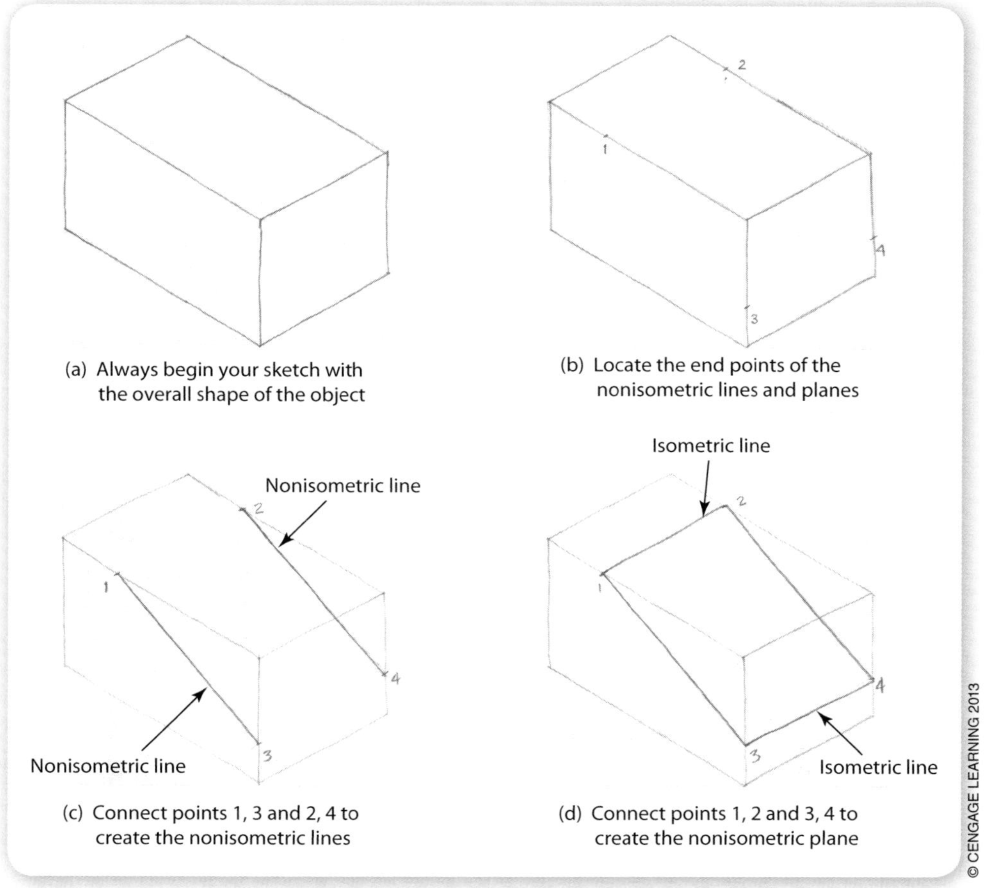

2
1
4
3
(a) Always begin your sketch with the overall shape of the object
(b) Locate the end points of the nonisometric lines and planes
Nonisometric line
Isometric line
Nonisometric line
Isometric line
(c) Connect points 1, 3 and 2, 4 to create the nonisometric lines
(d) Connect points 1, 2 and 3, 4 to create the nonisometric plane

FIGURE 5-12 **Development of nonisometric lines and planes (also called inclined lines and planes).**

Sketching Isometric Circles, Arcs, and Cylinders Circles and arcs look like **ellipses** when they are sketched in isometric. Simply put, an ellipse is a "flattened" circle that is developed with two equal large arcs and two equal smaller arcs. The easiest way to begin an isometric circle or arc is to begin with an isometric square. The circle or arc will appear slightly different in each view (front, top, and side) of the isometric shape. Figure 5-13 illustrates the steps for sketching each isometric circle. An arc, as you may recall from Chapter 4, is a segment or part of a circle. So, when you learn to sketch isometric circles, you also learn to sketch isometric arcs.

ellipse
An ellipse is a flattened circle.

Sketching isometric cylinders is the natural next step once you have learned to sketch isometric circles. Cylinders are solids with the ends shaped like circles. When you sketch the two opposite isometric circles, all you will need to do is connect the circles with two lines (the edges of the cylinder) tangent to the circles (remember, *tangent line* was defined in Chapter 4). Examine Figure 5-14 to see the steps you need to take to sketch isometric cylinders from the front, side, and top views.

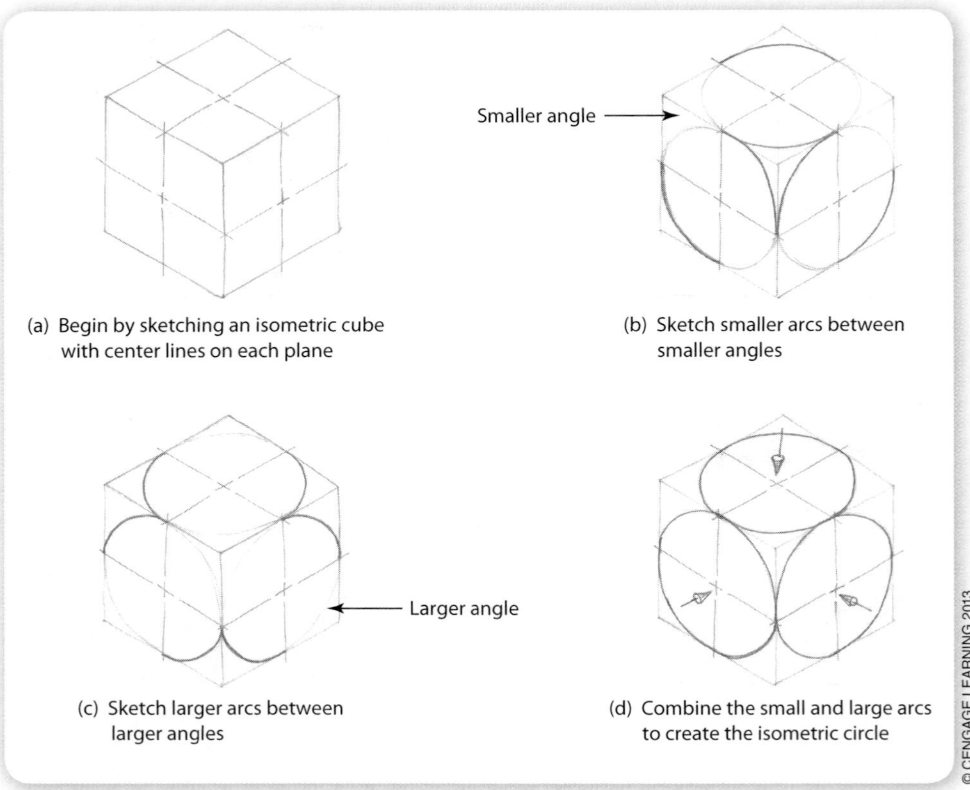

Smaller angle
(a) Begin by sketching an isometric cube with center lines on each plane
(b) Sketch smaller arcs between smaller angles
Larger angle
(c) Sketch larger arcs between larger angles
(d) Combine the small and large arcs to create the isometric circle

FIGURE 5-13 **Developing isometric circles and arcs.**

Sketching Isometric Holes Isometric holes can be thought of as negative cylinders. You can sketch an isometric hole in the same way you sketch an isometric cylinder. Instead of the cylinder being of solid material, it forms empty space. Figure 5-15 demonstrates how to sketch an isometric hole. If the length (also called *depth*) of the hole is less than the diameter of the circle, then you will be able to see part of the other end of the hole. However, if the depth is longer (deeper) than the diameter, you will not be able to see the end of the hole.

Sketching Complex Objects in Isometric So far you have learned to sketch simple isometric shapes such as rectangular solids, circles, arcs, and cylinders. You also learned to sketch inclined planes, or planes that are not perpendicular to the six normal planes (front, right side, top, left side, back, and bottom). Now you will take what you have learned and combine shapes to create recognizable objects.

All objects can be divided into simple isometric shapes. The best way to do this is visualize each part of the object as a rectangular solid. Once you have accomplished this, you can begin to add curved shapes such as circles, arcs, or cylinders. Figure 5-16 shows you a step-by-step process.

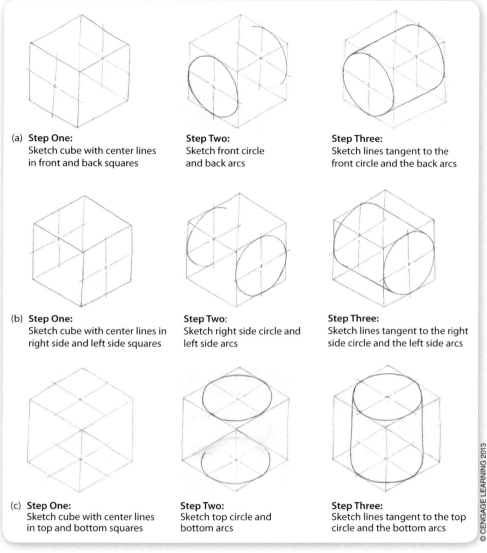

(a) **Step One:**
Sketch cube with center lines in front and back squares

Step Two:
Sketch front circle and back arcs

Step Three:
Sketch lines tangent to the front circle and the back arcs

(b) **Step One:**
Sketch cube with center lines in right side and left side squares

Step Two:
Sketch right side circle and left side arcs

Step Three:
Sketch lines tangent to the right side circle and the left side arcs

(c) **Step One:**
Sketch cube with center lines in top and bottom squares

Step Two:
Sketch top circle and bottom arcs

Step Three:
Sketch lines tangent to the top circle and the bottom arcs

FIGURE 5-14 (a) A front-facing cylinder; (b) a right-side-facing cylinder; (c) a top-facing cylinder.

perspective drawing or sketching
In perspective drawing or sketching, objects appear to become shorter and closer together as they move back and away from the viewer.

Perspective Drawing

Isometric pictorials provide a good representation of objects and are used often by engineers and designer, but perspective pictorials are the most realistic representations of objects. **Perspective drawing** or **perspective sketching** is used primarily in architecture and also in presentation drawings by designers and engineers (see Figure 5-17).

Concepts of Perspective Sketching When you look down your street or down a long straight highway, you should notice two things right away. First,

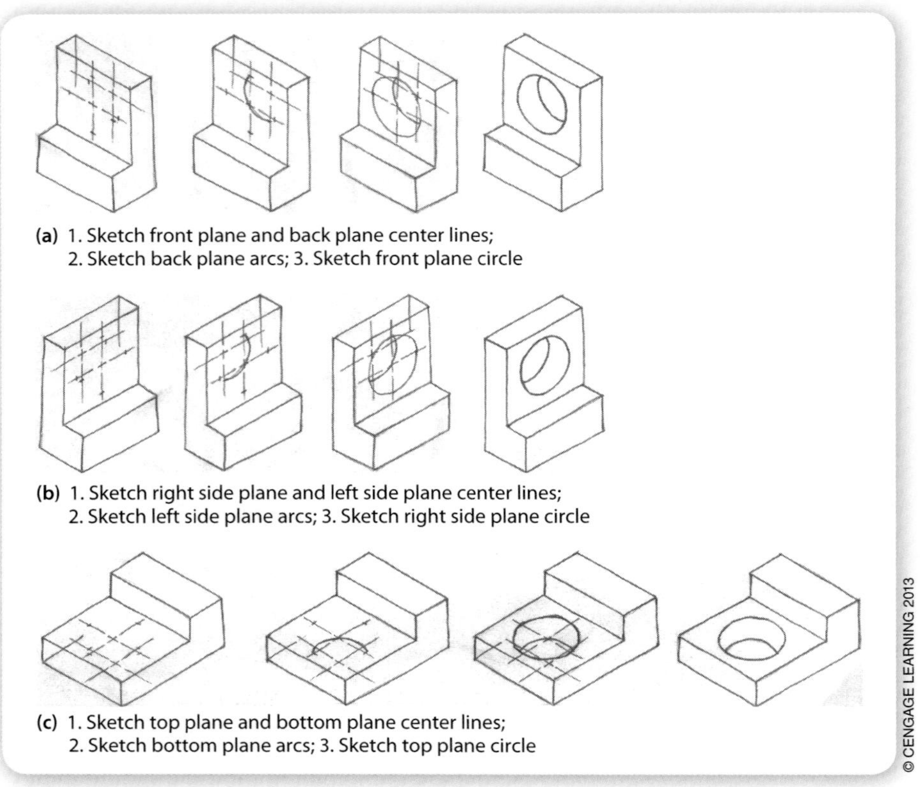

(a) 1. Sketch front plane and back plane center lines;
2. Sketch back plane arcs; 3. Sketch front plane circle
(b) 1. Sketch right side plane and left side plane center lines;
2. Sketch left side plane arcs; 3. Sketch right side plane circle
(c) 1. Sketch top plane and bottom plane center lines;
2. Sketch bottom plane arcs; 3. Sketch top plane circle

FIGURE 5-15 Sketching isometric holes. (a) Front-view-facing hole; (b) right-side-facing hole; (c) top-view-facing hole.

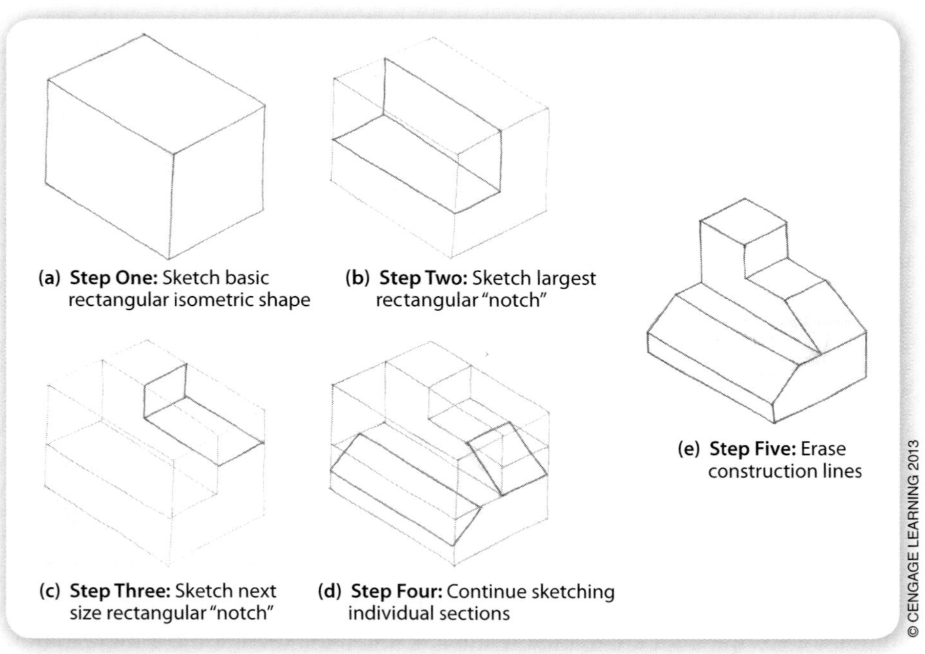

(a) Step One: Sketch basic rectangular isometric shape
(b) Step Two: Sketch largest rectangular "notch"
(c) Step Three: Sketch next size rectangular "notch"
(d) Step Four: Continue sketching individual sections
(e) Step Five: Erase construction lines

FIGURE 5-16 Developing complex isometric shapes.

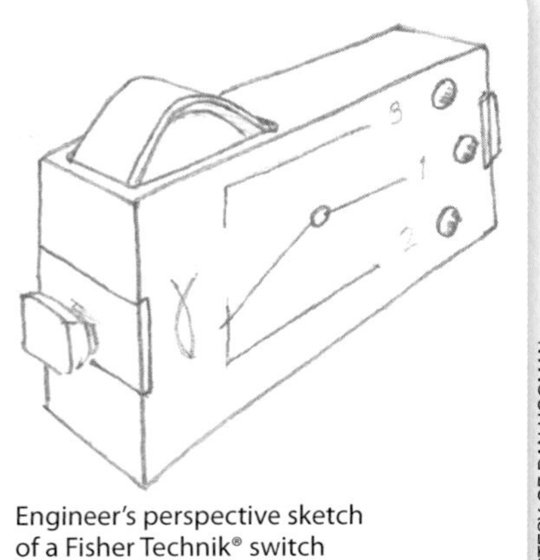

Engineer's perspective sketch of a Fisher Technik® switch

COURTESY OF DAN HOGMAN

FIGURE 5-17 Architects and engineers use perspective sketching.

as the street or highway gets farther away from you, the sides of the road seem to get closer together. Second, telephone and streetlight poles appear to get shorter and closer together. Other objects such as trees and buildings will appear the same way, getting shorter and closer together. This is **perspective**, or the way we see distance and depth in the things in our world (see Figure 5-18).

There are four basic parts to perspective sketches. The first is the *horizon line*, an imaginary line that represents the edge of how far we can see. The second part is the **vanishing point**, a location on the horizon line where

perspective
Perspective is the way we see distance and depth in the things in our world.

vanishing point
The vanishing point is the location on the horizon line where our line of sight merges.

Engineering Challenge

Engineers use quick freehand sketches to record their ideas for solving a design problem. Their sketches are drawings of rough ideas at first but become more detailed with pictorial sketches. Often an engineer will be asked to redesign an existing product to improve its function or appearance. This may require the engineer to make quick proportional sketches of the product to study the existing product and to sketch possible changes.

Your engineering challenge is to find a simple object such as a drinking-water bottle or a thumb drive (also known as a USB flash drive) and develop an isometric sketch. You may measure the product so you can sketch the object proportionally. After you have completed the sketch, redraw two isometric sketches of the product on the same sheet with different suggested improvements. Place notes on each improved product to indicate the changes you have made. Reminder: Be sure to use good sketching techniques when you develop your sketches. Once you have completed the challenge, share your concepts with your classmates.

FIGURE 5-18 Perspective is the way we see the world around us. (a) Notice the hall floor and ceiling become narrower and the lockers get shorter the farther they are from you. (b) Notice how the lines on the road become closer together and the poles get shorter.

our line of sight merges. The third part is the *ground line*, or the position a person is standing when viewing the object. The fourth part is the *true height line*, or the front edge of the object sitting on the ground line. Any measurement on this line will be true size to scale. All other lines will appear progressively smaller as they get closer to the horizon line. Figure 5-19 illustrates the three basic parts to a perspective sketch.

Perspective of the Viewer Perspective sketches are commonly developed in one of two different views. The choice will depend on how you want the viewer to see the object. The school hallway and the country

Science in Engineering

Perspective drawing is the process of representing 3-D objects on a 2-D flat surface. The scientific principles of elementary optics are applied to perspective drawing development.

Optics is the study of light, and it includes how humans see the world around them. Optically, objects appear to be smaller and less defined the farther away they are.

In this interior perspective view, objects farther away appear to be smaller and less defined.

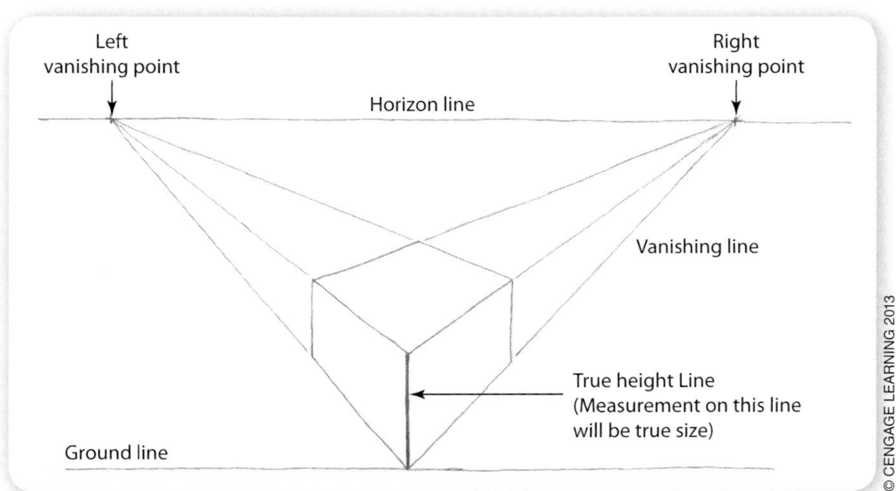

FIGURE 5-19 The basic parts of a perspective sketch.

optics

Optics is the study of light, including how humans see the world around them.

FIGURE 5-20 **This building view shows an example of two-point perspective.**

road shown in Figures 5-18a and 5-18b illustrate *one-point perspective*. If you were to extend all the horizontal lines of the objects, they would meet at one point.

two-point perspective
A two-point perspective view allows the viewer to see two sides of an object "vanishing" to two different points on the horizon line.

The second type of perspective view is the **two-point perspective**. This type of view allows the person viewing the object to see two sides of an object "vanishing" to different points on the horizon line. For example, the front of the object is vanishing to the *left vanishing point* and the right side view is vanishing to the *right vanishing point*. In Figure 5-20, observe that when the lines of the object are projected to the horizon line on the front view, they move to the left vanishing point. On the right side view of the object, the lines move to the right vanishing point.

Sketching in One-Point Perspective One-point perspective is used when you want to look at an object "head-on." Mathematically, we would say we are viewing the object *perpendicular* to our line of sight: if we were to draw a line from our eye to the face of the object, it would form a right angle.

The steps in developing a one-point perspective are simple. Study the steps in Figure 5-21 to understand how to sketch a one-point perspective.

Sketching one-point perspective circles, arcs, and holes is similar to sketching in isometric. Begin by developing a perspective square. Unlike isometric squares, for which all sides are equal, the height of the square

Did You Know?

A simple form of perspective drawing was first introduced in the fifth century B.C. in ancient Greece. In the early 1400s, Filippo Brunellschi, an Italian Renaissance architect, developed the geometric method of perspective drawing. This is the method used by artists and designers today.

Math in Engineering

We use the mathematical concept of proportion in perspective drawing to obtain the optical illusion of objects farther away from us (see Chapter 4). Objects are sketched proportionally smaller and smaller as they appear farther away.

Notice how the columns become shorter and closer together proportionally as they get farther away.

Station point Horizon line

Ground line

Step 1. Lay out the horizon line, station point, and ground line

Station point Horizon line

Ground line

Step 2. Sketch the front view of your object (This will be your perpendicular view)

Station point Horizon line

Vanishing line

Ground line

Step 3. Project the vanishing lines from each corner of the front view to the varnishing point

Station point Horizon line

Ground line

Step 4. Estimate the back side of the object. Each line will be parallel to the front view lines

FIGURE 5-21 Steps to sketch a one-point perspective cube.

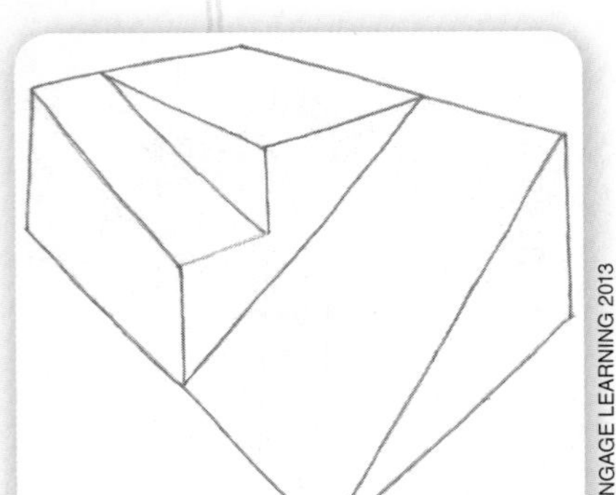

FIGURE 5-22 An engineering sketch in perspective. Compare this sketch with the isometric sketch in Figure 5-11.

is shorter on the back side, and the top and bottom edges are shorter in depth than the front height of the square.

Sketching in Two-Point Perspective Two-point perspective is used when you want to look at an object from a "corner" view. Your point of perspective view will allow you to see two sides of the object vanishing to the left and to the right (see Figure 5-22).

The steps for developing a two-point perspective are simple. Study the steps in Figure 5-23 to understand how to sketch a two-point perspective. Remember, objects or parts of an object such as holes, pins, slots, and braces (called **features**) will appear to be shorter and closer together the closer they get to vanishing points. Look at Figure 5-18 again to observe this perspective concept.

features

Features are objects or parts of an object such as holes, pins, slots, braces.

Sketching two-point perspective circles, arcs, and holes is similar to sketching in one-point perspective. The difference is that the front and side planes vanish to different vanishing points.

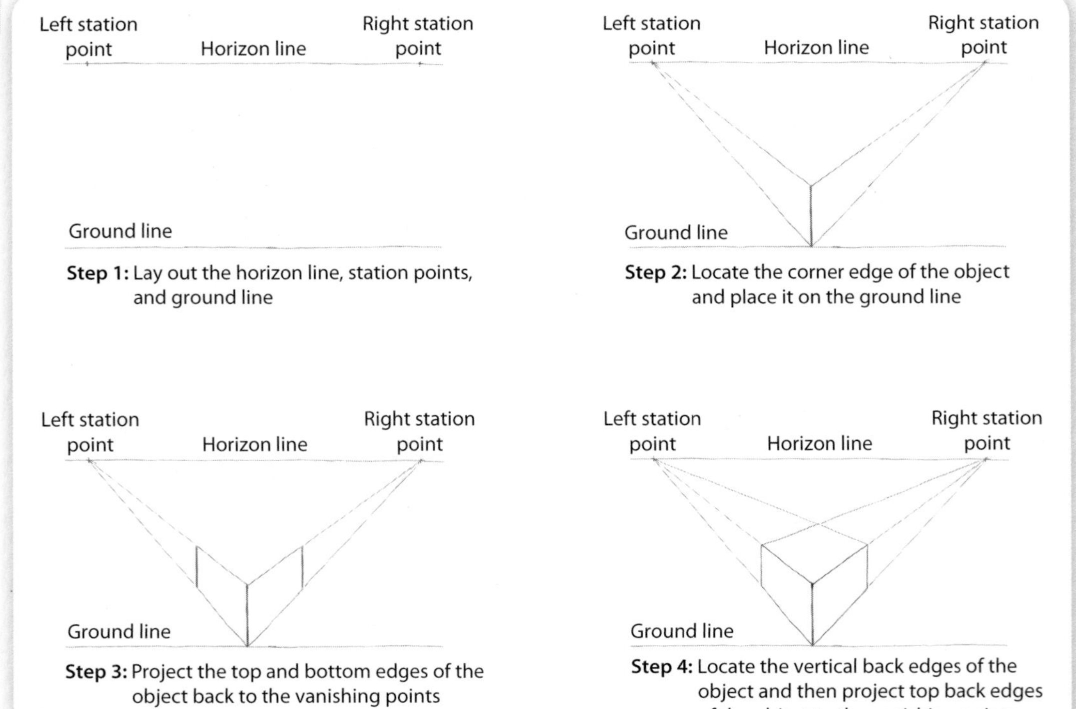
Left station point
Horizon line
Right station point
Ground line
Step 1: Lay out the horizon line, station points, and ground line
Left station point
Horizon line
Right station point
Ground line
Step 2: Locate the corner edge of the object and place it on the ground line
Left station point
Horizon line
Right station point
Ground line
Step 3: Project the top and bottom edges of the object back to the vanishing points
Left station point
Horizon line
Right station point
Ground line
Step 4: Locate the vertical back edges of the object and then project top back edges of the object to the vanishing points

FIGURE 5-23 Steps to sketch a two-point perspective cube.

Career Spotlight

Name:
Shane R. Sevo

Title:
Creative Designer, Superhouse Media

Job Description:
Sevo's job is to show people what a building will look like before it is built. He does this by making pictures on a computer.

First, Sevo gets plans from engineers and architects. The engineers decide on the function of the building, and the architects decide on the way it will look. Sevo combines their ideas on his computer, working with the same software that people use to make video games and movies. "I have to create a picture in a virtual 3-D world and make it look like it's real," he says.

Often, Sevo works with photographs. "I make changes that the engineers are proposing by removing things from the photo or adding new things to it."

When Sevo is finished, the architects, engineers, and client can look at the image he has created and decide if they like what they see. If not, they all work together to make changes, and Sevo makes the adjustments to his picture.

"The engineers and architects are serving the customer, and I develop a tool that helps them all communicate," he says.

Education:
Sevo received a bachelor's degree in mechanical engineering from Cedarville University. In his free time at school, he enjoyed creating graphics on his computer. After graduation, he found a way to combine his engineering background with his love of computer graphics.

"I did what I enjoyed, making pictures on the computer," he says. "But I did it for engineers because I understood the difficulties they face in communicating with clients and each other. So I was able to marry my two interests into a service. It gave me a good opportunity to use my education but also to enjoy my creativity."

Advice to Students:
Sevo urges students to pursue the subjects they love and then figure out a way to turn their interests into a career. "Study what interests you," he says. "Learn how to do the things you enjoy well enough so you can serve other people with those skills."

SECTION 2: PRESENTATION SKETCHES

Presentation sketches or renderings are used to give the engineer's or designer's sketches a more realistic look. Renderings help the nonengineer to better understand what the product will look like once it has been manufactured (see Figure 5-24). This type of sketching can also be used in advertising the product in magazines, journals, newspapers, catalogs, or even television advertisements.

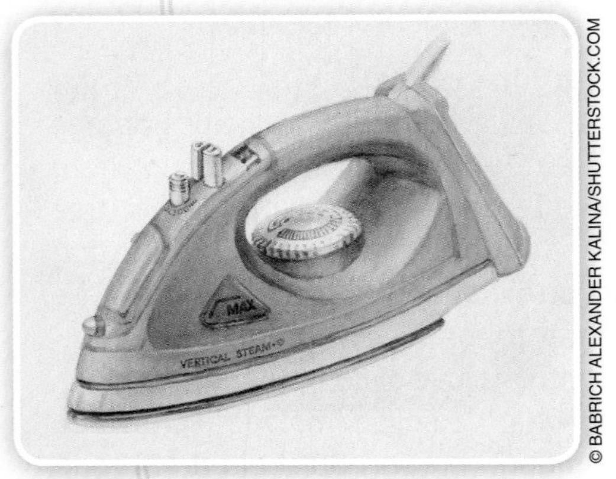

FIGURE 5-24 Rendering a sketch can help you give a more realistic view of your product.

Considerations When Rendering

You must consider three things when you render your sketch. The first is the direction from which light shines on the object. For example, does it come from behind the object, from the front, or maybe from the left or right sides of the object (see Figure 5-25)? Second, what is the intensity of light on a surface? In other words, how bright is the light? The third consideration

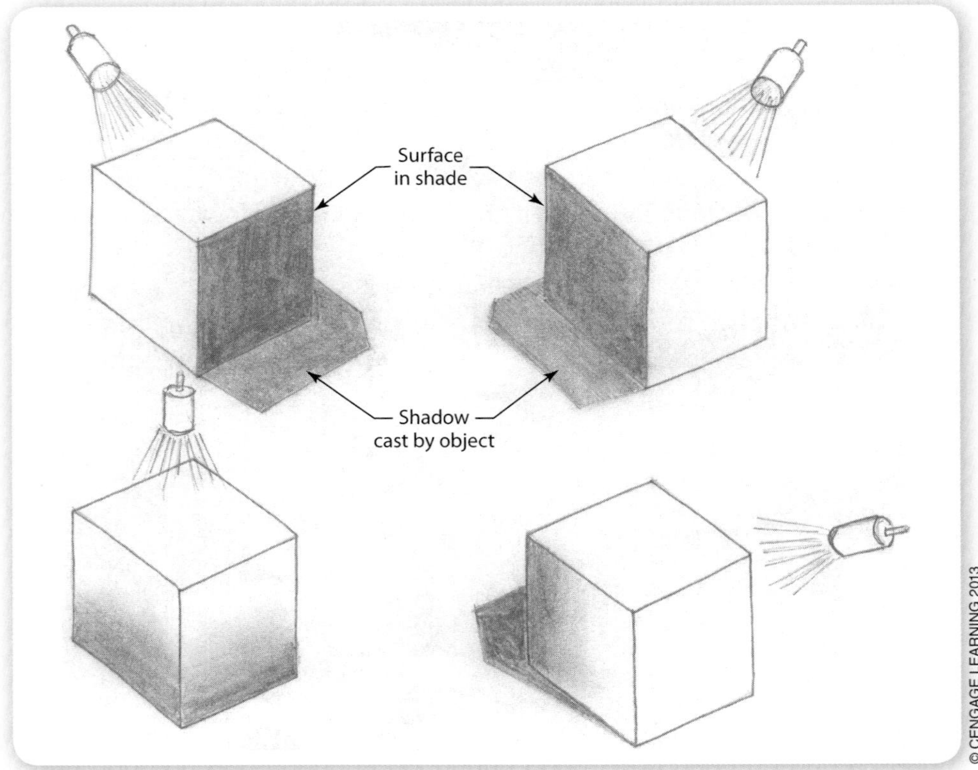

FIGURE 5-25 The location of shade and shadow is dictated by the direction of the light source.

is the type of surface on the object. The smoother the surface, the more light it will reflect. If the surface is rough, it will be less bright.

Rendering Techniques

Basic rendering of an object is done by using shading techniques. You can also increase the effectiveness of the rendering by using shadowing techniques. The amount of shade on an object is created by the amount of light present. The amount of shadow cast by an object is created by the amount of light present and the direction of the light source. Engineers and designers use shading and shadowing sparingly. Therefore, we will address simple shade and shadow techniques in this chapter. Although graphic artists use more techniques, we will explore only three rendering techniques. Remember, just as in freehand technical sketching, practice is necessary to develop the skill of rendering.

Smudge Shading and Shadowing You can create tones of very light to very dark with a smudging technique. This technique gives the closest representation to what actual surfaces might look like. You may use a very soft pencil (no. 2, H, or F grades will work) to develop the tones. Colored pencils may also be used for a more dramatic effect. Keep the pencil point dull. Use the edge of the pencil lead, moving the pencil back and forth. Figure 5-26 shows the results of such a motion. Be careful when using this technique because your sketch can become very messy. The best way to prevent

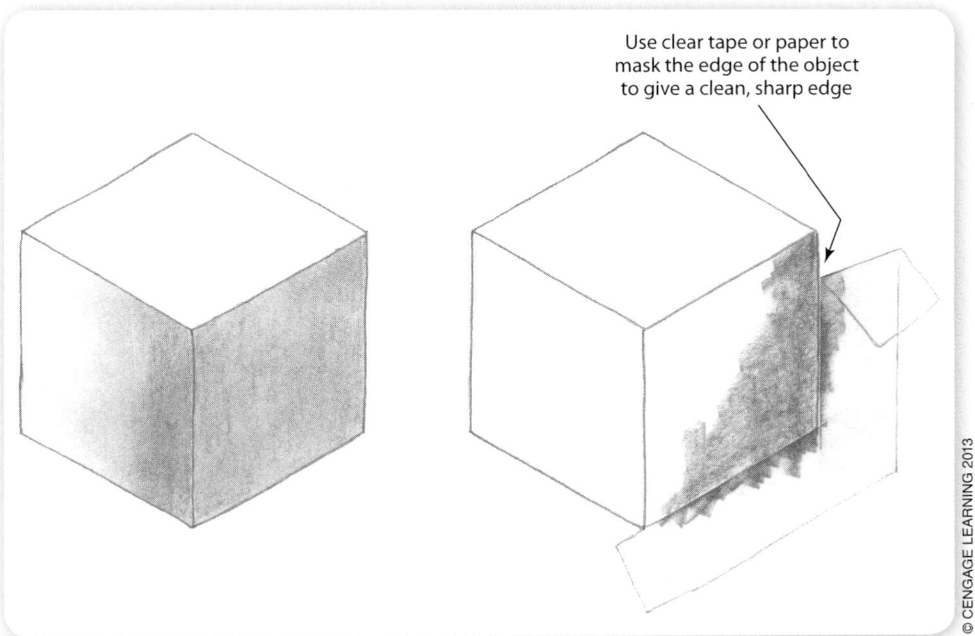
Use clear tape or paper to
mask the edge of the object
to give a clean, sharp edge

FIGURE 5-26 Smudging is one technique to create shade or shadow.

excess smudging is to mask off your sketch with additional paper or clear "easy-stick" cellophane tape.

Line Shading and Shadowing Like the smudging technique, you can create tones from very light to very dark using a line technique. Use a no. 2 or softer pencil with a slightly dull point. Colored pencils or colored markers also can be used in this technique. This technique is not messy like the smudge technique, so you do not have to take the same precautions. Shade or shadow is created by sketching parallel lines on the surface of the object. The closer the parallel lines are drawn, the darker the surface will appear. To create a lighter surface, sketch your lines farther apart (see Figure 5-27).

Stippling Shades and shadows of light to dark can be created by placing dots far apart or close together. This technique requires a soft pencil such as an H or F grade. As in the line technique, colored pencils or colored markers can be used for a more dramatic effect. A no. 2 pencil can be used, but it requires more time. To create dark tones, place the dots closer together. Placing dots farther apart will create a lighter tone (see Figure 5-28).

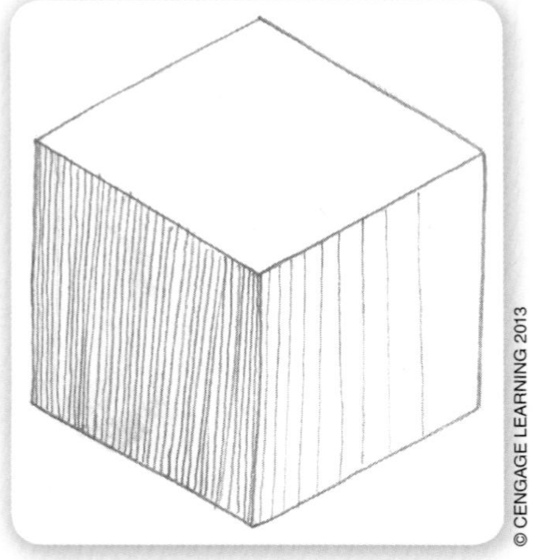

FIGURE 5-27 **Using line to create shade or shadow.**

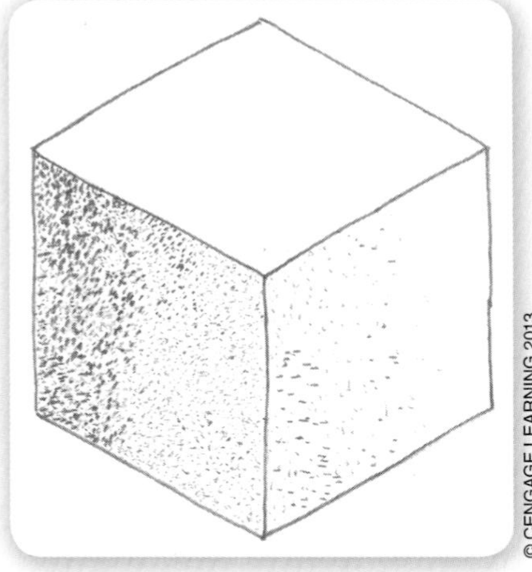

FIGURE 5-28 **Using stippling to create shade or shadow.**

SUMMARY

In this chapter you learned:

- Engineers use pictorial sketching because it represents how people naturally see the world.
- Your eyes see and your brain interprets the world around us in three dimensions: width, height, and depth. We referred to this as *three-dimensional* or *3-D*.
- Pictorial sketching is the process of graphically representing objects in three dimensions.
- Isometric pictorial (part of the axonometric group) is the most popular type used by engineers and designers.
- Perspective drawing, which can be divided into one-point and two-point, provides the most realistic view of an object.
- Isometric and perspective circles, arcs, and holes are represented by elliptical shapes.
- Developing complex pictorial drawings should begin with an object's overall basic rectangular shape.
- The concepts of perspective drawing are that objects appear to get shorter and closer together the farther they are from the viewer and that everything "vanishes" to the horizon line.
- Renderings are developed with the use of shading and shadowing techniques. Three common methods used are smudging, lines, and stippling.

VOCABULARY

Write a definition for each term in your own words. After you have finished, compare your answers with the definitions provided in the chapter.

Three dimensional (3-D) sketching
Pictorial sketching
Plane
Isometric sketching
Nonisometric planes
Ellipse
Perspective drawing or sketching
Perspective
Vanishing point
Optics
Two-point perspective
Features

STRETCH YOUR KNOWLEDGE

Provide thoughtful, written responses to the following questions and assignments.

1. Why is it important for engineers and designers to develop the skill of pictorial sketching?
2. What is the difference between isometric pictorials and perspective pictorials?
3. Who else uses axonometric drawings besides engineers, and what do they use them for?

Use sketch paper provided by your instructor to complete the following exercises. Your instructor will tell you how many of the exercises you will need to complete.

4. Using correct freehand technical sketching techniques, practice sketching isometric cubes. See Figure 5-29 for a practice example.

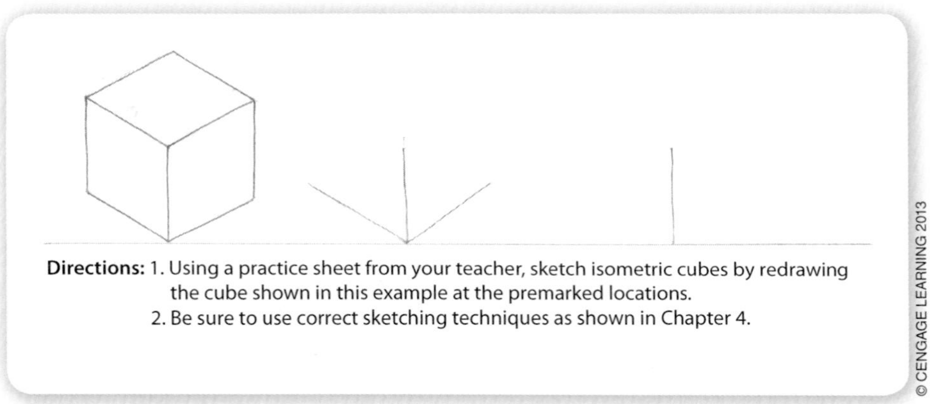

Directions: 1. Using a practice sheet from your teacher, sketch isometric cubes by redrawing the cube shown in this example at the premarked locations.
2. Be sure to use correct sketching techniques as shown in Chapter 4.
© CENGAGE LEARNING 2013

FIGURE 5-29 **Sketching practice: isometric cube.**

5. Using correct freehand technical sketching techniques, practice sketching isometric rectangular solids. See Figure 5-30 for a practice example.

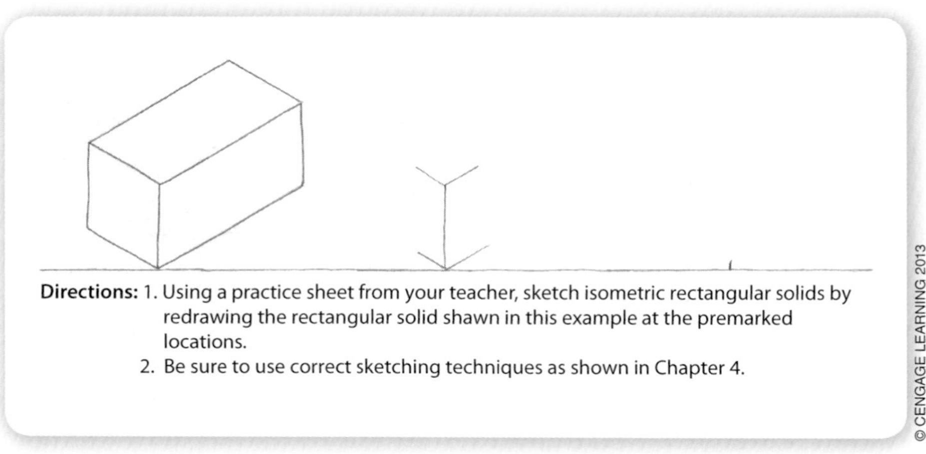

Directions: 1. Using a practice sheet from your teacher, sketch isometric rectangular solids by redrawing the rectangular solid shawn in this example at the premarked locations.
2. Be sure to use correct sketching techniques as shown in Chapter 4.
© CENGAGE LEARNING 2013

FIGURE 5-30 **Sketching practice: isometric rectangular solid.**

6. Using correct freehand technical sketching techniques, practice sketching complex rectangular solids. See Figure 5-31 for a practice example.

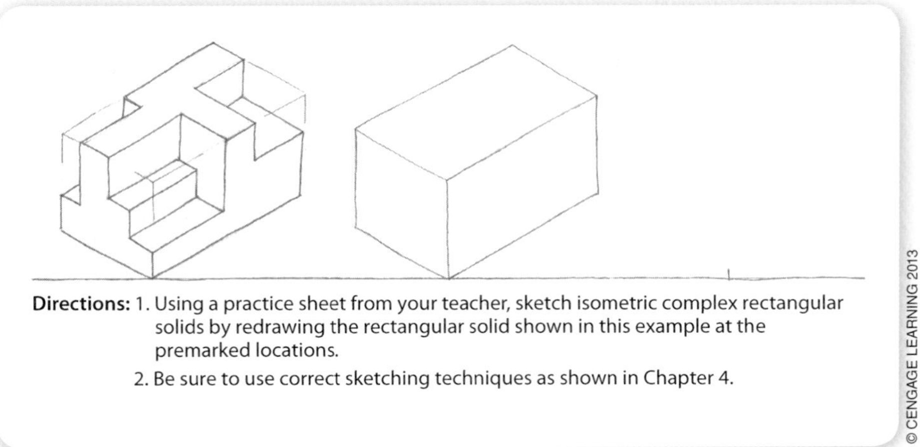
Directions: 1. Using a practice sheet from your teacher, sketch isometric complex rectangular solids by redrawing the rectangular solid shown in this example at the premarked locations.
2. Be sure to use correct sketching techniques as shown in Chapter 4.

FIGURE 5-31 **Sketching practice: isometric complex rectangular solid.**

7. Using correct freehand technical sketching techniques, practice sketching rectangular solids with inclined planes. See Figure 5-32 for a practical example.

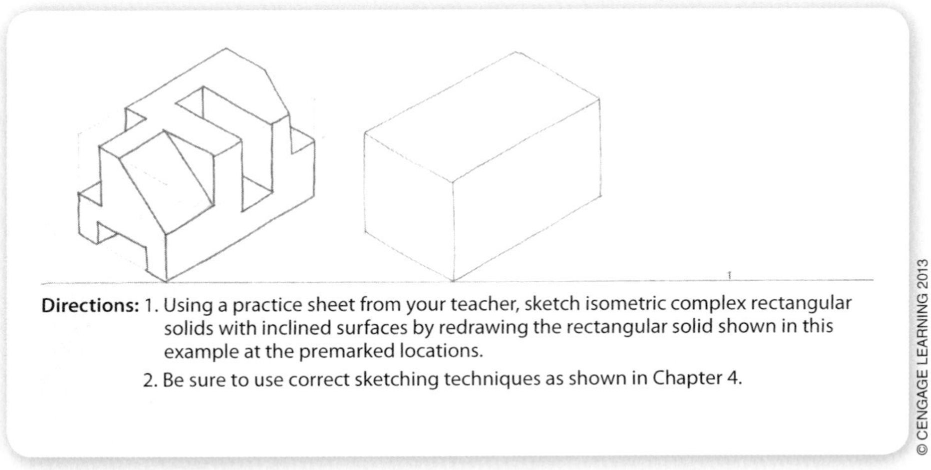
Directions: 1. Using a practice sheet from your teacher, sketch isometric complex rectangular solids with inclined surfaces by redrawing the rectangular solid shown in this example at the premarked locations.
2. Be sure to use correct sketching techniques as shown in Chapter 4.

FIGURE 5-32 **Sketching practice: isometric complex rectangular solid with inclined surfaces.**

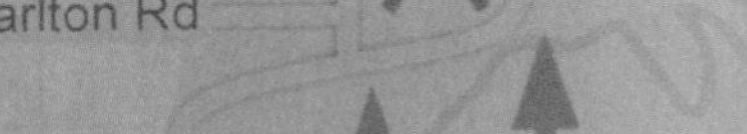
Carlton Rd

8. Using correct freehand technical sketching techniques, practice sketching isometric circles. See Figure 5-33 for a practical example.

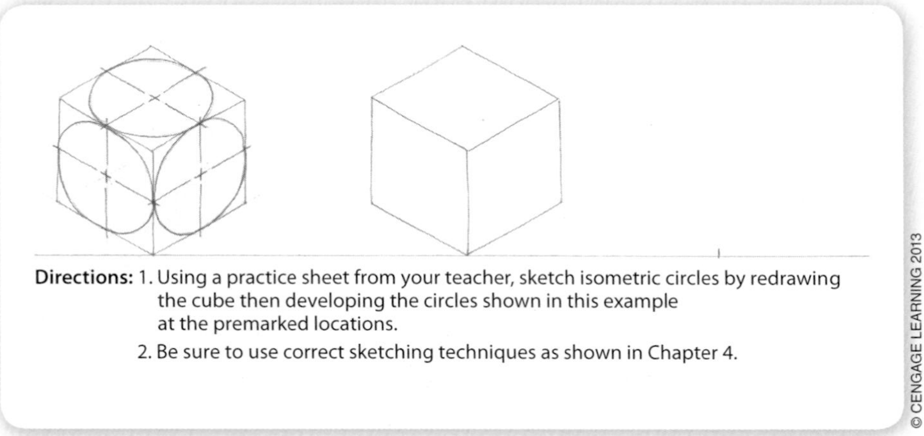
Directions: 1. Using a practice sheet from your teacher, sketch isometric circles by redrawing the cube then developing the circles shown in this example at the premarked locations.
2. Be sure to use correct sketching techniques as shown in Chapter 4.

FIGURE 5-33 **Sketching practice: isometric circles.**

9. Using correct freehand technical sketching techniques, practice sketching isometric holes. See Figure 5-34 for a practical example.

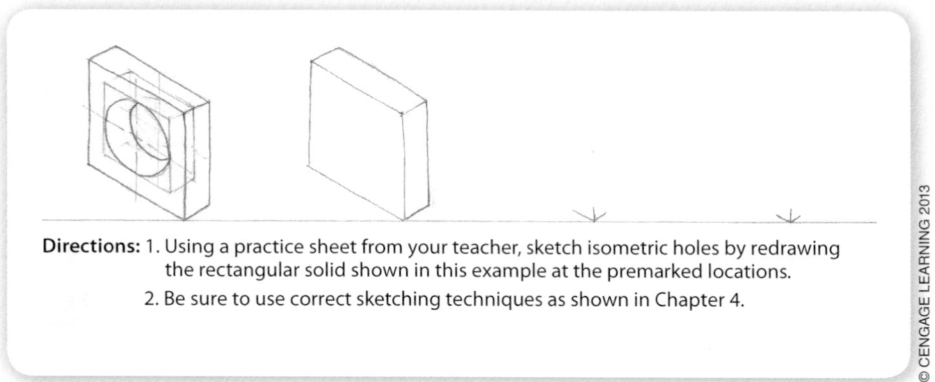
Directions: 1. Using a practice sheet from your teacher, sketch isometric holes by redrawing the rectangular solid shown in this example at the premarked locations.
2. Be sure to use correct sketching techniques as shown in Chapter 4.

FIGURE 5-34 **Sketching practice: isometric holes.**

10. Using correct freehand technical sketching techniques, practice sketching one-point perspective rectangular solids. See Figure 5-35 for a practical example.

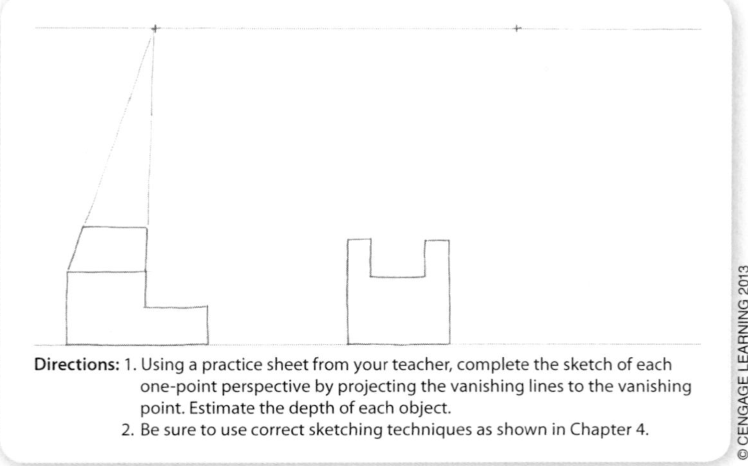

Directions: 1. Using a practice sheet from your teacher, complete the sketch of each one-point perspective by projecting the vanishing lines to the vanishing point. Estimate the depth of each object.
2. Be sure to use correct sketching techniques as shown in Chapter 4.
© CENGAGE LEARNING 2013

FIGURE 5-35 Sketching practice: one-point perspective.

11. Using correct freehand technical sketching techniques, practice sketching two-point perspective rectangular solids. See Figure 5-36 for a practical example.

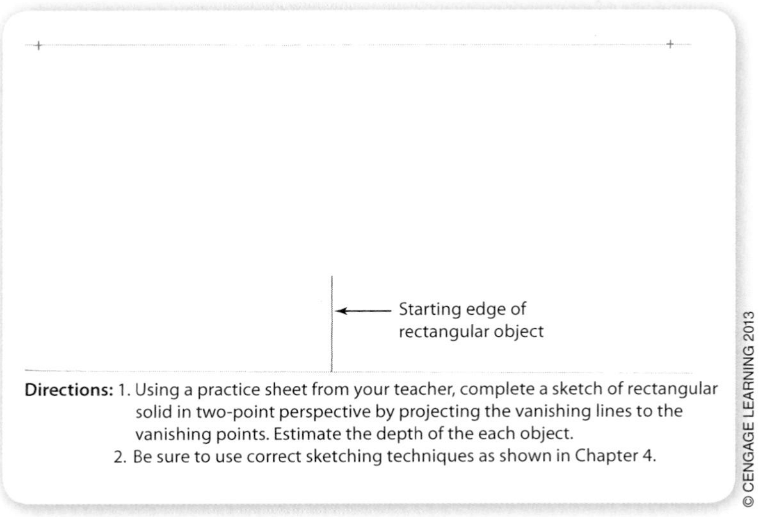

Starting edge of
rectangular object
Directions: 1. Using a practice sheet from your teacher, complete a sketch of rectangular solid in two-point perspective by projecting the vanishing lines to the vanishing points. Estimate the depth of the each object.
2. Be sure to use correct sketching techniques as shown in Chapter 4.
© CENGAGE LEARNING 2013

FIGURE 5-36 Sketching practice: two-point perspective.

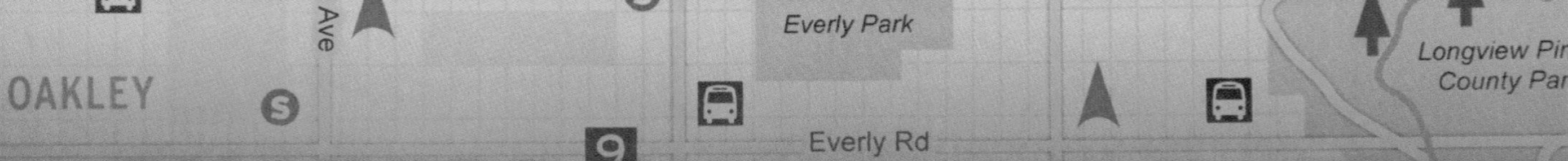

CHAPTER 6
Reverse Engineering

Menu

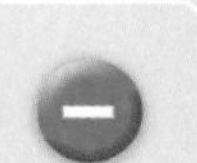

Before You Begin

Think about these questions as you study the concepts in this chapter:

1. What is reverse engineering?
2. Why do engineers use reverse engineering?
3. Why do engineers require multiview drawings?
4. What is orthographic projection?
5. Why are assembly drawings important?
6. What is an exploded view?
7. Why are dimensions required on engineering drawings?

© ISTOCKPHOTO/SKIP O'DONNELL

Engineering in Action

Do you have a younger sister or brother who has Legos? As a child, did you ever build things out of Legos? Were you or your sister or brother using Lego interlocking bricks or Mega Bloks? You may not have been able to tell the difference between these two types of blocks.

The Mega Bloks company used a process called *reverse engineering* to design its interlocking blocks so they would be compatible with Lego bricks. **Reverse engineering** is the process of measuring and analyzing an existing item. A designer makes technical drawings of the item during the process and uses them to re-create the item. Computers often generate these new drawings, which allows for computer modeling. We discuss this process in Chapter 7. Mega Bloks was legally able to perform this reverse engineering because the original 1958 Lego Group patent on the interlocking bricks had expired.

SECTION 1: REVERSE ENGINEERING

We can also use reverse engineering when the manufacturer of a product is no longer in business and the documentation of the product cannot be found. What if the antique sewing machine pictured in Figure 6-1 required a new part? An engineer could reverse engineer the part from a similar antique sewing machine and then produce the part for the broken sewing machine.

FIGURE 6-1 An antique sewing machine.

We can view reverse engineering as a logical process of steps. Figure 6-2 shows a six-step flowchart of the reverse-engineering process. First, we must identify the purpose of the investigation. In simple terms, why complete the reverse-engineering process? What question needs to be answered? Second, we must establish a hypothesis or answer to the question. Third, we must disassemble products with more than one part (but not a single part such as the Lego brick).

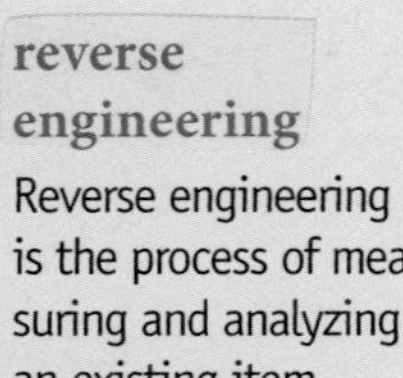

We analyze each part after disassembly (see Figure 6-3). In analysis, we identify the material, determine how the part was manufactured, and measure its size and features. There are many legitimate reasons to reverse

reverse engineering
Reverse engineering is the process of measuring and analyzing an existing item and then re-creating technical drawings of that item.

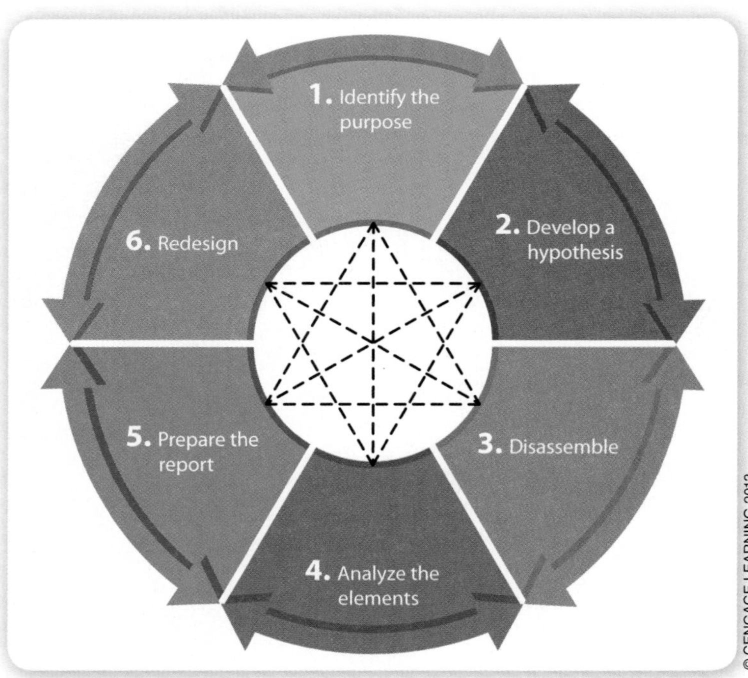

FIGURE 6-2 The reverse-engineering process flowchart.

Science in Engineering

Because it is sometimes difficult to identify the type of material a product is made from, engineers often turn to scientists for help. By experimenting, scientists can distinguish between different types of materials. In some experiments, we can determine what temperature a material melts at or what type of reaction occurs when a sample is dipped in different chemicals. Scientific inquiry is an essential part of the reverse-engineering process.

engineer a product, but it can also allow people and companies to infringe on the rights of a patent holder. We must consider both legal and ethical issues when using the reverse-engineering process.

In the fifth step, a written report details the findings from the disassembly process. This report should also include drawings of each part. Sixth, using the information obtained, we redesign the reverse-engineered product.

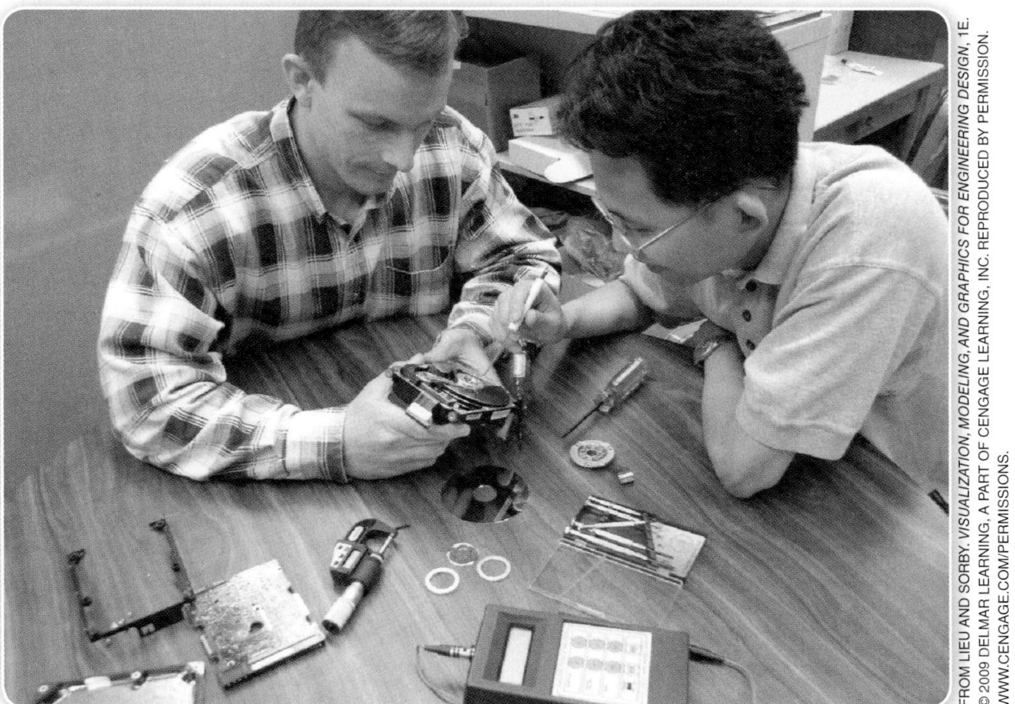

FIGURE 6-3 *Team members analyze disassembled components.*

Did You Know?

Ole Kirk Christiansen, a carpenter from Denmark, designed the first interlocking blocks in 1947. He called his first blocks Automatic Binding Bricks. Christiansen's workshop was the humble beginning of the now world-famous Lego Group. He coined the name Lego from the Danish phrase "Leg godt," which means "Play well." Did you know that the Lego bricks manufactured in 1963 will interlock with the Lego bricks manufactured today?

SECTION 2: MULTIVIEW DRAWINGS

For the designer to communicate the product's characteristics to the fabricator, they must produce a drawing of the item. Drawings can take the form of sketches, drawings done with drafting instruments, or drawings produced using a computer system. Drawings made with instruments or computers are termed **engineering drawings**.

Most engineering drawings contain more than one view of a product and are called **multiview drawings** (see Figure 6-4). Multiview drawings are required for the manufacturing process and also to permanently record the product's design.

Look at the block in Figure 6-5. How many sides does it have? Yes, it has six sides. Most objects, in fact, have six sides. The surface you see when looking

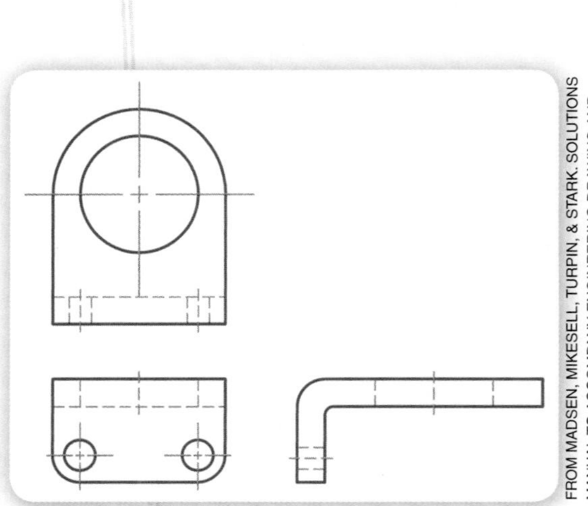

FIGURE 6-4 A multiview drawing.

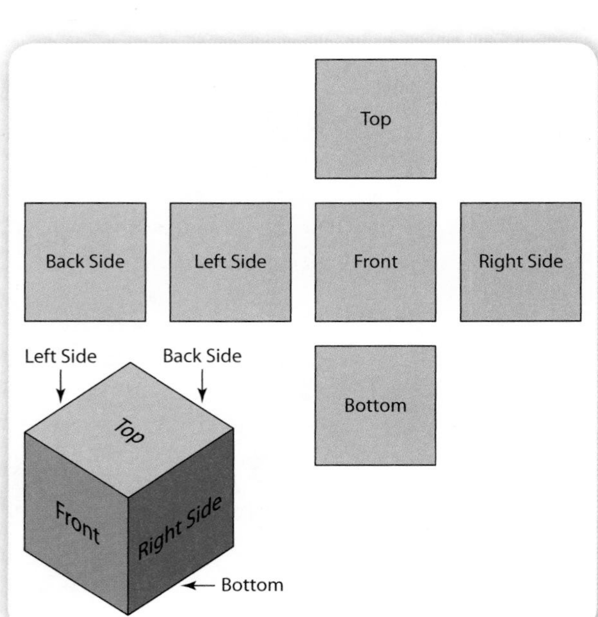

FIGURE 6-5 The six sides of an object.

at each side is called that side's *view*. Because objects have six sides, they have six views: (1) front view, (2) top view, (3) bottom view, (4) right side view, (5) left side view, and (6) back view. Even though objects have six views, multiview drawings contain only the number of views required to properly show the object. Typically multiview drawings contain the three views shown in Figure 6-4; the front view, the top view, and the right side view.

4

Orthographic Projection

To develop these six views, an engineer uses **orthographic projection** to project or transfer the view's features out from each side of the object. Figure 6-6 shows the orthographic projection of one surface. Projecting the view straight from the object shows the surface's true shape and size.

orthographic projection

An orthographic projection is a method of transferring the views of an object onto the planes that are perpendicular to each surface of the object.

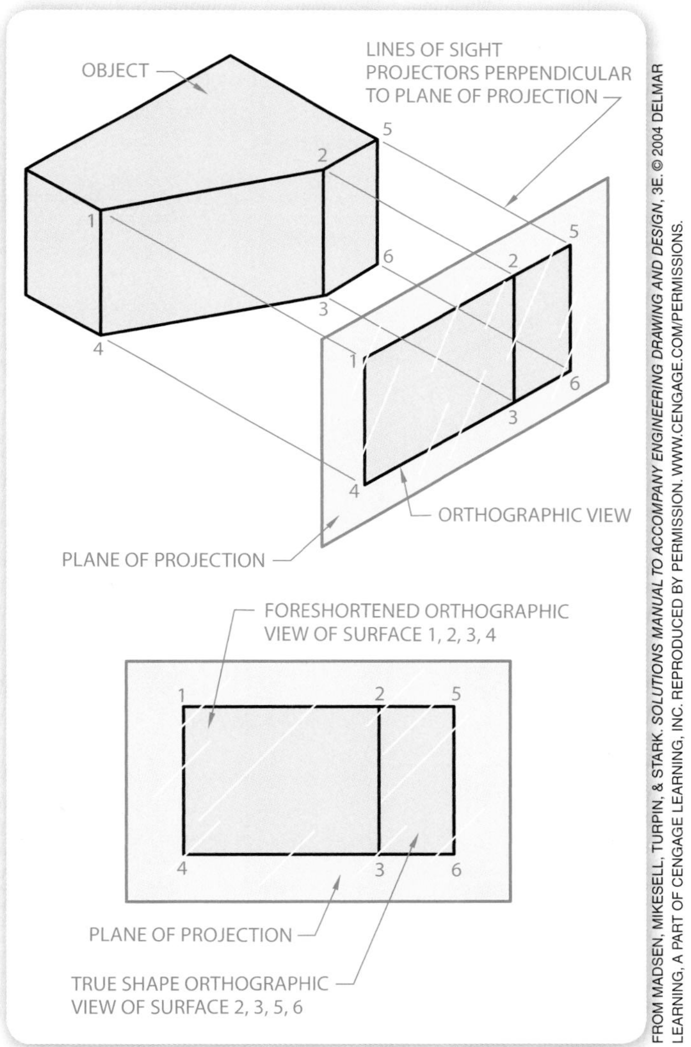

FIGURE 6-6 *Orthographic projection of a surface.*

Math in Engineering

Orthographic projection is derived from geometry, the area of mathematics that is concerned with the size, shape, and position of items in space. Geometry is one of the oldest areas of mathematics, dating back to 300 B.C. Ancient mathematicians who studied geometry used classical tools such as compasses and straightedges (rulers) to develop their discipline.

Look at the clock radio in Figure 6-7; its sides are not square. But if you look at the clock radio directly from the front, as in Figure 6-8, the front's true shape and size are visible. The **front view** shows the object's height and width. Figure 6-9 shows the clock radio's right **side view**, or its depth and height. Figure 6-10 provides the clock radio's **top view**, or its width and depth.

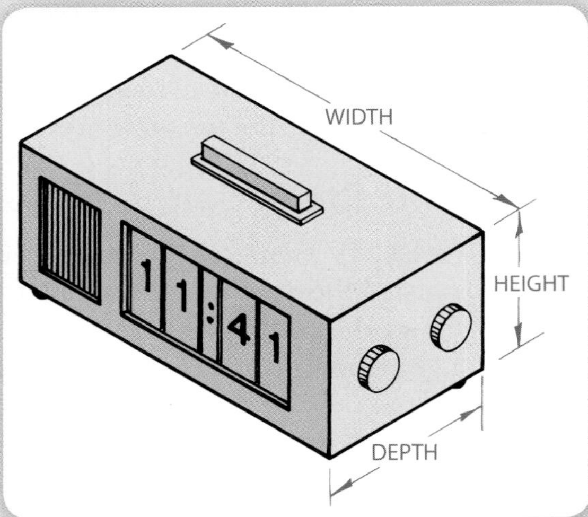

FIGURE 6-7 An isometric picture of a clock radio.

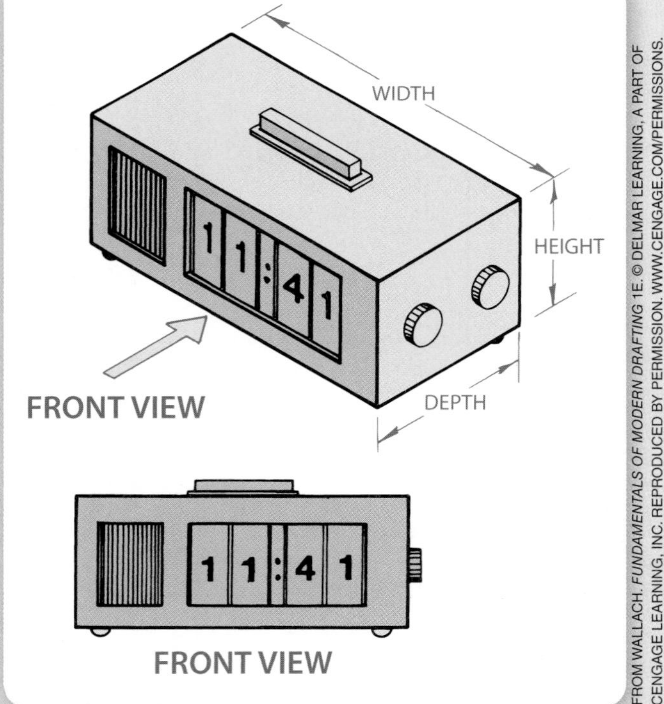

FIGURE 6-8 The front view shows the object's height and width.

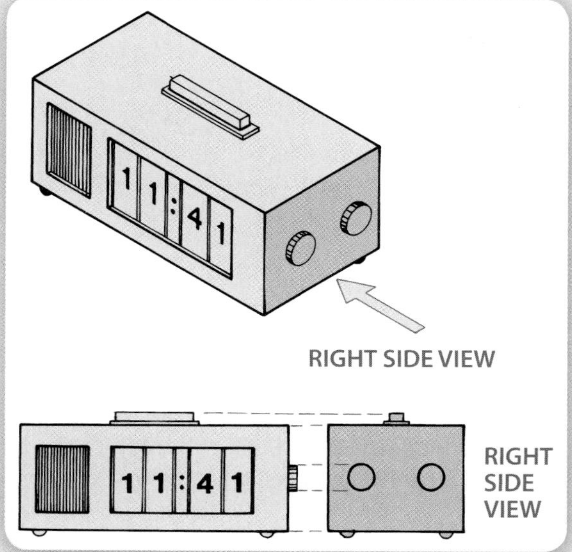

FIGURE 6-9 **The right side view shows the object's height and depth.**

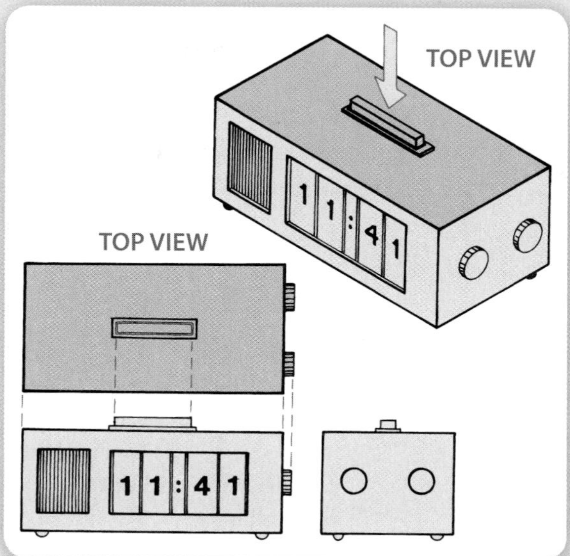

FIGURE 6-10 **The top view shows the object's depth and width.**

Figure 6-11 shows all six views of the clock radio. Which view shows the most detail about the clock radio? Yes, the front view. This was selected as the front view of an engineering drawing because it shows the most detail for the object.

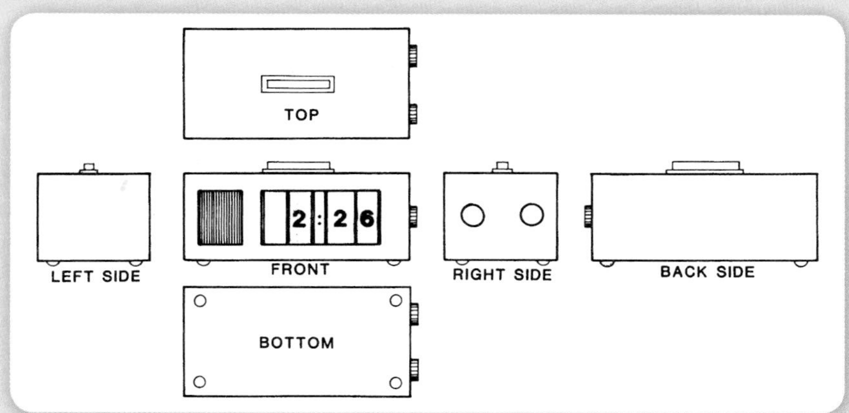

FIGURE 6-11 **The six views of the clock radio.**

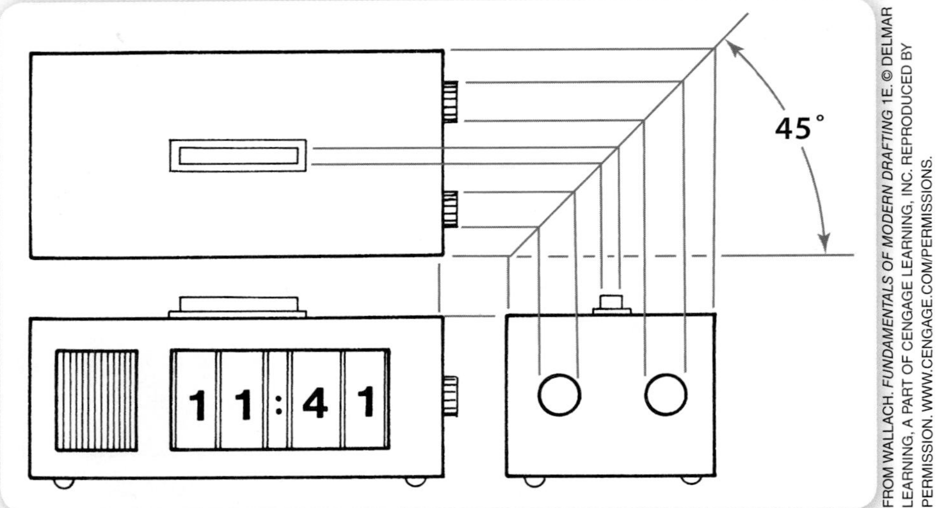

FIGURE 6-12 Using the 45°-angle projection technique to transfer locations.

Techniques

Look at the clock radio knobs shown in the right side view of Figure 6-10. These knobs are also visible in the top view. How did the designer know where to locate these knobs? Their size and location were projected or transferred using a 45°-angle technique (see Figure 6-12). This projection technique was also used to transfer the location of the handle for the top view to the right side view. We can also use the technique to transfer the location of the handle from the top view to the front view.

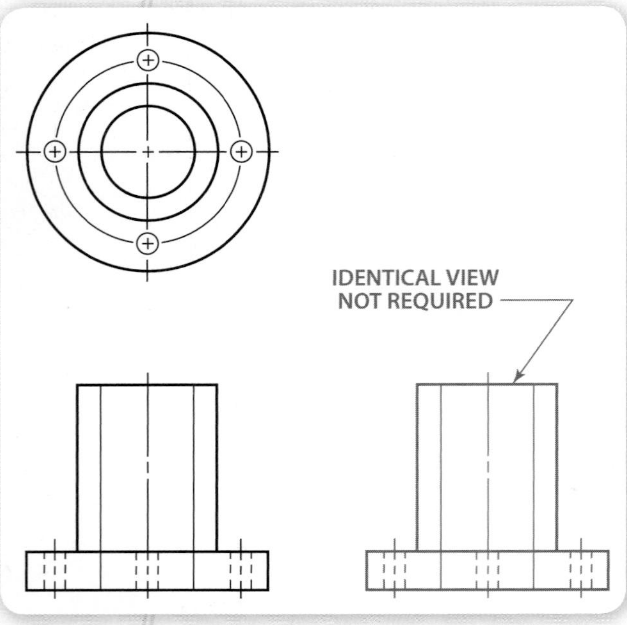

Even though the front view provides the most detail about the item, other views usually are required for a complete engineering drawing. We can add more views to provide details not seen in the front view. Most multiview drawings include three views and are thus called **three-view drawings**. Typically, the three-view drawings contain (1) the front view, (2) the top view, and (3) the right side view.

Some multiview drawings require only two views. Look at the round object in Figure 6-13. The front and right side views are identical, so only the front and top views are

FIGURE 6-13 A circular object may require only two views.

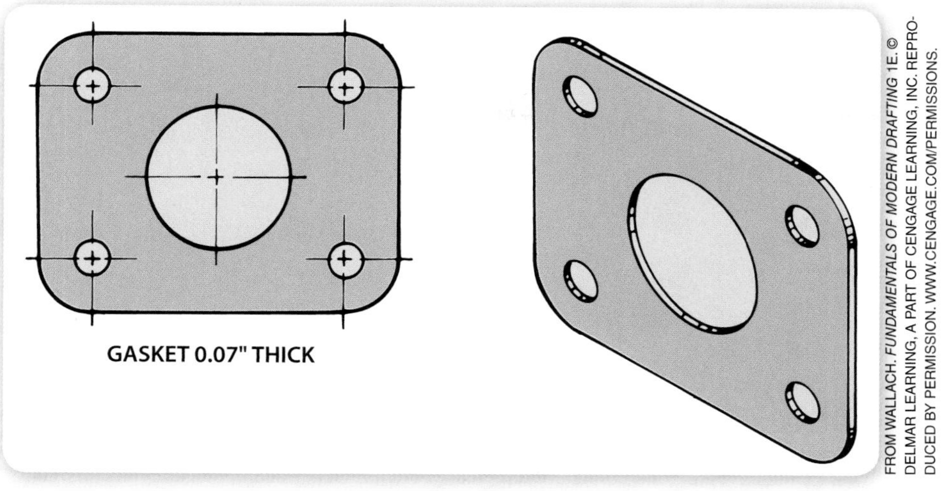

FIGURE 6-14 A flat gasket may require only one view.

required for this item. The gasket in Figure 6-14 requires only the front view. We could indicate the gasket's thickness by including a note on the drawing.

working drawing

A working drawing is a drawing from which the object can be fabricated.

assembly drawing

An assembly drawing is a pictorial drawing that shows the parts of an object placed in their correct locations.

SECTION 3: WORKING DRAWINGS

Working drawings provide all the information and details needed to make or fabricate the item. By including all the required information, we make sure the technician who makes the item will not have to guess any size or feature location. A common type of working drawing is the assembly drawing. Working drawings may also include exploded views of the assembled product.

Assembly Drawings

Most products contain more than one part, so engineers must show how the parts of the product will fit together. For this type of communication, engineers use **assembly drawings** that show the relationship of the parts and the assembled product. Figure 6-15 shows an assembly drawing of a pegboard toy. The drawing shows how the toy's sides fit onto the base and how the pegs fit into the toy. You may have seen assembly drawings in the instructions that come with a new bicycle or household item. Assembly drawings help consumers visualize how the product will look when it is assembled.

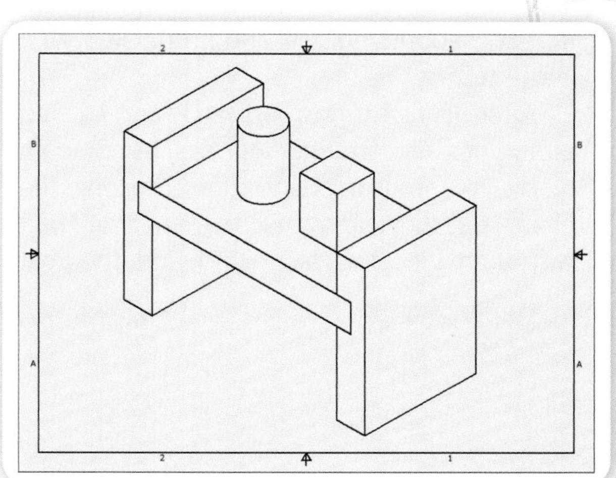

FIGURE 6-15 An assembly drawing of a pegboard toy.

Career Spotlight

Name:
Peter G. Hwang

Title:
User Experience Designer,
Hewlett-Packard Company

© CENGAGE LEARNING 2013

Job Description:
Hwang is the guy who makes connections between psychology and technology. His goal is to design exactly the kinds of computer printers that customers want. First, he works to understand the customers' needs and goals. He finds people where they work and live, observes them, and conducts extensive interviews. Back in the lab, he studies the data he has collected and brainstorms with other engineers. Then they create prototypes of computer printers and take them back to the people they interviewed.

"That allows us to get feedback into whether we're really solving their problems," Hwang says. "The idea is to keep the user as the focus at all times."

Hewlett-Packard gave Hwang a unique opportunity to observe customers in Japan for a month. "I got paid to walk the streets, spend time in people's homes, and immerse myself in the culture," he says.

After his first trip to Japan, Hwang created prototypes of computer printers using foam blocks. "We went back and showed people the blocks, but they interpreted them in a different way than we expected, which led to great design insight on our part," he says. "We were able to take all that back and launch new products we wouldn't have come up with if we hadn't had the deeper cultural knowledge."

Education:
Hwang earned a bachelor's degree in mechanical engineering from the University of California–Berkeley and a master's degree in product design from Stanford University. His master's degree was one-half engineering and one-half fine arts.

"Most people think you're either an artist or an engineer," he says. "But I found a happy medium where I could be both. There's a technical side to my job, but I do a lot of sketching and drawing as well. Being an engineer doesn't mean only that you're an egghead or a left-brained math/science person."

Advice to Students:
Hwang encourages students to work their passions into their studies and into their careers. "When I was in engineering school, I was looking for opportunities to draw and be creative, and now I do that in my work," he says. "Students with a passion for surfing, for example, can become engineers who design surfboards. Students who love sports can become engineers who design sports equipment. It makes your career that much more meaningful."

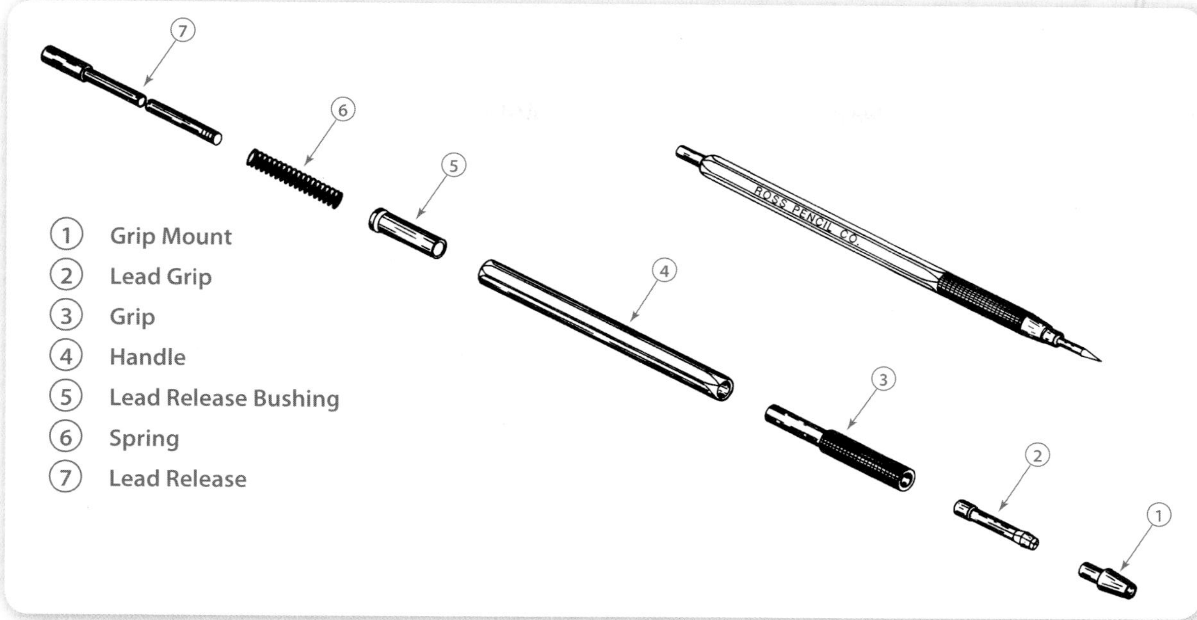

FIGURE 6-16 An exploded view of a pen with its parts labeled.

exploded view
An exploded view is part of a pictorial drawing that shows the parts of an object disassembled but in relation to each other.

Exploded Views

An **exploded view** shows how the parts of a product separate from one another. The parts may look as if they are floating in space. It would be impossible to see the internal components of the pen pictured in Figure 6-16 without the exploded view. We commonly see exploded views in assembly manuals, and it is also common for the different parts to be labeled, as they are in the exploded view of the pen.

SECTION 4: DIMENSIONING

Rationale

dimensions
Dimensions are the sizes and notes placed on a mechanical drawing that record an object's linear measurements such as width, height, and length, as well as the location of the object's features.

To make a part, a technician must know all the details of the part. We can communicate some details through pictures. Other details must be communicated by sizes and notes placed on an engineering drawing: these are called **dimensions**. Dimensions can define both the size of the object and the location of features such as holes and angles.

Look at Figure 6-17a. What can you tell about the part? Now look at Figure 6-17b. This is the same part, but dimensions have been added. The dimensions communicate that the part is 5.75 inches wide and 3.50 inches high. You can also see that a 1.50-inch diameter hole is to be placed in the part. You also know that there will be a 45° angle cut in its left side. Dimensions are an essential component of the engineering design process.

Math in Engineering

Engineers do not overdimension a drawing. In other words, they show only the minimum number of dimensions that will allow someone read the drawing. Therefore, engineers and technologists must use their mathematical skills to calculate distances and features that are not dimensioned.

Techniques

The rules for dimensioning are established by the American National Standards Institute (ANSI). According to ANSI, dimensions can be in inches (U.S. customary) or in metric or in dual dimensions that list both U.S. customary and metric. The position of the dimensions on an engineering drawing can be either unidirectional or aligned (see Figure 6-18).

Figure 6-19 shows the correct placement of dimensions on a drawing. The numerical dimension is centered between its **dimension lines**. At the end of each dimension line is an **arrowhead** that contacts a line that has been extended for the drawing. This line is called an **extension line**. The extension line does not touch the object lines of the drawing. Both the extension line and the dimension line are thinner than the object lines on the drawings. (See Chapter 4 for more information on the types of lines.)

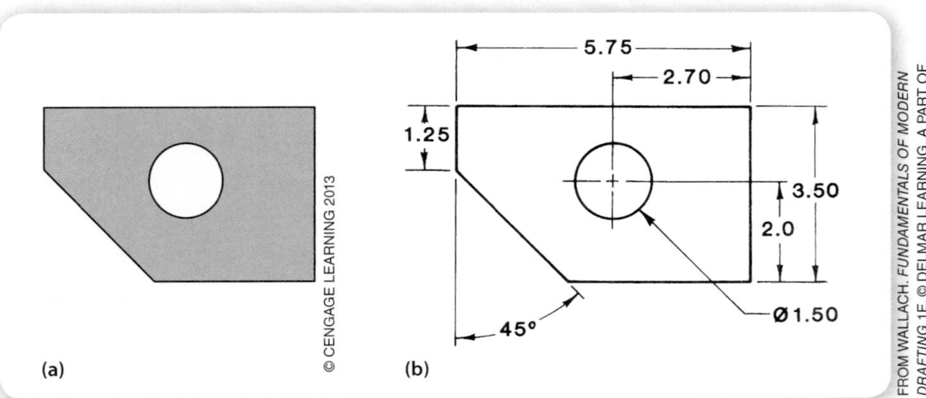

FIGURE 6-17 Drawing of a block (a) without and (b) with dimensions.

ENGINEERING CHALLENGE 1

Four parts are pictured. In addition to a picture of each part, there is a front view engineering drawing of the corresponding part. Each front view has the appropriate dimensions provided. Using your knowledge of mathematics for calculating distance and area, answer each of the following questions for each part. This will require you to interpret the dimensions from each engineering drawing.

Part 1:

1. What is the distance from point A to point B?
2. What is the surface area of the part's front view?

Part 2:

1. What is the surface area of the part's front view?

Part 3:

1. What is the distance from point A to point B?
2. What is the distance from point C to point D?

Part 4:

1. What is the surface area of the part's front view?

Interpreting dimensions.

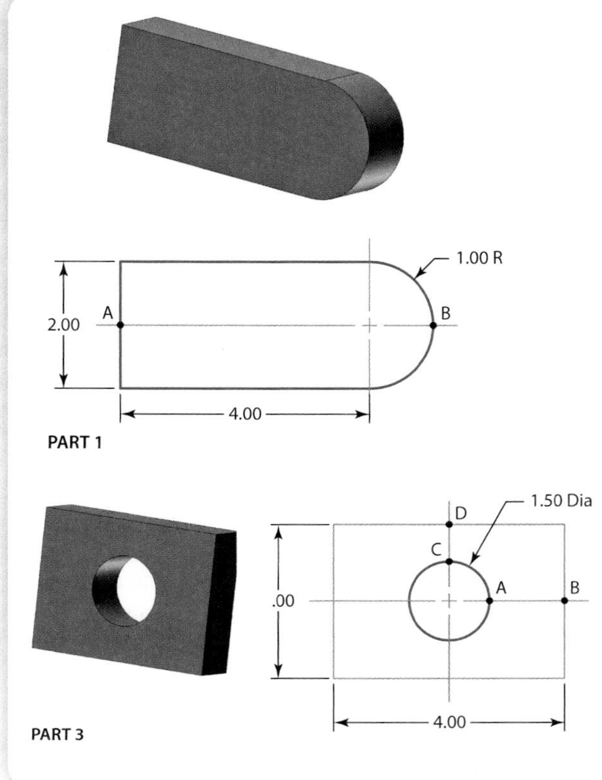

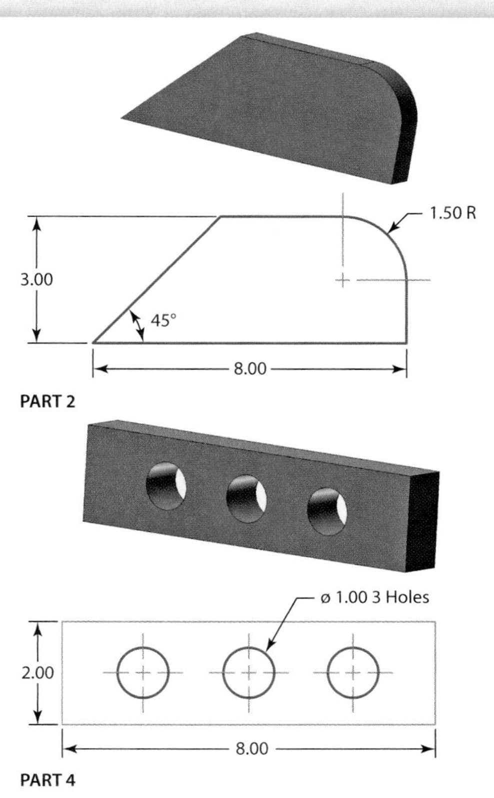

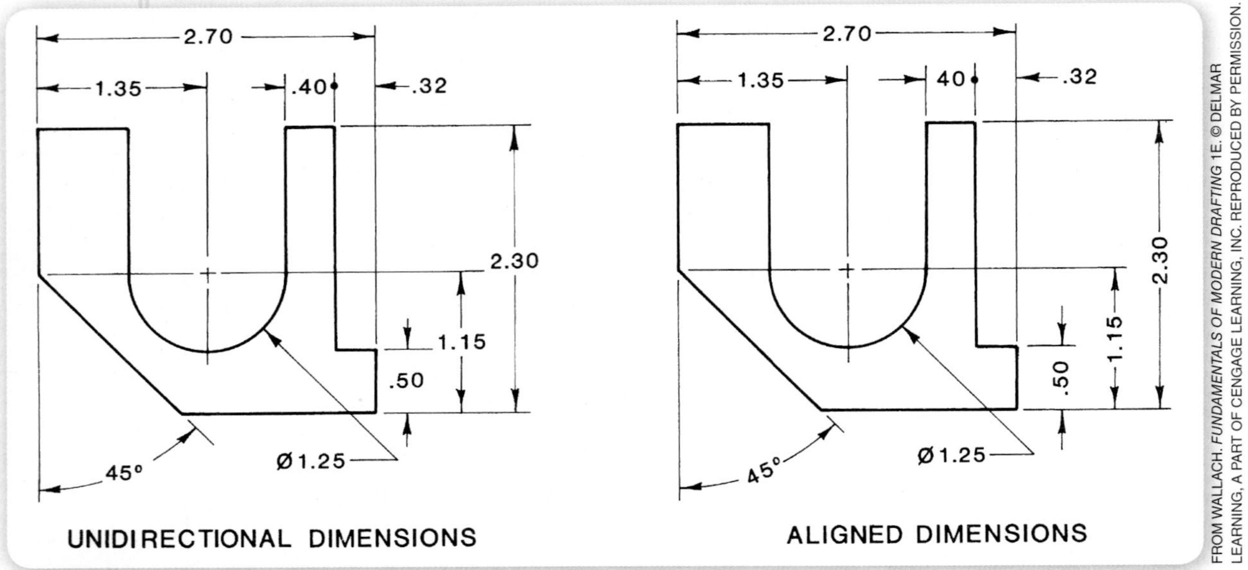

FIGURE 6-18 Unidirectional and aligned dimensions.

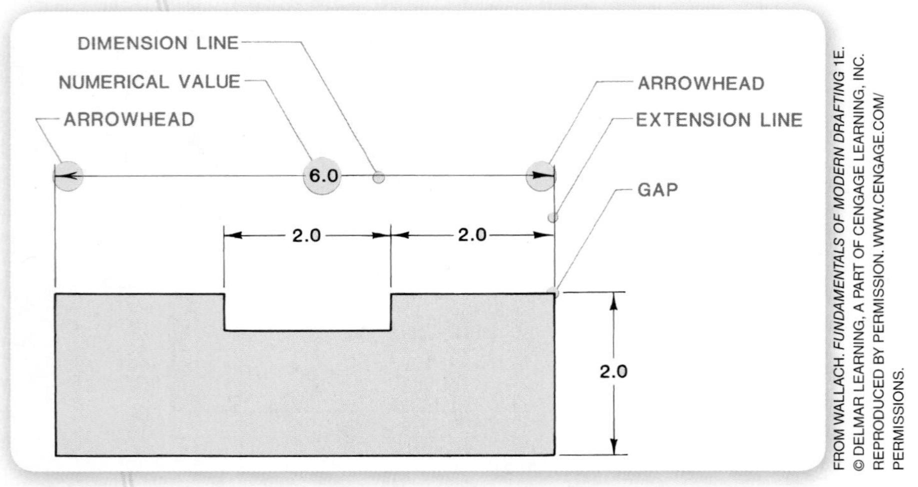

FIGURE 6-19 The correct placement of dimensions on a drawing.

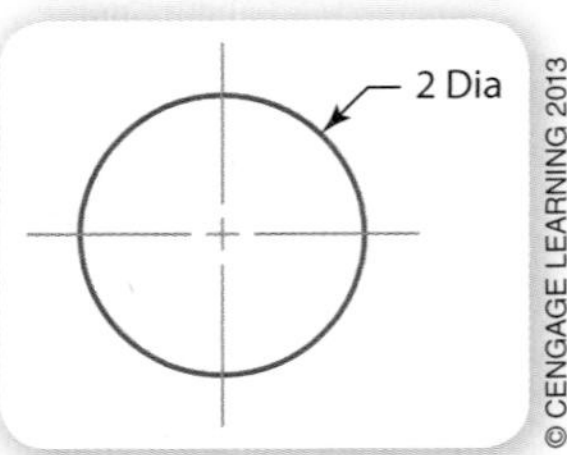

FIGURE 6-20 Dimensioning a circle with center lines and a leader.

To locate a circle or a hole on an engineering drawing, the engineer uses **center lines**. These lines cross at a 90° angle in the center of the circle or hole (see Figure 6-20). The dimension of the hole's diameter is placed at the end of a **leader**. The leader points to the center of the circle but only extends to the circle's outer edge.

ANSI has established a complete set of guidelines for dimensioning an engineering drawing. The following are some common guidelines to follow. Figure 6-21 also shows some guidelines for dimensioning.

- Place the dimension on the view that shows the true shape of the object.
- Group dimensions together on a view.
- Do not place dimensions inside the object's views.
- Place dimensions between the views.
- Do not cross dimension lines with extension lines.
- Do not duplicate dimensions.

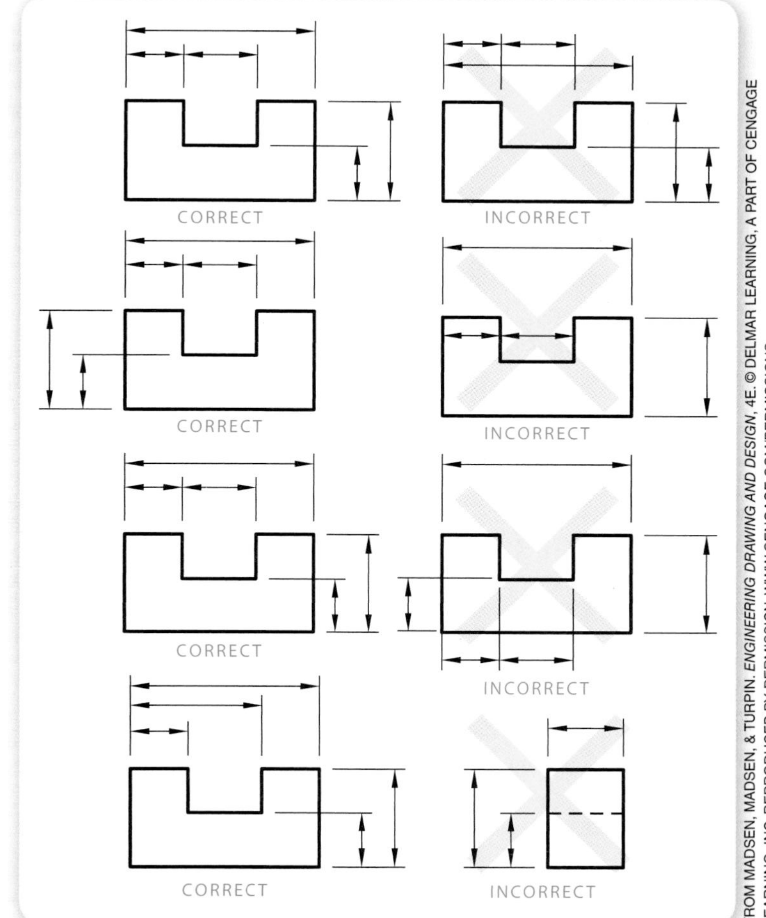

FIGURE 6-21 *The guidelines of dimensioning.*

Engineering Challenge

ENGINEERING CHALLENGE 2

Your team has been asked to reverse engineer one of the pictured products. (Your teacher may provide you with a different set of products.) Follow the six steps of the reverse-engineering process. Document your findings in your engineering notebook. Then make the appropriate engineering drawings needed to re-create the product.

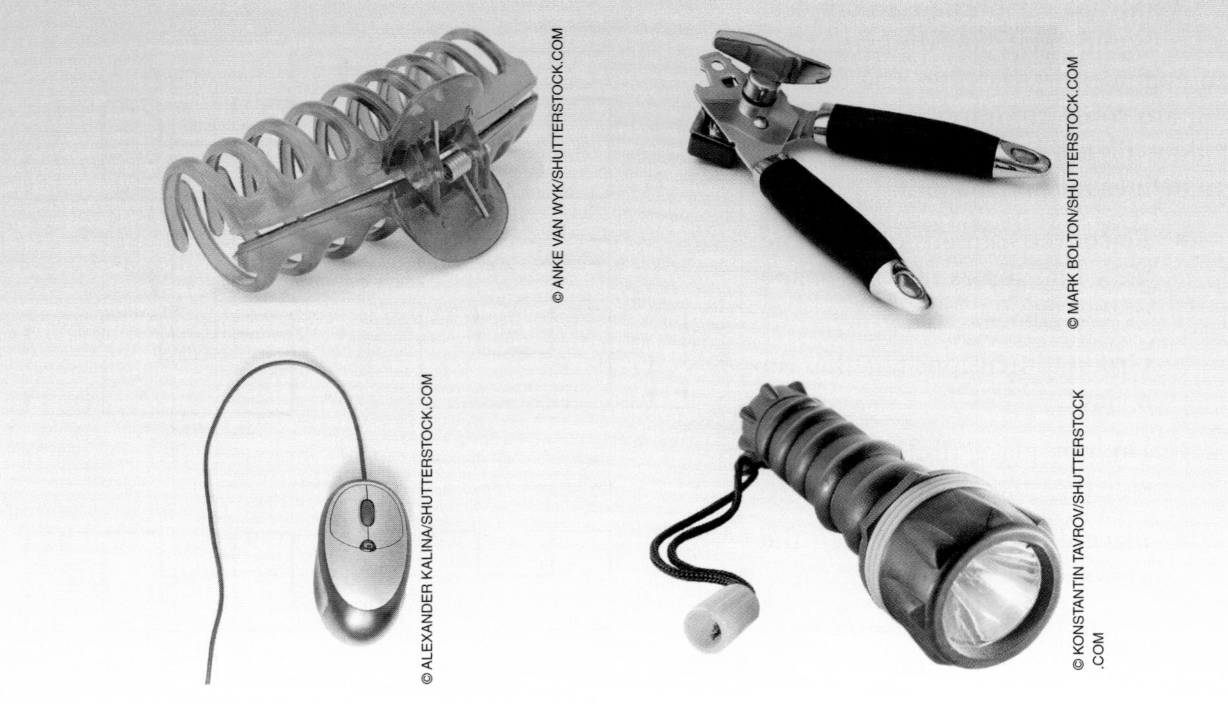

Products to reverse engineer.

SUMMARY

In this chapter you learned:

- Reverse engineering is a process used to copy an existing product.
- Orthographic projection is a technique used to develop views of a part that show the part's true shape and size.
- Most multiview drawings consist of three views: front view, top view, and right side view.
- Assembly drawings provide a picture of how the component will look with all of its parts assembled.
- Exploded views allow someone to see how the parts of a product fit together.
- Dimensions communicate the size and feature locations from the engineer to the person who makes the part.

VOCABULARY

Write a definition for each term in your own words. After you have finished, compare your answers with the definitions provided in this chapter.

Reverse engineering
Orthographic projection
Working drawing
Assembly drawing
Exploded view
Dimensions

STRETCH YOUR KNOWLEDGE

Please provide thoughtful, written responses to the following questions.

1. What would be an appropriate and ethical use of reverse engineering?
2. While considering the globalization of the world's industries, describe some possible negative impacts of reverse engineering.
3. Why is it important that engineering drawings be drawn accurately and contain accurate dimensions?
4. Describe how a company might enhance the assembly drawings of a product for household consumer use.
5. Describe why it is important to follow the ANSI dimensioning guidelines.
6. What determines whether the dimensions are placed in U.S. customary units, metric units, or both?

Onward to Next Destination ▶

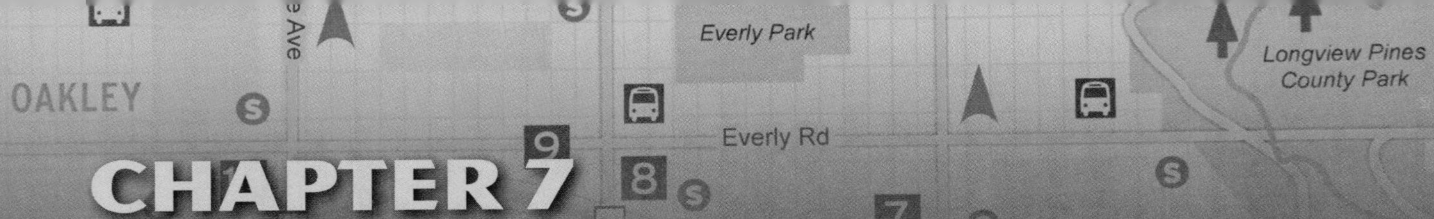

CHAPTER 7
Parametric Modeling

Menu

Before You Begin

Think about these questions as you study the concepts in this chapter:

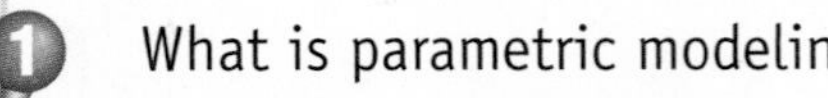

1. What is parametric modeling?

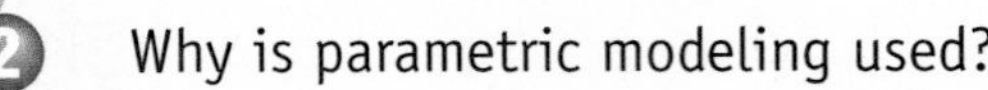

2. Why is parametric modeling used?
3. Who uses parametric modeling?
4. How does parametric modeling differ from nonparametric computer-aided drafting or design?
5. What are the fundamental steps for developing an object in parametric modeling?
6. What is the difference between a sketch, a working drawing, and a 3-D model?

A rapid prototype of a computer mouse design.

Engineering in Action

Historically, designing products for everyday use was a process that often took months or even years to complete. Engineers now use parametric modeling software to develop and test product ideas in just days. This software allows an engineer to create the shape as well as test the strength of a chosen material. The engineer can make decisions about the design without having to manufacture the product by simulating the product using parametric modeling software. This saves time and money.

CAD
CAD is the acronym for *computer-aided drafting* (or *design*).

parametric modeling
Also known as *feature-based solid modeling,* parametric modeling is a 3-D computer drawing program.

profile
A profile is the two-dimensional outline of one side of an object.

SECTION 1: INTRODUCTION TO PARAMETRIC MODELING

Modern **CAD**—an acronym for *computer-aided drafting* or *design*—has its roots in the early 1960s. Aerospace, automotive, and electronics companies first used CAD. These large companies were the only ones who could afford computers that could perform the massive calculations needed for CAD. As personal computers became affordable and CAD software became available for this size of computer, CAD became a universal application in design, manufacturing, and construction (see Figure 7-1).

John Walker founded Autodesk in 1982 and introduced two-dimensional (2-D) CAD (AutoCAD) for use on personal computers. Another company, Pro/ENGINEER, introduced feature-based parametric modeling (three-dimensional, or 3-D, CAD) in 1988. Autodesk Inventor, a parametric modeling CAD program, became available to the public in 1999 (see Figure 7-2). Several 3-D modeling programs are available today. Two- and three-dimensional CAD programs are among the primary tools used to design products and structures today.

What Is Parametric Modeling?

Parametric modeling, also known as *feature-based solid modeling*, is a 3-D computer drawing program. Engineers and designers draw simple 2-D geometric shapes called **profiles** to begin their design concepts. Once a basic 2-D shape has been developed, the designer can force the lines to be parallel, perpendicular, horizontal, or vertical. The designer also assigns exact dimensions to each line of the geometric shape. This process is known as *applying constraints* to the drawing (see Figure 7-3).

As defined in Chapter 3, constraints are limitations that restrict the position or relationship between features of the geometric shape. Some examples of constraints are making two lines parallel or perpendicular to each other. We call these **geometric constraints.** We call the process of dimensioning the length of a line or the diameter of circle *dimensioning constraints* (see Figure 7-4).

The word *parametric* comes from the word *parameter*. A **parameter** is any set of physical properties whose values determine the characteristic of something. In the case of

FIGURE 7-1 Large mainframe computers were required to run the first CAD software programs. Powerful personal computers have now made CAD accessible to large and small companies.

parametric modeling, the "something" is drawn geometric shapes such as circles, rectangles, and triangles or any combination of shapes. The shape and size of drawn geometric shapes are the physical properties. So, we can say that parametric modeling is the process of determining the shape and size of the drawn or sketched geometry with parametric constraints.

Remember, parametric modeling gives engineers the ability to sketch their design ideas on a computer without having to worry about the sizes or locations of features such as holes, notches, and braces (see Chapter 5). Once we have completed the basic designed shape or profile, we assign constraints to the features of the object. We discuss the process of applying constraints later in this chapter (see Figure 7-5).

© RAGMA IMAGES/SHUTTERSTOCK.COM

FIGURE 7-2 Powerful personal computers have made CAD accessible to large and small companies.

The Need for Parametric Modeling

In the past, an engineer designed products independent of other engineers and those involved with manufacturing and marketing (selling) products. This process could involve a great deal of time, effort, and money. The engineer would complete the design before talking with those who were going to make the product or sell it. This discussion would often result in changes to the product because of

geometric constraint

A geometric constraint establishes a fixed relationship between the features of a drawing such as lines and shapes—for example, constraining two lines to be parallel or perpendicular.

parameter

A parameter is a physical property whose values determine the characteristic of something.

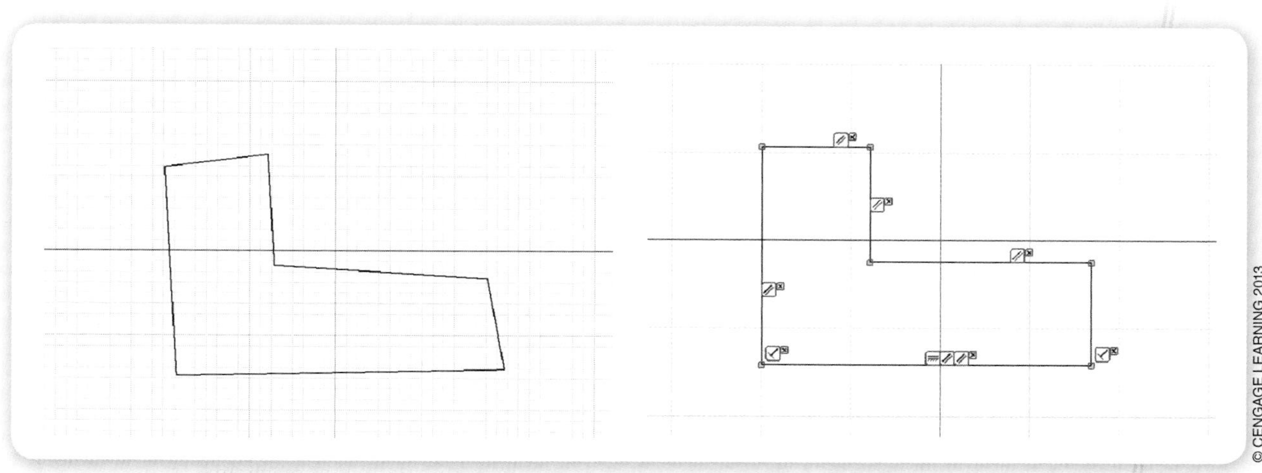

© CENGAGE LEARNING 2013

FIGURE 7-3 Sketched geometry before and after constraints have been applied.

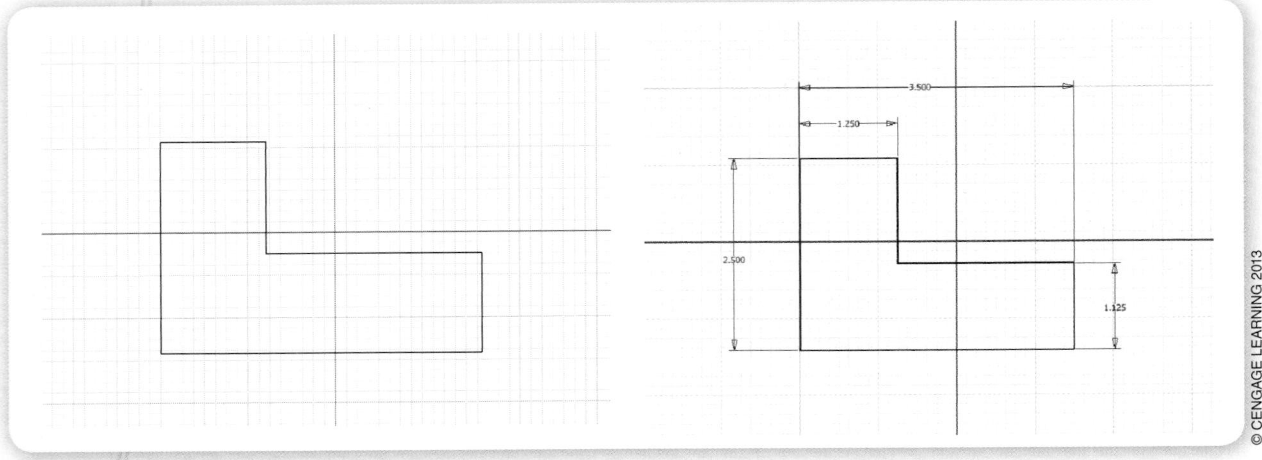

FIGURE 7-4 Sketched geometry before and after dimensional constraints have been applied.

concurrent engineering
Concurrent engineering is the process of involving everyone responsible for the design and manufacturing of a product right from the beginning of the process.

problems with manufacturing or trying to market (sell) the product. Engineering drawings would often have to be completely redrawn when design changes were made. Time and money would be lost because of the needed changes.

The concept of involving everyone from the beginning of the process revolutionized the design and manufacturing world. We save time and money because all engineering, marketing, manufacturing, and financing departments are involved in the design from the beginning of the process. In most cases, a better product is designed and produced. Designing a product this way is known as **concurrent engineering** (see Figure 7-6).

The use of CAD and the Internet improved the process of connecting engineers with other engineers. This did not help much, though, when changes to engineering design drawings had to be made. It was still time consuming to change the drawings. The introduction of parametric modeling changed all of that. Parametric (dealing with parameters) drawing allows the engineer to easily change an object's parameters such as width, height, and depth without having to redraw the shape of the object. Even the sizes of holes and their locations can be changed easily without having to redraw. We often refer to this process as *dimension-driven* or *geometry-driven* drawing.

We introduce the basic steps of creating an object in parametric modeling later in this chapter.

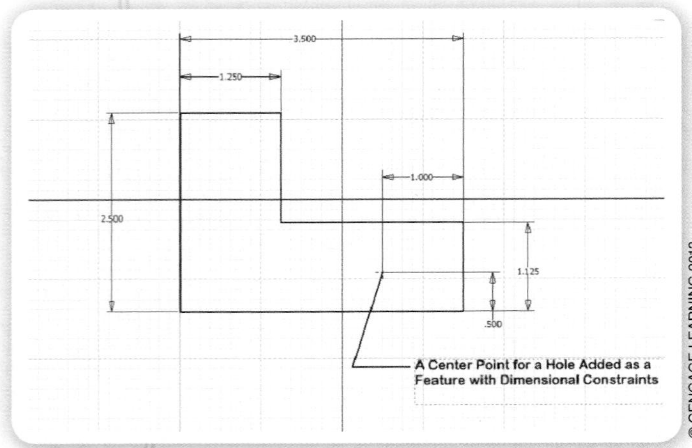

FIGURE 7-5 Geometric features added to the profile.

Science in Engineering

Stress analysis is an engineering process that uses scientific knowledge to determine the stress capacities of materials used in structures such as bridges, auto bodies, auto frames, and even bicycle frames. The aim of this scientific process is to determine whether a structure can safely withstand certain forces or loads. We analyze forces such as compression, tension, and torsion to examine the strength of the structure.

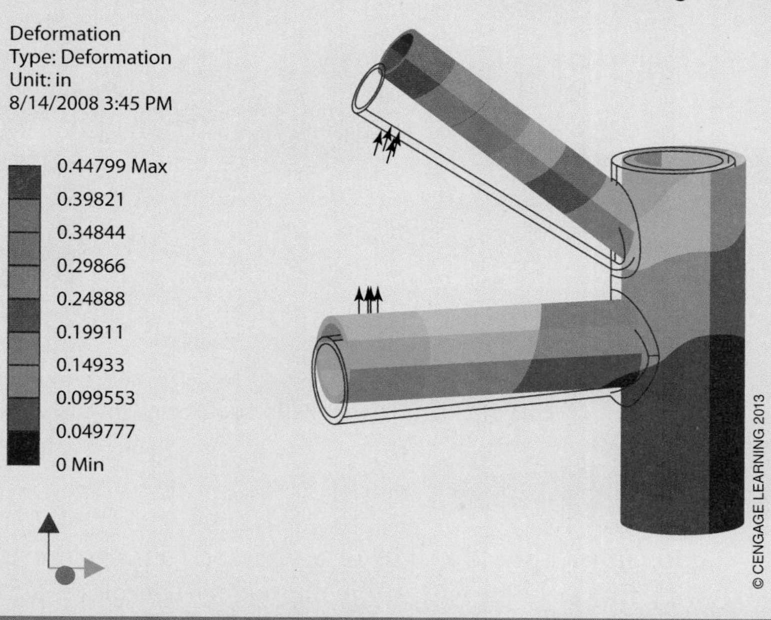

3

Who Uses Parametric Modeling?

So far, we have mentioned only engineers and designers as users of parametric modeling. They use this type of drawing program to design products and analyze the structural and mechanical characteristics of their designs. By using parametric modeling, engineers can test the strength of their designs or the mechanical movement of parts within their designs before the products are actually built. This **engineering analysis** process can save time and money (see Figure 7-7).

People other than engineers also use parametric modeling. People who work in computer graphics and animation are now using this type of program. Artists, sometimes called

FIGURE 7-6 **Engineers, manufacturers, sales representative, and even accountants make up the concurrent engineering team.**

engineering analysis
Engineering analysis is the process used by engineers to test the strength of their design or the mechanical movement of parts within the design before making the product.

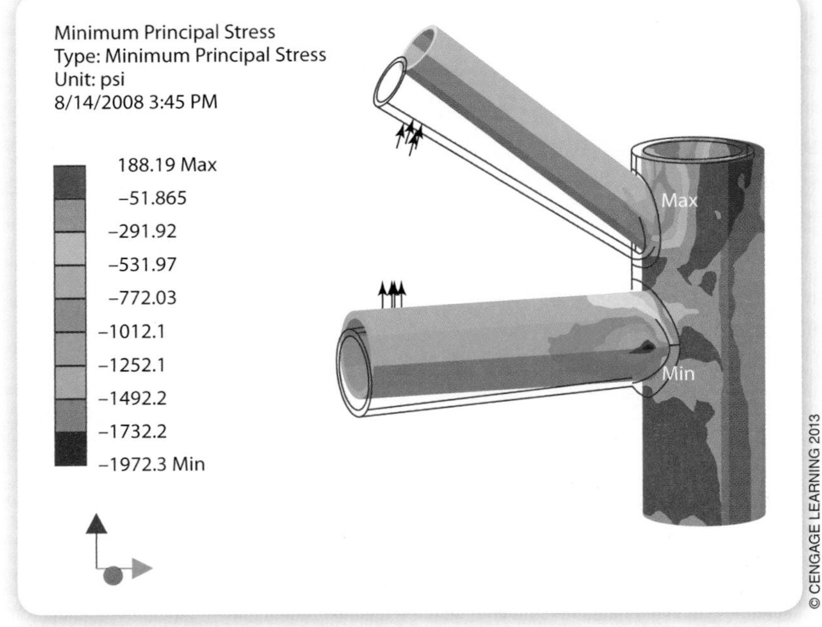

FIGURE 7-7 *One of the many analysis tools used in Autodesk Inventor.*

animators, draw the skin of a character and parametrically associate it with the skeleton within the character. Therefore, as the skeleton moves (is animated), the skin moves with it (see Figure 7-8).

FIGURE 7-8 *In addition to engineering design, parametric modeling is also used in animation.*

Medical solid modeling also uses parametric modeling. Magnetic resonance imaging scanners use this modeling technology to create 3-D internal body features. Engineers use medical information in conjunction with parametric modeling to design prosthetics (artificial limbs) and body parts such as replacement hip and knee parts (see Figure 7-9).

Engineers also use parametric modeling to create model parts when using rapid prototyping machines. Rapid prototyping is a method of "printing" in 3-D. Different rapid prototyping machines use different materials to produce the model, such as plastic or starch-based materials (see Figure 7-10).

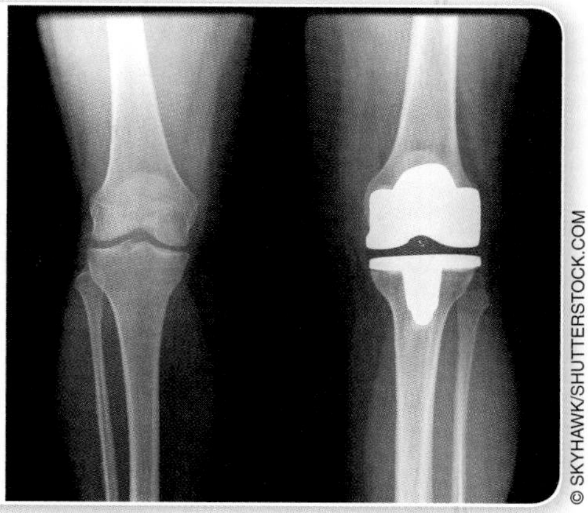

FIGURE 7-9 Parametric modeling is used to design artificial limbs and replacement knee and hip parts.

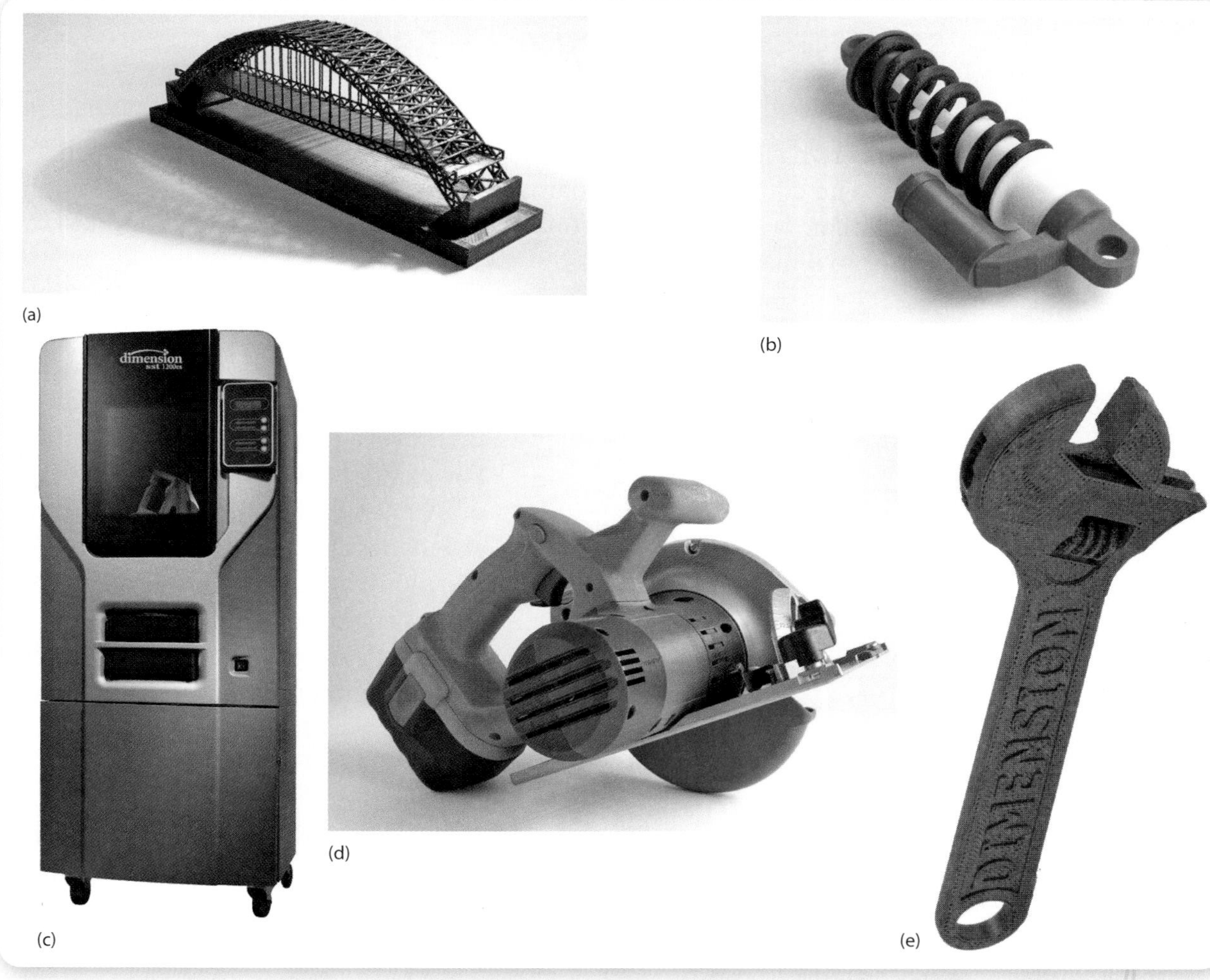

(a)

(b)

(c)

(d)

(e)

FIGURE 7-10 After designing a part in parametric modeling, a part model can be quickly "printed" using a rapid prototyping machine, which is sometimes called a 3-D printer. Shown here: (a) model bridge; (b) spring mechanism; (c) Rapid Prototype Machine; (d) circular saw model; (e) adjustable wrench model.

Advantages of Parametric Modeling

Traditional CAD drawing requires an engineer to draw geometric shapes to exact sizes and shapes. If the size or the shape of a design were to change, then the engineer would have to redraw the shape as well as change the dimensions of the shape. Parametric modeling requires only the dimension to be changed. The shape is automatically resized once the constrained dimension has been changed.

Editing a simple or complicated geometric shape in parametric modeling will automatically change all the views associated with the shape. For example, when the diameter and depth of a hole is changed in one view of an object, the other views will also change. Not only do the dimensions change but also the geometric shapes change.

Finally, the engineer has the ability to view a created part or produce an unlimited number of views.

SECTION 2: DEVELOPING AN OBJECT USING PARAMETRIC MODELING

Engineers have several parametric modeling software programs to choose from. The concepts and fundamental processes in each of these programs are similar. We cover these concepts in this section by using Autodesk Inventor to illustrate the process.

Drawing in Traditional CAD

Drawing in a traditional CAD program requires that lines and shapes be drawn to the exact measurements required. For example, a 4-inch square must be drawn at 4 inches. If a square is drawn at 3 inches and labeled with a 4-inch dimension, the square would still actually be 3 inches. However, a square can be drawn at any size in parametric modeling then dimensioned to 4 inches and the square will automatically change to 4 inches. (see Figure 7-11).

Changing the size of the square in CAD would require you to redraw the square and change the dimensions. Changing the size of the square in parametric modeling, on the other hand, requires you to change only the dimension. Drawing basic shapes is often referred to as sketching.

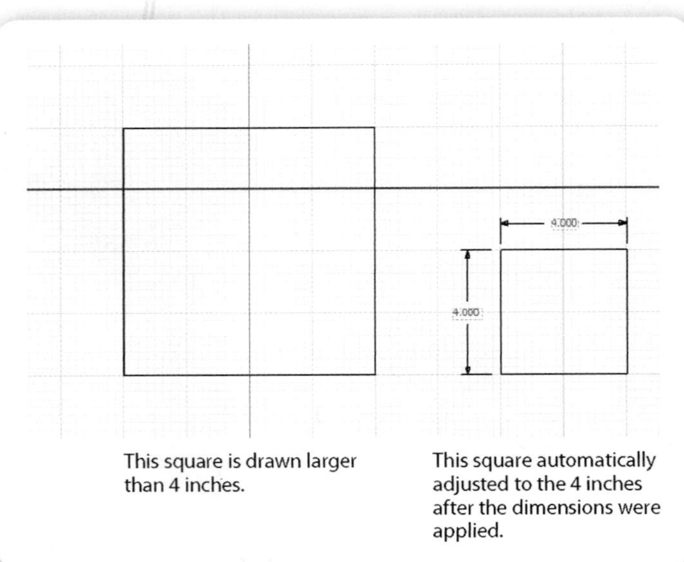
4.000
4.000
This square is drawn larger than 4 inches.
This square automatically adjusted to the 4 inches after the dimensions were applied.
© CENGAGE LEARNING 2013

FIGURE 7-11 In parametric modeling, a square can be drawn at any size and then dimensioned to 4 inches; the square will automatically change to 4 inches. However, it does not matter what size you draw the shape, once you place dimensions, the shape will adjust to the given size.

Geometric Shapes

It is important to be familiar with basic 2- and 3-D geometric shapes before you attempt to develop parametric drawings. You must be able to computer sketch 2-D rectangles, circles, arcs, or triangles before you can convert them to 3-D geometric solids such as cubes, spheres, and pyramids. A review of Chapter 4 will provide a refresher in understanding geometric shapes.

Cartesian Coordinate System

Two- and three-dimensional drawing in CAD programs is based on the mathematical **Cartesian coordinate system**. This system provides a method for locating points on the drawing screen. It is important to understand and be able to work with the coordinate systems.

Cartesian coordinate system
The Cartesian coordinate system is the graphical mathematical system used to locate points in two- and three-dimensional space.

So how does the Cartesian coordinate system work? Very simply, the 2-D system uses two lines, called *axes*, that are perpendicular (at right angles) forming a plane, referred to as an *xy*-plane. We call the horizontal axis the ***x*-axis** or *x*-coordinate (also known as the abscissa of the point) and the vertical axis the ***y*-axis** or *y*-coordinate (also known as the ordinate of the point) (see Figure 7-12). We call the point where the two axes intersect the **origin** and label it O (for origin) or 0,0. The *x*-axis is horizontal, and the *y*-axis is vertical. We divide the coordinate system into four quadrants and label them with Roman numerals: I (+,+), II (–,+), III (–,–), IV (+,–) (see Figure 7-12). The quadrants are normally labeled counterclockwise, with number I beginning in the upper right quadrant. The value of each coordinate number is either negative (–) or positive (+). The *y* values are positive if they are above the *x*-axis (horizontal) and negative if they are located below this line. The *x* values are positive if they are to the right of the *y*-axis (vertical) and negative if they are to the left of this line. We can see this in Figure 7-12.

origin
The origin point (0 or 0,0) is where the *x*- and *y*-axes intersect.

CAD uses only quadrant I (the upper right one). This means that *x* and *y* values will always be positive. We can draw a line by identifying two points in this quadrant, for example, *x*2, *y*3 and *x*5, *y*3 (usually stated 2, 3 and 5, 3). Begin by locating the first point 2 units along the *x*-axis to the right of the origin and 3 units directly up from that point and parallel to the *y*-axis. The second point is located 5 units along the *x*-axis to the right of the origin and 3 units directly up from that point, parallel to the *y*-axis. By connecting these two points, we create

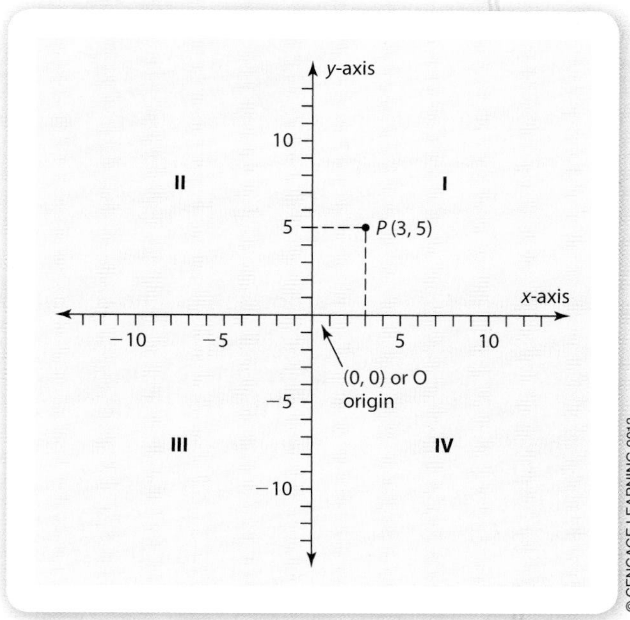

FIGURE 7-12 The *x*- and *y*-axes, 0,0 or O (origin), and four quadrants.

Math in Engineering

The Cartesian coordinate system has many mathematical applications, including robotics, computer-controlled-machining, and computer-aided drafting, mentioned in this chapter. However, the most common system used today is the world mapping concept of longitude, latitude, and height or altitude. The prime meridian is the originating meridian for measures of longitude and represents the *y*-axis. The equator is the originating line for latitude and represents the *x*-axis. Any distance above Earth from sea level represents the height (altitude), or the *z*-axis. We use the mathematical concept of this 3-D Cartesian coordinate system to develop maps and operate the global positioning system (more commonly called GPS).

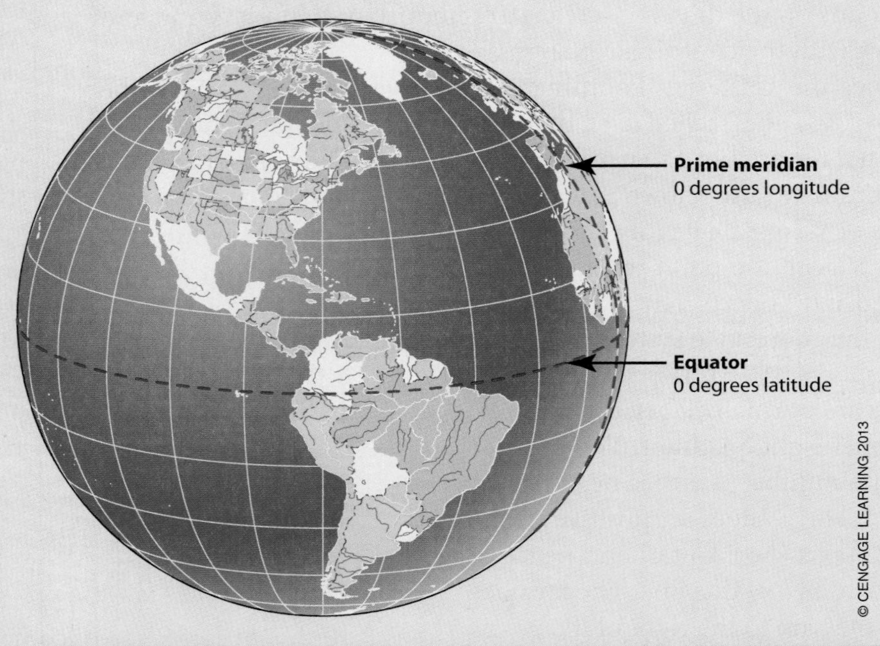

plane
A plane is a flat surface that has no thickness, sort of like a piece of paper that extends forever.

a horizontal line 3 units long (see Figure 7-13). All points and lines in the 2-D system are located on a single plane formed by the *x*- and *y*-axes. A **plane** is a flat surface with no thickness, similar to a piece of paper that extends forever.

The 3-D Cartesian coordinate system uses three axes: *x*, *y*, and *z*. Look at Figure 7-14. The three axes are perpendicular to each other. This system allows the viewer to visualize any point in space. Take a look at your classroom. Look over to one corner of the room where the two walls meet the floor. The line that is formed by the floor and one wall creates one axis, while the line where the other wall and floor meet creates another axis. The third axis, which is vertical, is formed by the two walls meeting. The three axes form three planes—*xy*, *yz*, and *xz* (see Figure 7-15). Understanding this will be important when you begin drawing in parametric modeling.

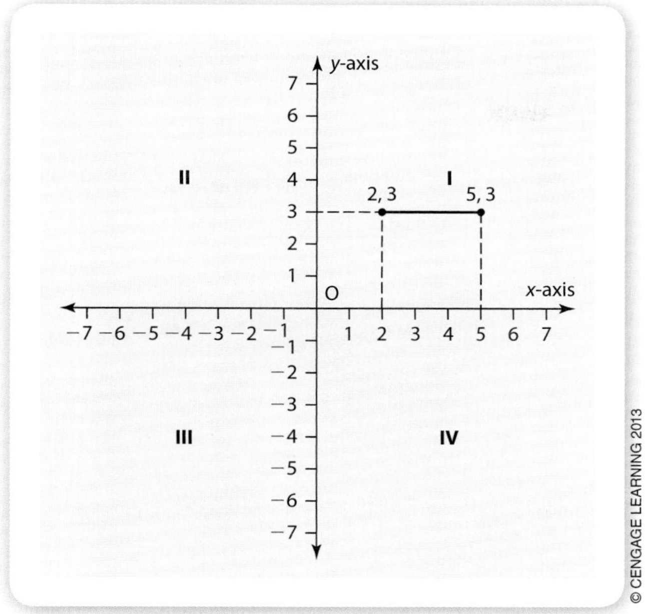

FIGURE 7-13 Plotting a line.

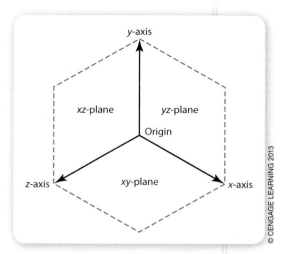

FIGURE 7-14 3-D Cartesian coordinate system showing *xy*-, *xz*-, and *yz*-planes.

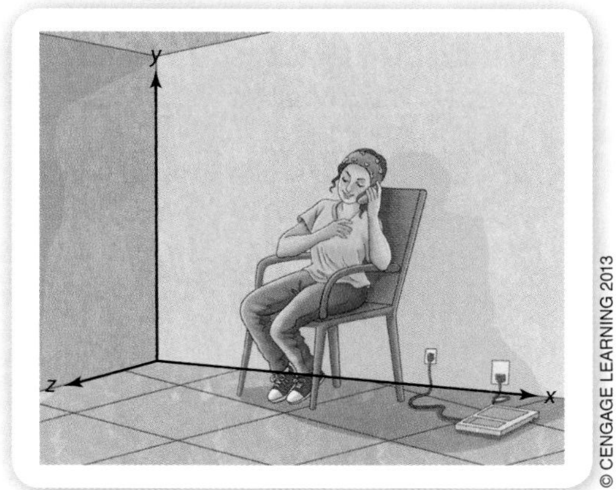

FIGURE 7-15 The walls and floor form a 3-D Cartesian coordinate system.

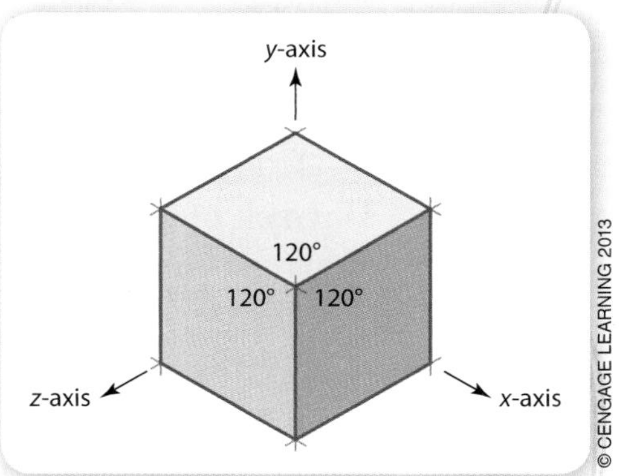

FIGURE 7-16 Isometric shape showing the 3-D Cartesian coordinate system relationship.

Isometric drawings are an example of using the 3-D Cartesian coordinate system. Remember in Chapter 5 that you learned about sketching isometric shapes. Figure 7-16 demonstrates the relationship of the 3-D Cartesian coordinate system to isometric drawing.

Six Basic Steps

5

To create a parametric part model, follow six basic steps:

1. Develop a 2-D profile sketch of the basic shape of the part (sketch geometry).
2. Place geometric and dimension constraints on the 2-D sketch.

Did You Know?

French mathematician and philosopher René Descartes and French lawyer and mathematician Pierre de Fermat developed what we now call the Cartesian coordinate system separately in 1637.

Pierre de Fermat

Rene Descartes

3. Create a solid 3-D feature by extruding, sweeping, or revolving the 2-D sketch.
4. Continue part development by adding additional parametric features such as holes, slots, and bevels.
5. Assemble the parts.
6. Create working drawings for manufacturing production.

We next discuss each step separately.

sketch geometry
Sketch geometry is the process of drawing geometric shapes in a parametric modeling program.

Step 1: Sketch the Profile Geometry We often refer to drawing geometric shapes in a parametric modeling program as **sketch geometry**. The first step is to sketch the shape of the profile view (also known as a two-dimensional view of one side of an object). This may be a square, a circle, or a more complex shape. We will use more complex shapes in this example (see Figure 7-17). Remember, drawing the exact size or shape of the profile is not necessary here and will be taken care of in the next step.

Step 2: Apply Geometry and Dimensional Constraints The next step is to constrain the parts of the sketched geometry, such as edges (lines), circles, and center points, and their sizes and locations (dimensions). Some of the types of geometric constraints are parallel, perpendicular, tangent, and concentric (see Figure 7-18). (See the section on Geometric Language in Chapter 4 for more information about these terms.) These constraints will become more familiar as you use parametric modeling software more often. Dimensions are of two types: (1) size (e.g., the length of a line, the size of a circle or an arc) and (2) location (e.g., the location of the center of a hole from two edges of the object) (see Figure 7-19).

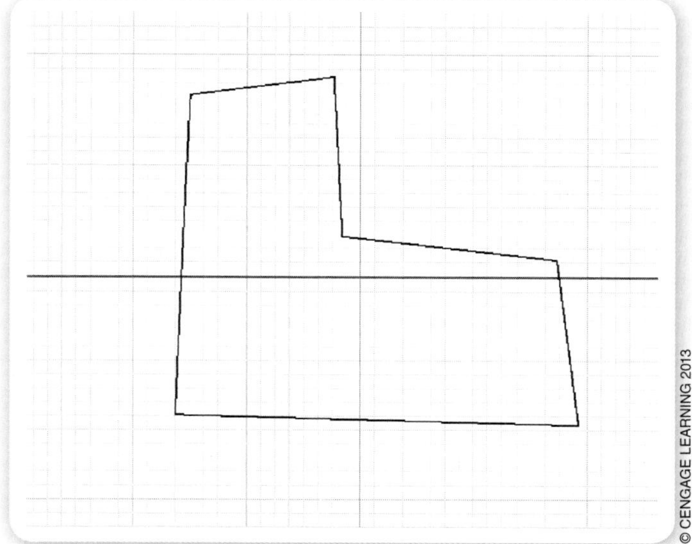

FIGURE 7-17 Step 1: Sketch the profile geometry.

Step 3: Create a Solid 3-D Figure Now that the sketched geometry and the dimensions are constrained, we must convert the 2-D drawing into a 3-D sketch. We do this by changing the 2-D profile view to a 3-D view (in this case, an isometric view) (see Figures 7-20 and 7-21). We have to switch the 2-D sketching toolbar to the 3-D development toolbar. Most modeling software programs have a tool command to extrude the isometric profile into a 3-D solid shape. **Extruding** is the process of adding depth to the geometric

extruding
In parametric modeling, extruding is the process of adding depth to the geometric shape.

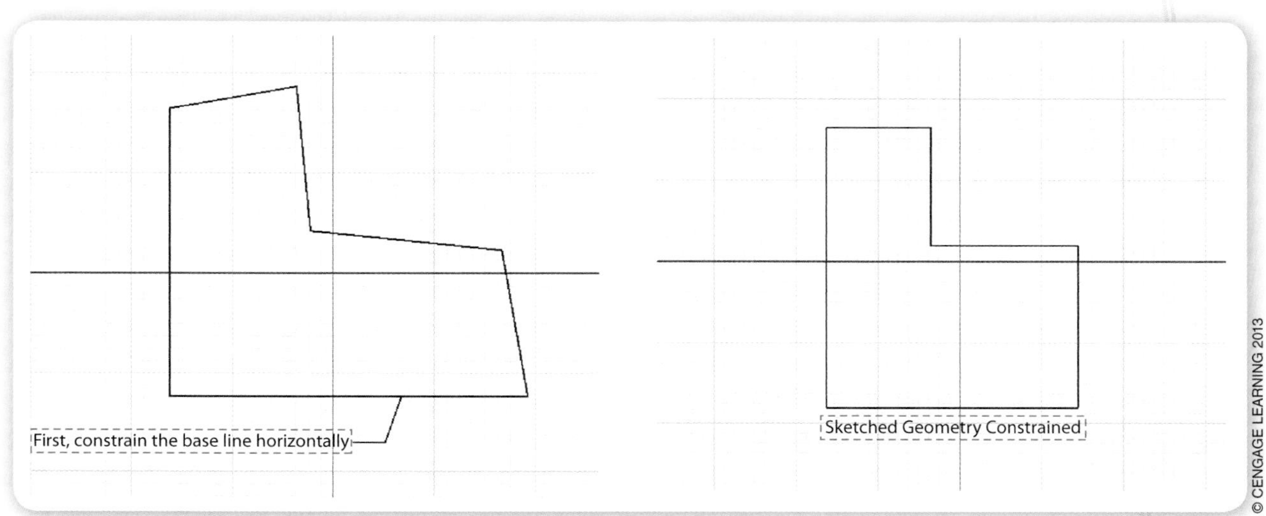

FIGURE 7-18 Step 2: Apply geometric constraints.

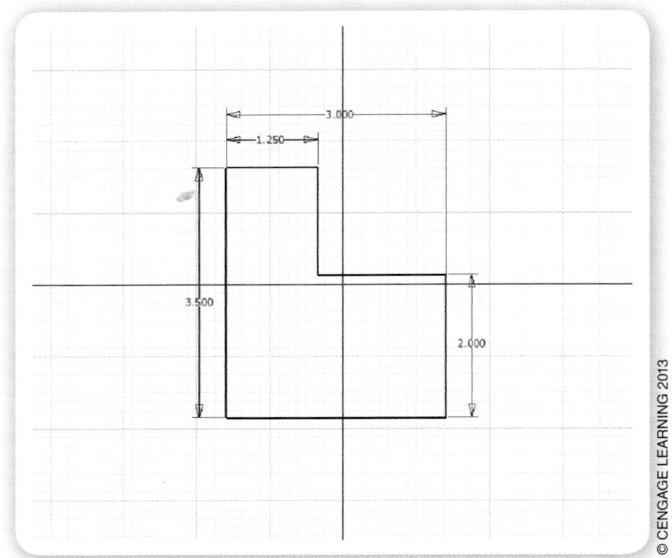

FIGURE 7-19 The second part of Step 2, constrain the dimensions of the profile.

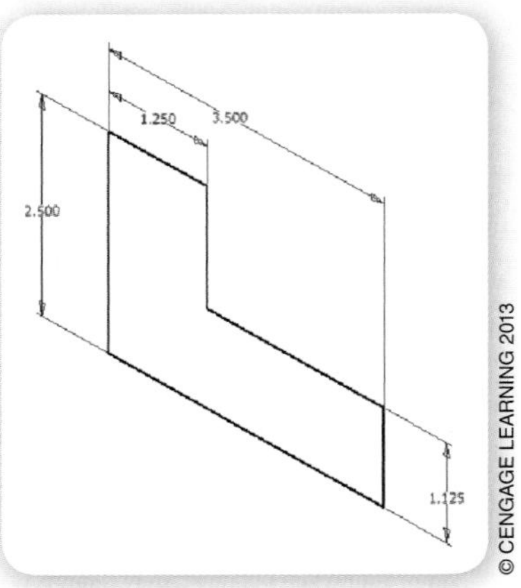

FIGURE 7-20 Two-step process to convert the 2-D sketch into an isometric (3-D) sketch.

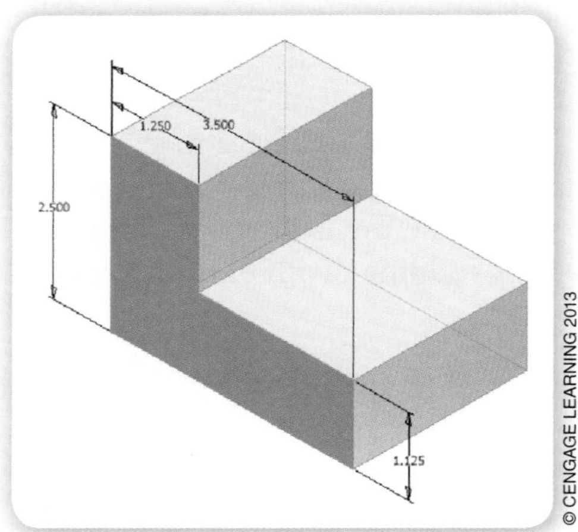

FIGURE 7-21 Sketched geometry in isometric view.

FIGURE 7-22 Extrude the profile to form 3-D solid model.

shape. Modeling programs allow the profile to be extruded in front of or behind the profile. The recommended direction is behind the profile, making the profile the front view of the object (see Figure 7-22).

Step 4: Add Part Features Product parts are usually more complex than a simple block or L-shape. Parts may have holes, slots, braces, or raised ridges, just to mention a few possible features. Once we have created the basic geometric shape, we can add such features.

For this example, we will place a hole in the object. Return to the 2-D sketch toolbar. We must select a *sketch plane* to draw on the surface of the object. In parametric modeling software, a **sketch plane** is the surface on which a geometric feature is drawn. Select the sketch plane on the surface where the hole will be placed. Next, select the center point geometric tool; then click anywhere on the sketch plane to create a center point for the hole (see Figure 7-23). Apply dimension constraints to locate the center point at the desired location on the object surface (see Figure 7-24).

Switch back to the 3-D toolbar and select the hole tool; then select the diameter of the hole and indicate how deep the hole will be in the object. Set the hole diameter to 0.5 inches and the depth to go all the way through the object (see Figures 7-25 and 7-26).

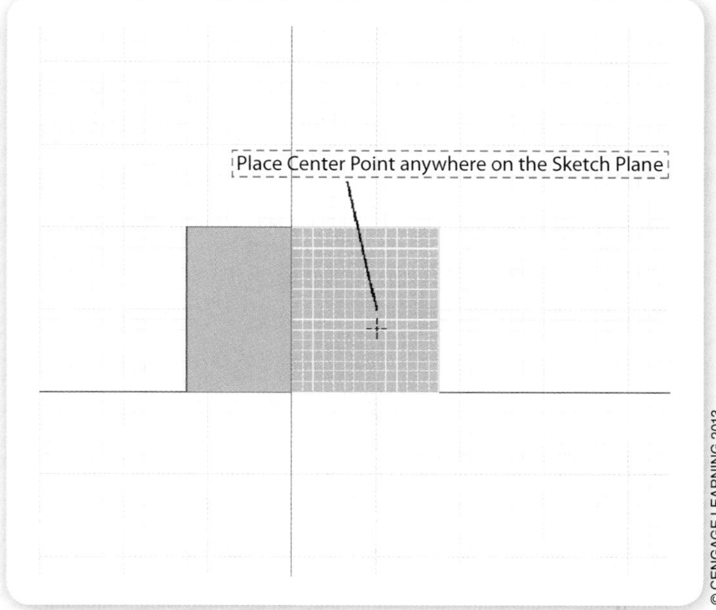

FIGURE 7-23 Select the sketch plane and place the center point.

sketch plane
In parametric modeling software, a sketch plane is the surface on which a geometric feature is drawn.

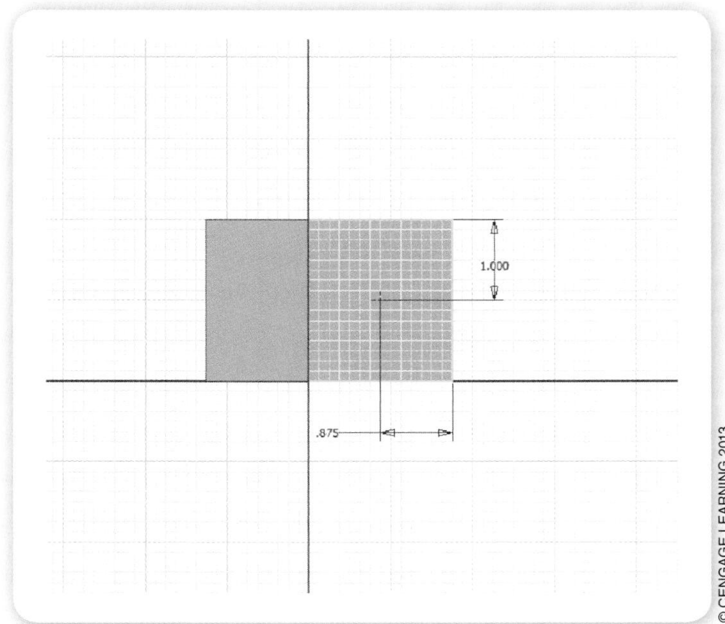

FIGURE 7-24 Apply dimension constraints.

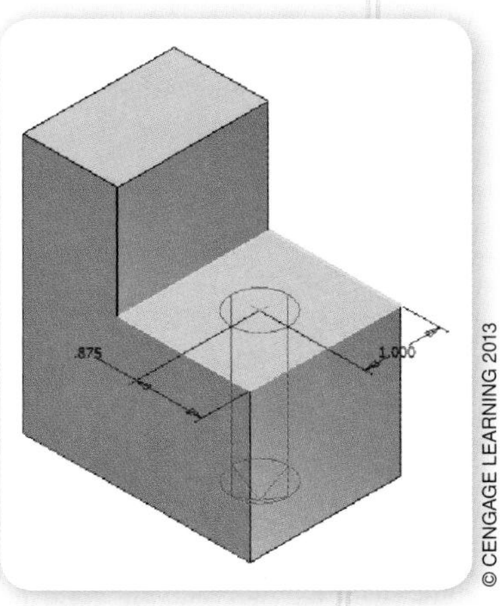

FIGURE 7-25 Place the hole feature.

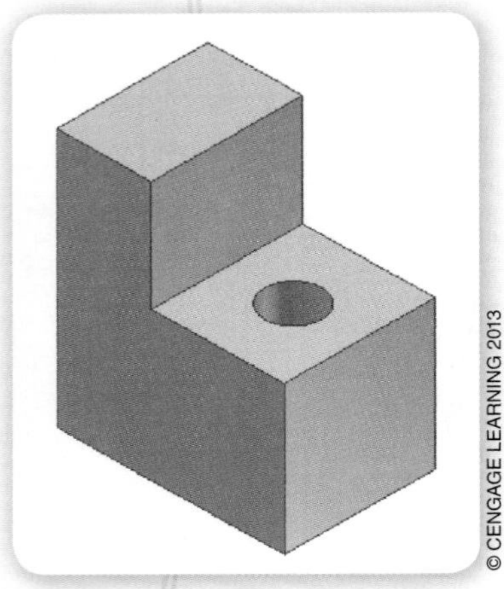

FIGURE 7-26 Completed geometric shape.

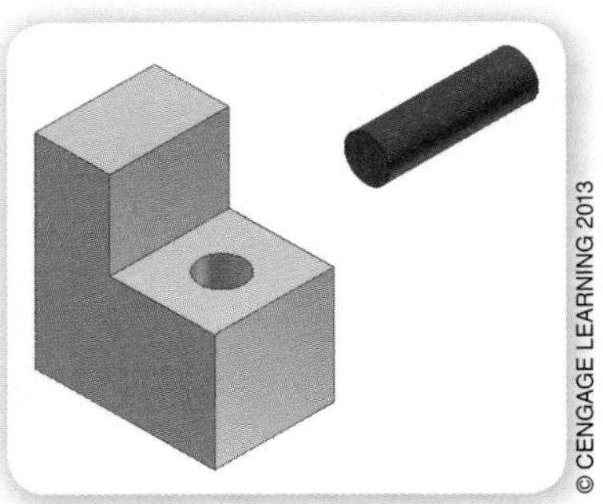

FIGURE 7-27 Assemble the components.

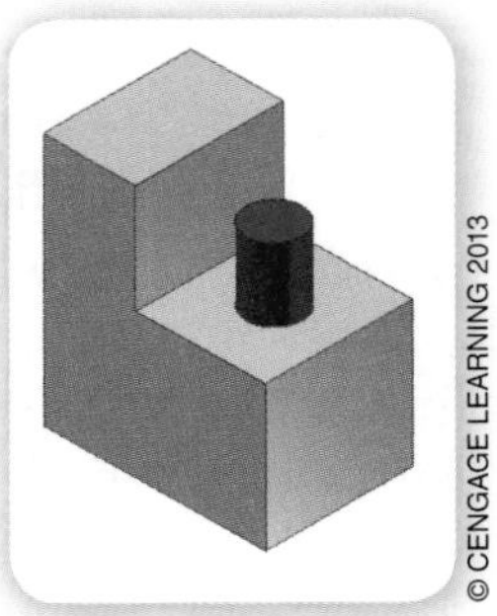

FIGURE 7-28 Assembled components.

working drawing
A working drawing is a 2-D drawing of individual parts that shows different views (e.g., front, side, and top views) with dimensions and annotations (notes).

Step 5: Assemble the Parts Most designed products have more than one part and must be assembled (see Figure 7-27). As part of the design process, we have to make sure parts can be assembled without any problems, so the designer must be able to demonstrate how well the parts fit. This can be done without having to manufacture or construct the product. Parametric modeling gives the designer the drawing tools necessary to produce a computer model of the assembled product.

Each parametric modeling software program accomplishes the product assembly using a slightly different procedure. However, the fundamental process is to open an assembly file (Standard.iam in the case of Autodesk Inventor) to create the assembled model. Each part of the product is imported and constrained in the correct position and orientation to the overall product (see Figure 7-28).

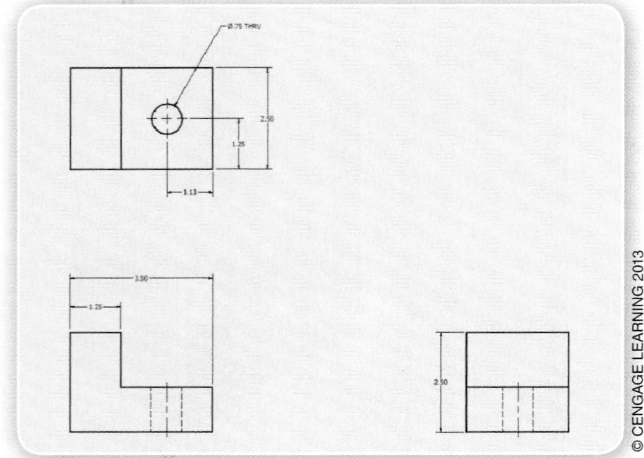

FIGURE 7-29 Working drawing for manufacturing (also called a shop drawing).

Step 6: Create Working Drawings for Manufacturing Pictorial (3-D) drawings of the product help the engineer and others involved in the design and selling of the product to understand how the product works. Those who are involved with manufacturing the product must have very detailed 2-D drawings to be able to correctly manufacture the product. The pictorial drawing cannot provide the needed detail. A **working drawing** (sometimes called a *shop drawing*) is needed (see Figure 7-29).

As noted in Chapter 6, working drawings are 2-D drawings of each part showing different views (i.e., front, side, and top) with dimensions and annotations (notes) (see Figure 7-30). Parametric modeling software makes it easy to create working drawings.

We can create working drawings using Autodesk Inventor file format Standard.idw. Once we have created the drawings, we can place appropriate dimensions and notes on each view.

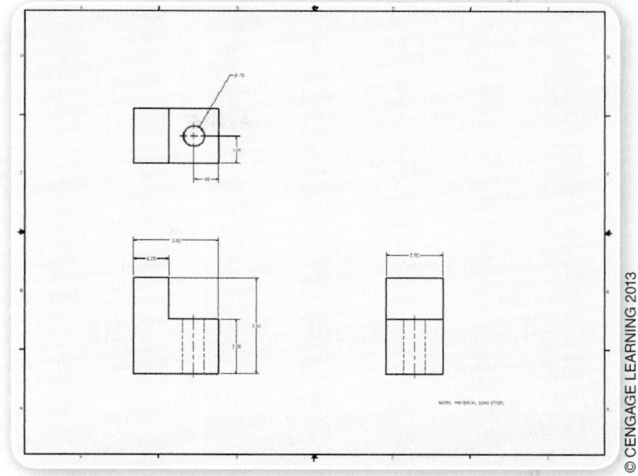

FIGURE 7-30 Working drawing of parts for manufacturing.

SUMMARY

In this chapter you learned:

- Parametric modeling software is used by engineers, designers, animation artists, and even engineers in the medical field.
- Parametric modeling has improved efficiency and shortened the time needed to design a product by using the software with concurrent engineering practices.
- An understanding of 2-D and 3-D geometric shapes is necessary to use parametric modeling successfully.
- The Cartesian coordinate system is used by 2-D and 3-D CAD programs to identify points in drawings.
- Editing (changing, correcting) in parametric modeling is one of the biggest advantages over conventional CAD.
- We can combine basic geometric shapes to create complex shapes in parametric modeling.
- Parametric modeling software is based on the principle of the geometric coordinate system.

VOCABULARY

Write a definition for each term in your own words. After you have finished, compare your answers with the definitions provided in the chapter.

CAD
Parametric modeling
Profile
Geometric constraint
Parameter
Concurrent engineering
Engineering analysis
Cartesian coordinate system
Origin
Plane
Sketch geometry
Extruding
Sketch plane
Working drawing

STRETCH YOUR KNOWLEDGE

Provide thoughtful, written responses to the following questions.

1. Compare and contrast the differences between traditional CAD and parametric modeling.
2. Discuss which type of software program—CAD or parametric modeling—would be used in concurrent engineering design and explain why.
3. Outline the steps taken to develop a cube with a hole through the center using a parametric modeling program.
4. Explain why applying geometric and dimensional constraints to an object in parametric modeling can assist in the design process.
5. Create a new use for parametric modeling other than the uses mentioned in this chapter.

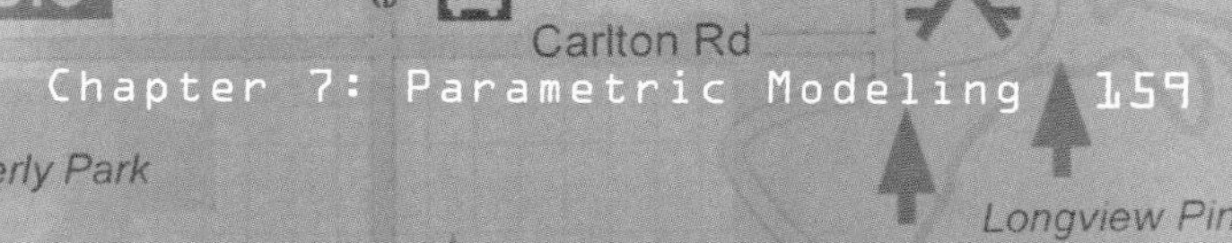

Engineering Challenge

Design engineers, along with production engineers, marketing specialists, and others, create design solutions every day. When they work together, it is called concurrent engineering. Imagine you are a design engineer. Your challenge is to pick two other students to serve with you on your concurrent engineering team and solve the following design problem. Make sure your team includes a production engineer (a person who designs a way to make the product) and a marketing specialist (a person who designs an advertizing campaign). Your team must work together to develop a successful product that will sell.

Design Problem: Create a small picture frame that will hold a 3″ × 5″ photograph and give you the time of day and the current temperature.

Solution: You must create your solution using a parametric modeling program. Provide an isometric view, an assembled 2-D mechanical drawing, and individual mechanical drawings of each part. Your team should provide a list of steps to make the picture frame (not the clock or thermometer) and an advertising flyer to promote the product.

Onward to Next Destination

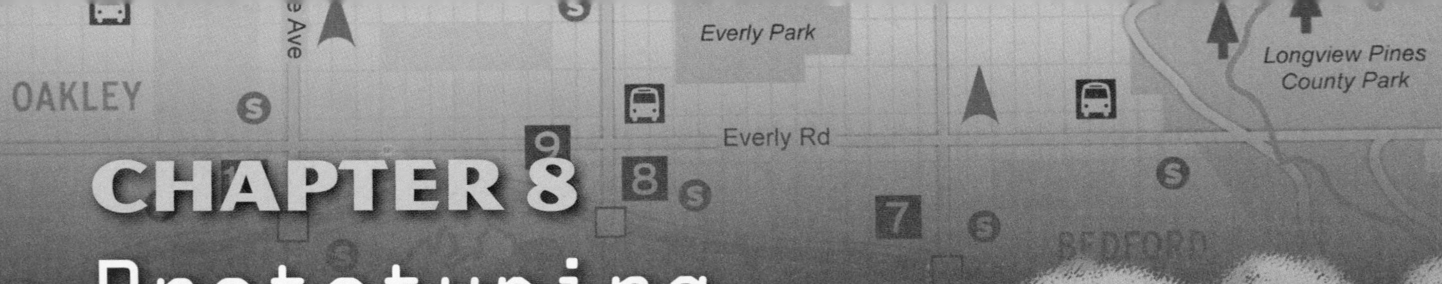

CHAPTER 8
Prototyping

Menu

Before You Begin

Think about these questions as you study the concepts in this chapter:

1. What is a prototype?
2. How is a prototype different from a model?
3. How do engineers classify materials?
4. What are a material's mechanical properties?
5. Why should engineers understand the mechanical properties of materials when designing a product?
6. Why should engineers understand the mechanical properties of materials when fabricating a prototype?
7. What external forces can be applied to a material?

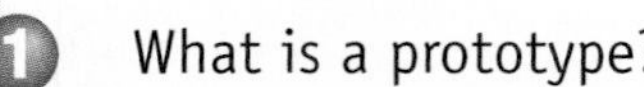

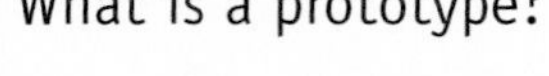

Thomas Edison's "speaking machine."

Engineering in Action

Your ability to listen to music from portable electronic devices such as a radio, an MP3 player, or an iPod would not be possible without engineers fabricating and testing prototypes more than 100 years ago.

Thomas A. Edison built the first prototype of his "speaking machine" in 1877. At first, Edison did not think his phonograph would record music. But after he tested the prototype and used his results to make changes, his design was successful. Eventually, people heard the nursery rhyme "Mary Had a Little Lamb" coming from Edison's machine.

Edison understood the importance of testing designs. Eventually, people began to call this successful inventor the "Wizard of Menlo Park." Today, you can thank Edison for the technologies that let you hear your favorite tunes whenever and wherever you want.

Did You Know?

Thomas A. Edison was granted his first patent in 1869 (no. 90,646) for an electronic voting machine. However, in 1869 no one was interested in purchasing this machine. He resolved then never to invent a product that would not sell.

SECTION 1: WHAT IS A PROTOTYPE?

After engineers have designed a product, they must physically test its design. Before that is done, modern computer technology allows engineers to test their designs using computer-assisted design (CAD) software in a process called *parametric modeling*. Then the design is physically tested before the product is put into production. Two types of product representations are common to engineering: models and prototypes.

Models

model
A model is a three-dimensional representation of an existing object.

A **model** is a three-dimensional representation of an existing object. Models are not full size. They are typically smaller versions of items that were previously built or are being planned for construction. Think about a model airplane kit you may purchase, a globe (a model of the Earth), or the model house in Figure 8-1. These items already exist. The model allows you to

FIGURE 8-1 Architects often create model buildings.

bring the smaller version into your room. Some models may be a larger representation of a smaller item. An example would be a model of the human eye used in a science class. Engineers do not test models.

Prototypes

As noted in Chapter 3, a **prototype** is a full-scale working model used to test a design concept. Through observations, testing, and measurements, engineers can find out whether the design concept will solve the design problem. A prototype is built full size to allow for testing. If you were designing a new bicycle, someone would have to ride the bicycle to test the new design. This would not be possible with a small model. Engineers use the information obtained through prototype testing to revise their design and improve the product. Figure 8-2 shows a parametric model, a mock-up, and a prototype of a new bicycle design.

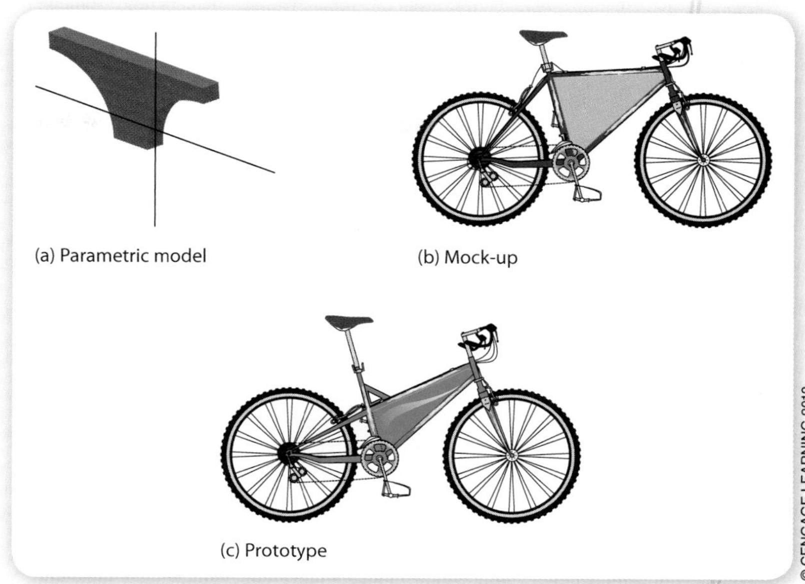

FIGURE 8-2 A parametric model for a bicycle leads to mock-up and prototype testing.

2

prototype
A prototype is a full-scale working model used to test a design concept by making actual observations.

fabrication
Fabrication is the process of making or creating something.

SECTION 2: MATERIALS

Engineers and technologists must use various types of materials to fabricate both the prototype and the actual product. **Fabrication** is the process of making or creating an object. We can classify materials used to fabricate a product into two basic types: natural and synthetic.

Natural Materials

Natural materials are naturally produced by the Earth or living organisms. Examples include iron ore, granite, and wood. We can easily perform basic fabrication processes on wood. Therefore wood, or lumber, is the most commonly used natural material for prototypes. We can also use various types of metals to fabricate prototypes.

3

We classify lumber into two types: hardwoods and softwoods. **Hardwoods** come from deciduous trees, or trees that shed their leaves each year. Hardwoods include oak, maple, and walnut. They are typically used in furniture construction because of their appearance and strength. **Softwoods** are products of conifer or evergreen trees that have needles and bear seed cones (see Figure 8-3).

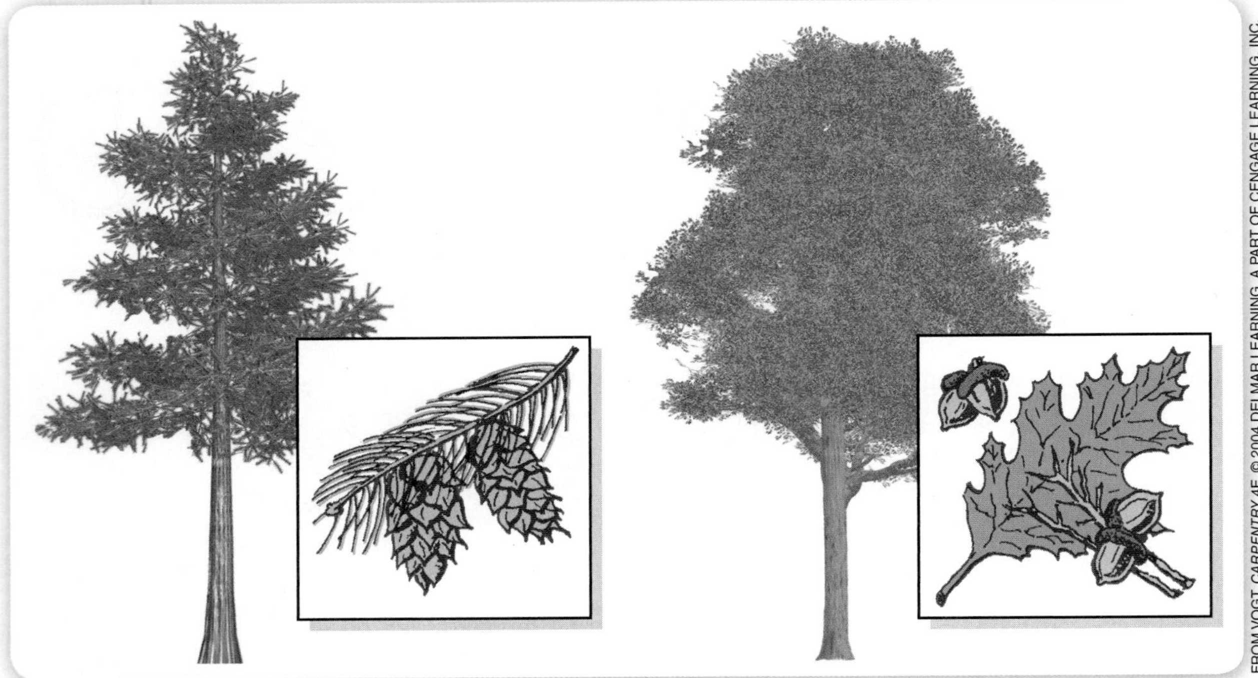

FIGURE 8-3 Softwoods come from cone-bearing trees. Hardwoods come from leaf-bearing trees.

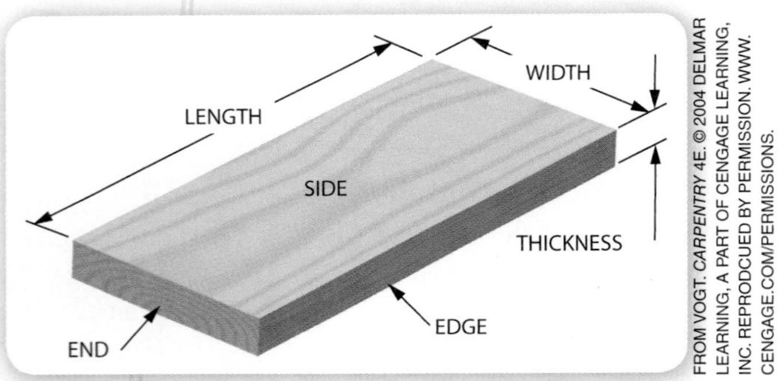

FIGURE 8-4 We distinguish lumber surfaces with specific names.

Because deciduous trees grow slowly, their woods cost more than softwoods. As the name *softwood* implies, these woods do not have the strength of hardwoods. We commonly use them for prototyping because they are easier to fabricate. The most common softwood is pine.

Lumber size is given by its length (distance with the grain), width (distance across the grain), and thickness as shown in Figure 8-4. The grain in lumber is actually the yearly growth rings of the tree. If you have ever tried to split firewood, you know that the strength of lumber lies in its grain. Engineers have used this knowledge to develop another lumber product called plywood. As shown in Figure 8-5, plywood has cross-bonded layers of wood that help develop a much stronger product. We call this layering of materials **lamination**. Engineers use plywood in fabrication and prototyping when larger flat surfaces are required.

ferrous metals
Ferrous metals contain iron.

nonferrous metals
Nonferrous metals do not contain iron.

Metal is another type of naturally produced material used in prototyping. **Metal** is durable, opaque, a conductor, reflective, and heavier than water, and it can be melted, cast, and formed. We classify metals as either those that contain iron (**ferrous metals**) or those that do not contain iron (**nonferrous metals**).

Ferrous metals include both iron and steel. Nonferrous metals include copper, aluminum, gold, and silver. We can form an **alloy** by combining two or more metals. Engineers develop alloys to produce a metal with the desired properties. For example, steel is an alloy of iron and carbon. The carbon makes the steel alloy stronger than pure iron. Brass is an alloy of copper and zinc.

Glass, clay, and ceramics are also natural materials, but they are not typically used in the fabrication of prototypes.

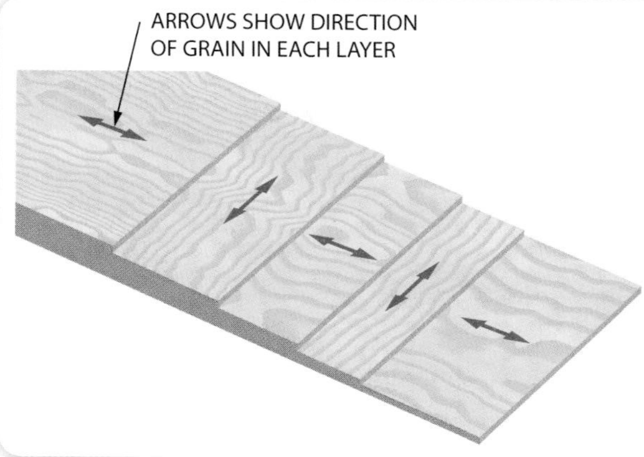

ARROWS SHOW DIRECTION OF GRAIN IN EACH LAYER

FIGURE 8-5 Plywood is constructed of cross-bonded layers.

Synthetic Materials

Synthetic materials are produced by humans. Humans take natural materials such as petroleum and chemically change their structures to make materials with desired properties. We call the most common synthetic materials **plastics**, meaning that they are synthetic materials with large and heavy molecules that are capable of being formed by heat or pressure (or both). We produce them primarily from petroleum products. We can divide plastics into two types: thermoplastics and thermosetting plastics. Thermoplastics may be heated and shaped many times. Thermosetting plastics can only be formed by heat one time.

Common thermoplastics include polyethylene, polystyrene, acrylic, polyvinyl chloride, and polypropylene (see Figure 8-6). We can easily identify a plastic by looking at the recycling triangle symbol code found on each product. Polyethylene is made from natural gas. We use polyethylene

alloy
An alloy is a combination of two or more metals into a single compound.

plastics
Plastics are synthetic materials that have large, heavy molecules, can be formed by heat, and are primarily produced from petroleum products.

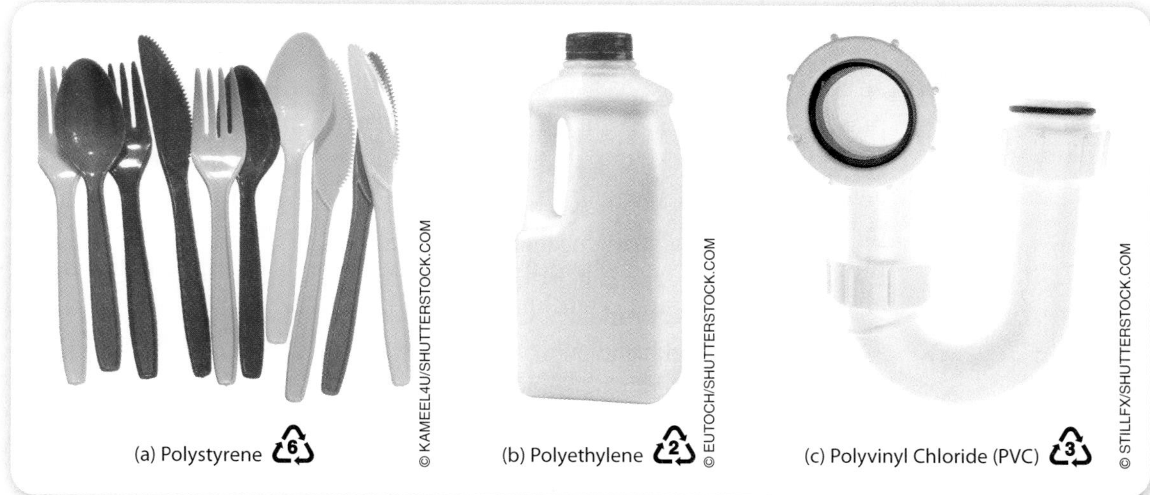

FIGURE 8-6 Three examples of common thermoplastics.

Science in Engineering

Most plastics are made from petroleum products, which are produced from fossil fuels. Fossil fuels are not renewable resources, which is one reason why plastics should be recycled. Today, Americans recycle only some 5 percent of all plastics that are produced. Discarded plastics, especially water bottles, have become a major environmental concern (see Figure 8-7).

Why do Americans recycle so few plastics? We have no simple answer. One reason might be that each plastic has different characteristics, so that different types of plastics must be recycled separately. To help consumers and recyclers identify different plastics, the Society of Plastics Engineers has developed a symbol code (see Figure 8-6). Currently, 39 states require these codes on plastic products. Scientists are continuing to research better methods to recycle plastics.

© HUGUETTE ROE/SHUTTERSTOCK.COM

FIGURE 8-7 **Plastic-water-bottle waste builds up.**

© ISTOCKPHOTO/PATRICK HERRERA

FIGURE 8-8 **An electric switch plate is made from thermosetting plastics.**

for a variety of products such as milk cartons. Polystyrene provides an inexpensive and easily fabricated synthetic material. We make the plastic silverware typically used on a picnic from polystyrene. Acrylic is a durable synthetic material that comes in sheets and may be colorless or tinted. It is easily formed it into clear containers, from fish tanks to salad bowls. We can find polyvinyl chloride, or PVC, in many plumbing systems. Polypropylene is an extremely strong plastic, and it also is used to insulate winter clothing.

The two types of thermosetting plastics commonly used are polyester and phenolic. We can use polyester wherever a strong material is required. We can use phenolics, sometimes called Bakelite, for electrical components (see Figure 8-8). Thermosetting plastics require special equipment for their fabrication, so we do not typically use them in prototype fabrication. Most thermosetting plastics also cannot be recycled.

Mechanical Properties

Engineers use the term **mechanical properties** to describe a material's characteristics. Hardness, toughness, elasticity, plasticity, brittleness, ductility, and strength are the most common characteristics described. Through years of research and study, engineers have devised measurements to describe each mechanical property.

mechanical properties
Mechanical properties are descriptions of a material's characteristics and include hardness, toughness, elasticity, plasticity, brittleness, ductility, and strength.

Hardness refers to a material's ability to resist permanent indentation by another object. Hardness indicates to the engineer how resistant the material is to wear. Diamonds are extremely hard, and we sometimes use them to cut metal. A material such as pine will not withstand even the force of a nail being driven into it, so we consider pine a soft material.

Toughness is the ability of a material to withstand a force applied to it. It indicates the ability of a material to absorb energy. A softball glove is tough. It absorbs the energy from a ball every time you catch it. The leather gives a little bit, but it does not break.

Whenever a material has a force applied to it, it typically changes shape. **Elasticity** is the ability of a material to return to its original shape once the force is gone. A product that demonstrates high elasticity would be a spring. When force is applied, the spring compresses. The spring returns to its original shape once the force is removed. Elasticity is only a temporary change in the shape of the material. **Plasticity**, on the other hand, is the ability of a material to be deformed and then remain permanently in the new shape.

Brittleness refers to a material that is not flexible and is easily broken. Brittleness is the opposite of plasticity. A brittle material will break before it is deformed. Glass is an example of a brittle material. Glass will not absorb the force of a softball. It will break.

Ductility is the ability of a material to bend, stretch, or twist without breaking. Most metals are very ductile. You can bend them into shape. Lumber, like pine, may be soft, but it is not ductile. You cannot bend a piece of pine without breaking it.

Strength is the ability of a material to withstand a force without changing shape or breaking. We can apply four types of force to a material: tension, compression, torsion, and shear. Engineers calculate and describe the strength of a material using these four forces. Engineers then select the material for use in the prototype based on the type of force that the component will be receiving.

Tension is a force that pulls the material from each end. Figure 8-9 shows

FIGURE 8-9 The rope in a game of tug-of-war is under *tension*.

Math in Engineering

S T E M

When engineers study the strength of a material, they must first know its area. Mathematics provides the engineering formulas to calculate the area of any surface. Check your mathematics textbook and identify the formulas to calculate the area of a square, a rectangle, a triangle, and a circle.

Engineering Challenge

ENGINEERING CHALLENGE 1

Your engineering team has been asked to research five different materials and rate the mechanical properties of the materials. Perform the following three tests. Decide how best to report the findings. Then prepare a technical report to submit your findings.

Materials to be tested: Pine lattice (or wooden yardstick), a piece of oak (1 inch × 2 inches), an old license plate (or 22-gauge aluminum), angle iron, and acrylic (1/16-inch thick).

Tools required: Scratch awl and mallet.

- Hardness test: Place each material on a workbench. Place the point of the scratch awl on the material and then strike the awl with the mallet. Repeat the test on each material.
- Toughness test: Place each material on a workbench. Place the point of the scratch awl on the material and then draw the awl across the material as if you were writing your name. Repeat the test on each material.
- Elasticity and plasticity test: Place each material on a workbench with one-half of the material hanging over the edge of the bench. Holding the material on the bench, apply pressure to the overhanging section and then release the pressure. Repeat the test on each material.

Math in Engineering

Have you studied how money grows in a savings account based on the percentage interest rate paid? Engineer use this same type of mathematical calculation when they determine the percentage a material gains under tension or loses in compression. Both bankers and engineers must be able to calculate and understand percentages.

two teams in a tug-of-war. The rope between each team is under tension. **Compression** is a force that pushes or squeezes the material. The stool in Figure 8-10 is under compression from the weight of the elephant. **Torsion** is a force that twists the material. Before flying the wind-up airplane in Figure 8-11, you apply torsion force to its rubber band. The rubber material, because it turns easily, is not a material with strong torsion strength. **Shear** refers to a splitting set of forces, one on each side of the material. The sheet of paper in Figure 8-12 is under shear force from the scissors. Forces on both sides of the paper are attempting to shear the paper.

FIGURE 8-10 An elephant standing on a stool puts compression on the stool.

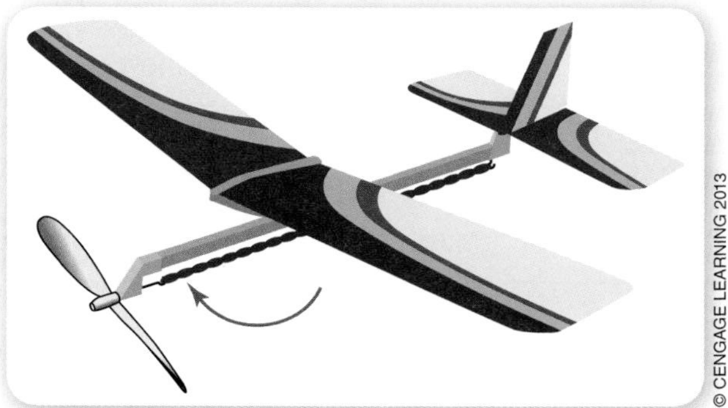

FIGURE 8-11 When you twist a rubber band on a toy airplane, you apply *torsion*.

FIGURE 8-12 Scissors cutting through paper show *shear* force.

ENGINEERING CHALLENGE 2

Engineers use mathematical calculations when studying the strength of materials. One method of testing a metal's strength is the *tensile test*. The test involves pulling on a metal sample (applying tension) and measuring the change in its length. Tension is applied slowly and is increased until the sample breaks. We call the tension applied to the sample *stress*. We call the stretching of the sample *strain*. Engineers measure stress as force (pounds) divided by the sample's area (square inches) or pounds per square inch (psi). We describe strain as the increase in the sample's length as a percentage of its original length. Let us look at an example.

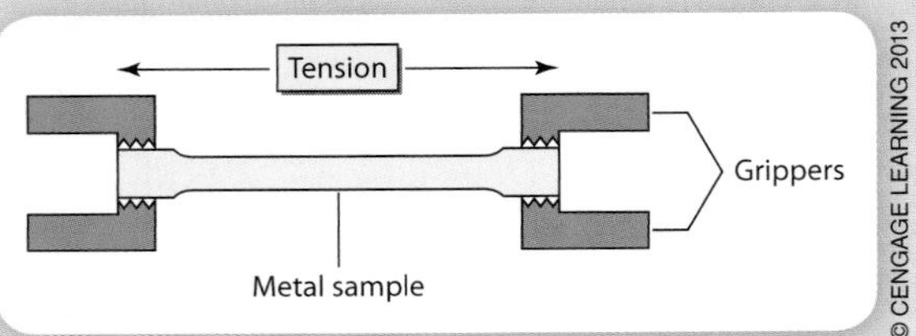

Tensile testing.

The metal sample has a diameter of 1 inch. The grippers are pulling with a force of 100 pounds when the sample breaks. The stress can be calculated as:

$$\text{Stress (psi)} = \frac{\text{Force (pounds)}}{\text{Area (in.}^2\text{)}} \qquad \text{Stress (psi)} = \frac{100 \text{ pounds}}{(3.14 \times 0.5 \text{ in.} \times 0.5 \text{ in.})} \qquad \text{Stress} = 126 \text{ psi}$$

If the sample stretches from 2 inches long to 2.02 inches long, the strain can be described as:

$$\text{Strain (\%)} = \frac{[\text{New length (in.)} - \text{Original length (in.)}]}{\text{Original length (in.)}} \times 100\%$$

$$\text{Strain} = \frac{(2.02 \text{ inches} - 2 \text{ inches})}{2 \text{ inches}} \times 100\%$$

$$\text{Strain} = 1\%$$

Calculate the breaking stress and the strain for the three test samples in the following table.

	Stress		Strain	
Sample	**Diameter (inches)**	**Breaking tension (pounds)**	**Original length (inches)**	**Stretched length (inches)**
A	1	2000	2	2.001
B	1	500	2	2.050
C	1	1000	2	2.020

- Which material is the strongest?
- Which material stretched the most?
- Why do engineers use samples with the same diameter?
- Why do engineers use a set distance to measure the sample's stretching?

Math in Engineering

When you describe a hill on a road, you might say that the road slopes up. If the hill is tall, you would say the hill has a steep slope. If the hill rises only a little, you would say it has a low slope. Engineers also use the term *slope*, but they use a mathematical calculation to define it. Engineers define *slope* as the rise (the upward distance) over the run (the horizontal distance).

Engineering Challenge

ENGINEERING CHALLENGE 3

Engineers graph these measurements of stress and strain on a stress–strain diagram. The shape of the graph indicates the sample's mechanical properties to the engineer. Mathematicians call the angle of the graph its slope. The end of the graph line indicates the breaking point of the sample. The stress–strain diagram in graph A has a steep slope. The slope of the stress–strain diagram in graph B is low. The steep slope indicates the metal does not stretch. The diagram shows the metal can withstand high tension but breaks without warning. The second sample stretches under a smaller force before it breaks. What comparison can you make between these two samples?

- Which sample, A or B, is stronger?
- Which sample, A or B, is more ductile?
- Would the stress–strain diagram of a strong material have a steeper or lower slope?
- Would the stress–strain diagram of a ductile material have a steeper or lower slope?
- How can this knowledge help engineers in designing a prototype?

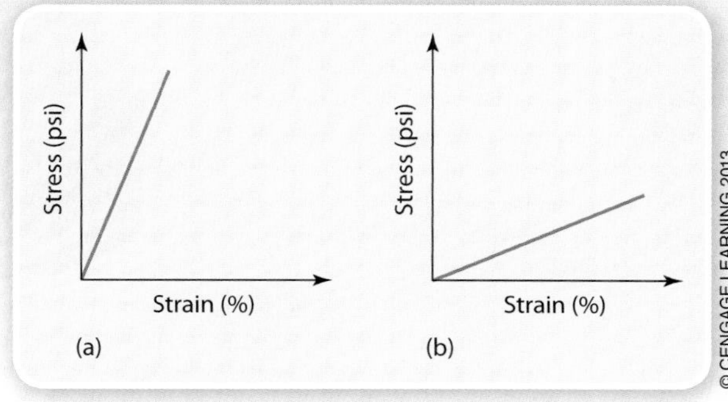

Stress–strain diagrams.

SECTION 3: TOOLS

Tools are objects that help humans perform work. They can be very simple like a hammer or very complex like a computer-controlled milling machine. The simplest tools used by humans during prototyping are common hand tools. Power equipment performs other, more complex processes.

Hand Tools

Hand tools are designed and built based on the properties of simple machines (see Chapter 10). Hand tools use the mechanical advantages of simple machines to help us perform various tasks. Before performing some of these tasks, a technician uses measuring tools to lay out the prototype.

Layout Tools The first step in prototype fabrication is transferring measurements from the drawing and laying them out on the material. Figure 8-13 shows common hand tools—a steel rule and a tape measure—used for measuring and layout. These measuring tools are typically graduated in 1/16-inch measures. Try squares or combination squares are used to transfer perpendicular lines to the material. For lumber, the transfer is marked with a pencil. If the material is metal, a scratch awl is used.

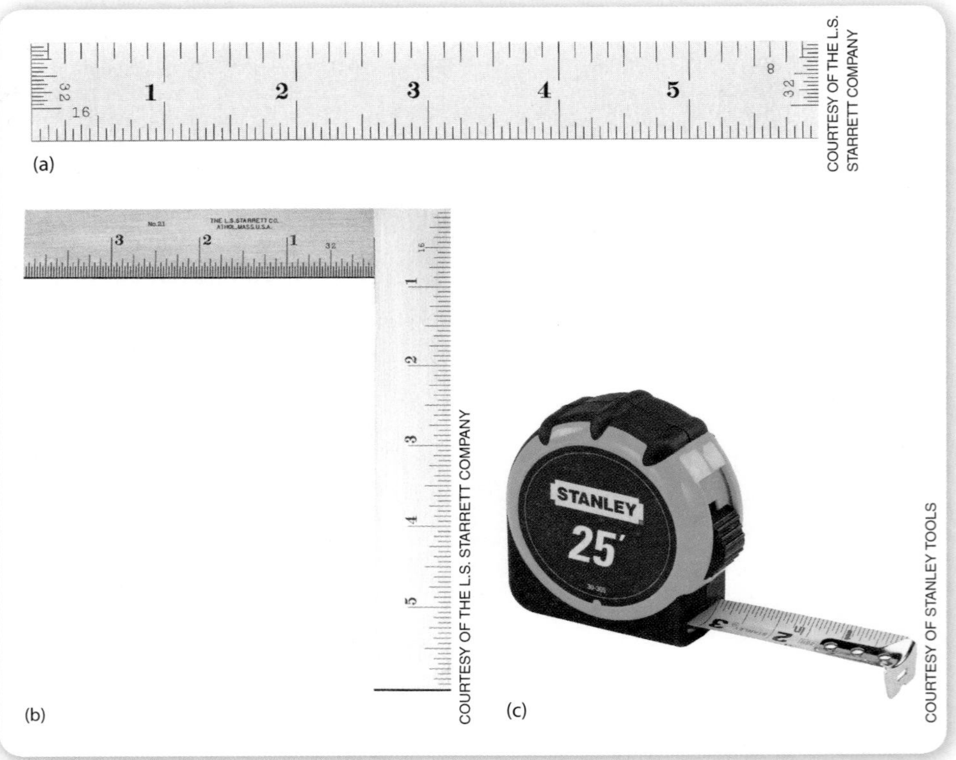

FIGURE 8-13 **Common measuring and layout tools: (a) a steel rule, (b) a steel try square, and (c) steel tape measure.**

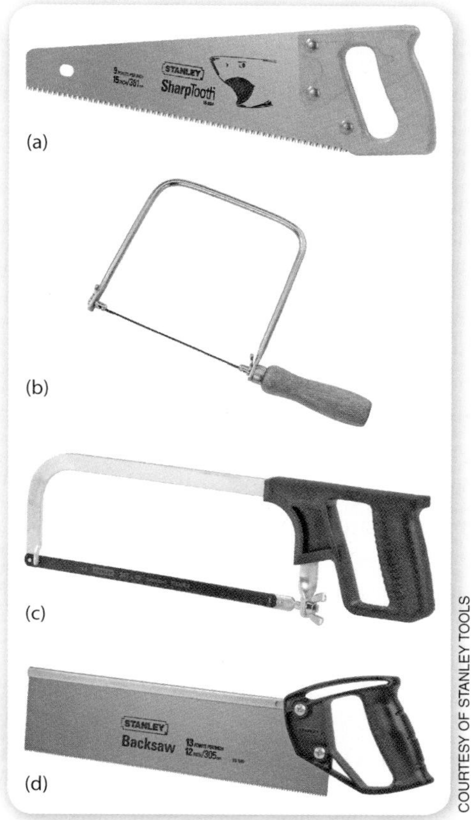

COURTESY OF STANLEY TOOLS

FIGURE 8-14 Common separating tools for sawing: (a) a handsaw, (b) coping saw, (c) a hack saw, and (d) a back saw.

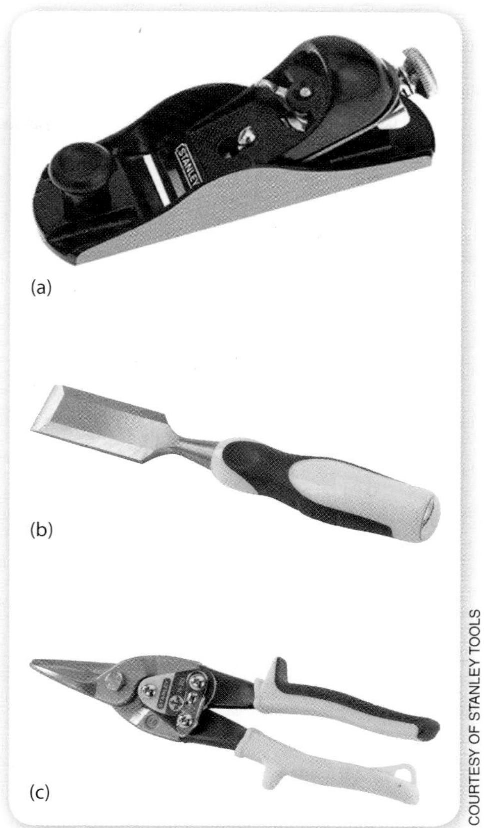

COURTESY OF STANLEY TOOLS

FIGURE 8-15 Common separating tools for shearing: (a) block plane, (b) wood chisel, and (c) aviation snips.

Separating Tools Once layout is complete, the material must be separated. Common hand separating tools either remove chips (for example, a saw) or shear the material (for example, a pair of scissors) (see Figure 8-14). Common saws include the handsaw, the backsaw, the coping saw, and the hacksaw. We can cut thick metal with a hacksaw, and we can shear thin metal with aviation snips. We also can remove soft material such as lumber with a chisel. The block plane shears material from the edge surface of a piece of lumber (see Figure 8-15).

Figure 8-16a shows three common types of files: flat, tapered, and rat-tail. We use the flat file to remove small amount

COURTESY OF STANLEY TOOLS

FIGURE 8-16 Common files: (a) flat, tapered, and rat-tail; (b) file card (cleaner).

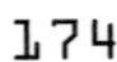

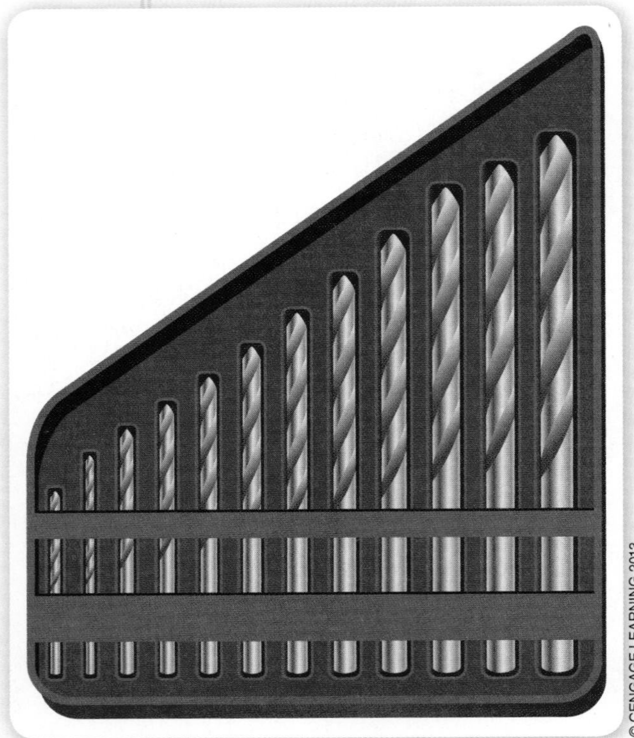

FIGURE 8-17 A set of high-speed drill bits.

of material from a flat surface. We can use the tapered file, or triangular-shaped file, to remove material from corners and so on. The rat-tail file is round and is used on circular areas. When a file is being used, its teeth can become full of the excess material. The file card is a tool designed to clean the file's teeth (Figure 8-16b).

To make a hole in a material, engineers use drill bits. The high-speed drill bits in Figure 8-17 are the most common, but there are various types of specialty drill bits for specific applications. Wood screws require that two different holes be drilled. The shank hole is slightly larger than the screw. The pilot hole is smaller than the diameter of the screw to allow for the screw threads to engage the material. When a flat-head screw is installed, the countersink is used to perform the countersinking process before installing the screw (see Figure 8-18).

Combining Tools We use different hand tools during combining processes. We use screwdrivers to install and remove screws. Screwdrivers have tips that match the type of screw on which they are used. We use slotted screwdrivers on slotted-head screws and the Phillips screwdriver on Phillips-head screws. Some screws have square drives, hex drives, or torx drives. Figure 8-19

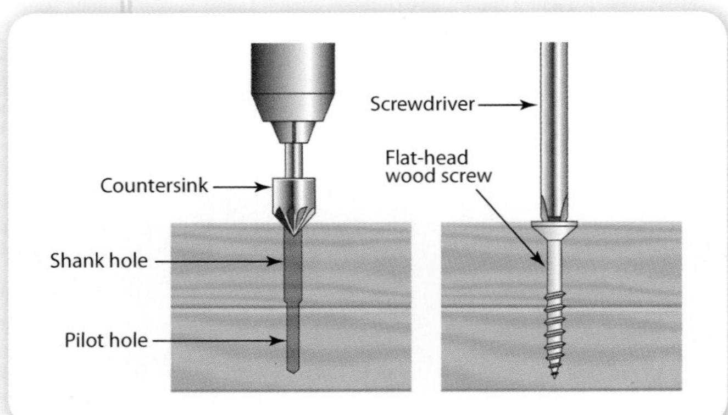

FIGURE 8-18 Countersinking a hole and installing a flat-head screw.

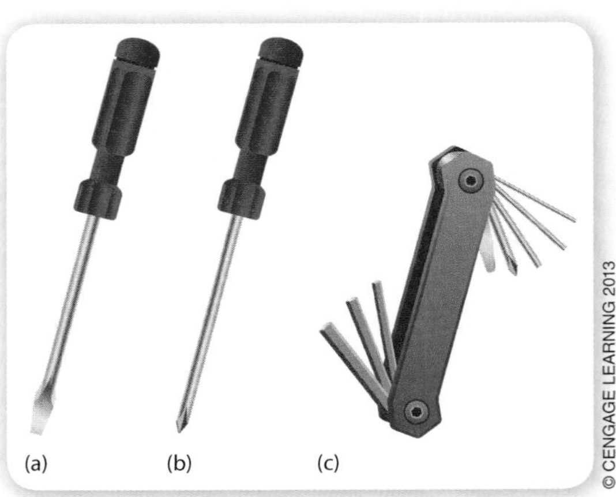

FIGURE 8-19 Common screwdrivers: (a) slotted-head, (b) Phillips-head, and (c) hex.

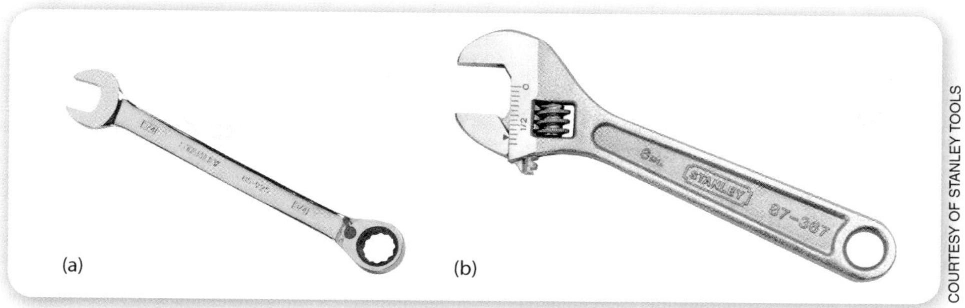

FIGURE 8-20 **Common wrenches: (a) combination and (b) adjustable.**

shows various types of screwdrivers. It is important to select the appropriate size and type of screwdriver when installing or removing a screw.

Machine screws (or bolts) require special tools. The combination wrench is one type of hand tool used with machine screws. The adjustable wrench is more versatile, allowing the technician to use it with different sizes of machine screws (see Figure 8-20). We may also use pliers, second-class levers, during the combining process. Pliers can be found in various types (see Figure 8-21). Slip joint pliers allow material to be held firmly. Needle-nose pliers can hold smaller items. Diagonal pliers (cutters) are actually a shearing tool for wire.

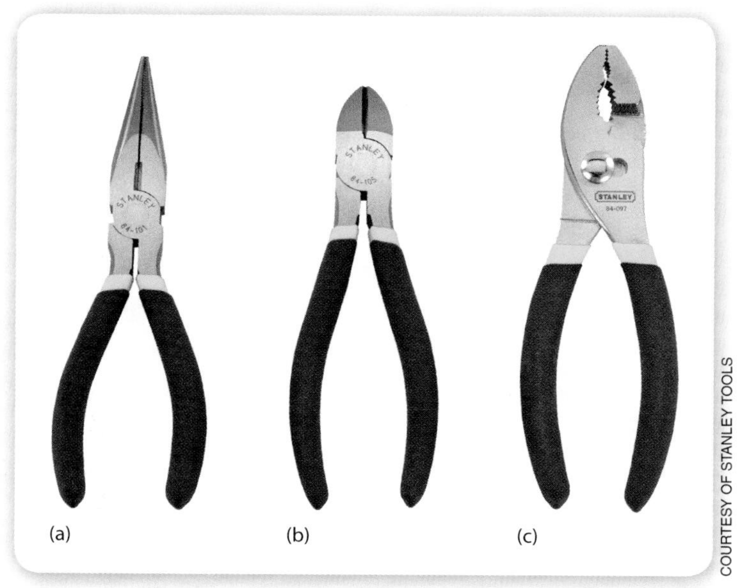

FIGURE 8-21 **Types of pliers: (a) needle-nose, (b) diagonal, and (c) slip joint.**

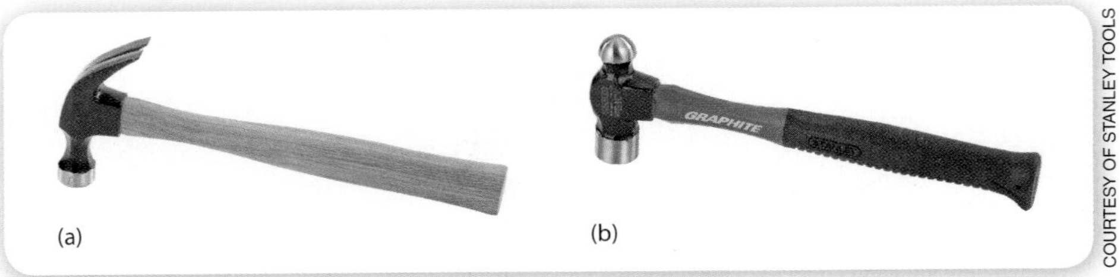

FIGURE 8-22 Common hammers: (a) claw and (b) ball-peen.

FIGURE 8-23 There are many types of mallets, including plastic and rubber.

We use the claw hammer to install or remove nails. We can use the ball-peen hammer in metal processing (see Figure 8-22). Mallets—plastic, rubber, or rawhide—help in the combining process (see Figure 8-23). When materials are glued together, they must remain under pressure until the glue cures. We use the clamps pictured in Figure 8-24 (C-clamp, hand-screw clamp, and spring clamp) for this purpose.

Power Equipment

Some processes are better performed using power equipment. Power equipment transfers electrical energy to replace the human energy used in hand tools. Two types of power equipment used in

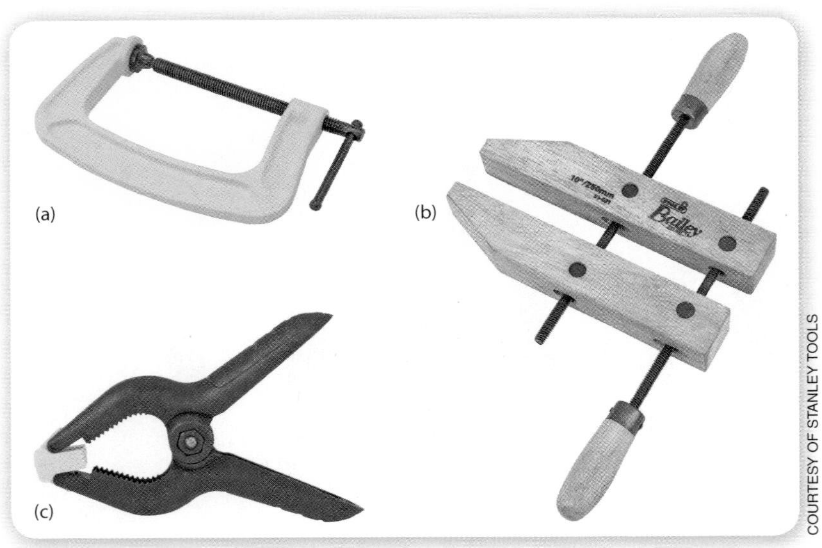

FIGURE 8-24 Types of clamps: (a) C-clamp, (b) hand-screw clamp, and (c) spring clamp.

the school laboratory are the scroll saw and the drill press. Both tools perform separating processes on wood, metal, and plastic.

The scroll saw is a motorized coping saw (see Figure 8-25). Its blade reciprocates up and down to move its saw teeth against the material. We use a drill press (Figure 8-26) to drill holes with high-speed drill bits. We clamp the material to be drilled to the drill press table and then lower the revolving drill bit into the material. The drill press can also be used with a variety of specialty drill bits. The drill press can accommodate fixtures to assist in the fabrication processes. As shown in Figure 8-27, we can use **fixtures** to align the material so the process is performed accurately.

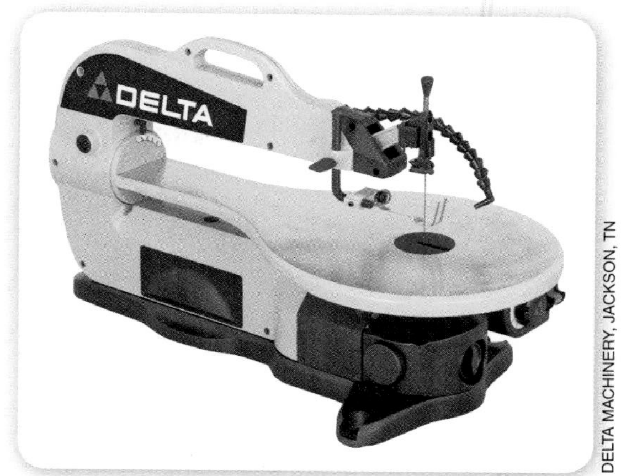

DELTA MACHINERY, JACKSON, TN

FIGURE 8-25 The scroll saw.

DELTA MACHINERY, JACKSON, TN

FIGURE 8-26 The drill press.

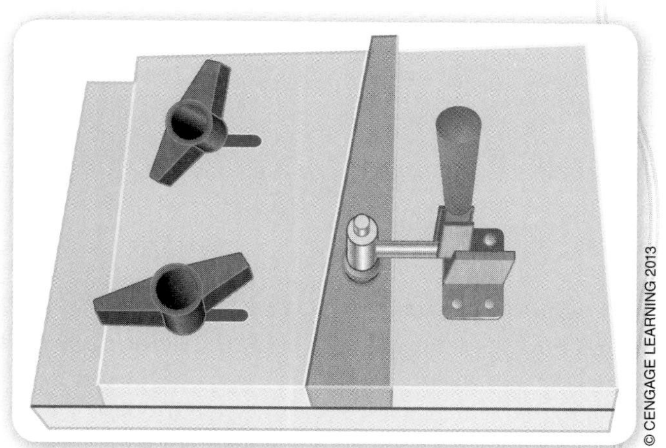

FIGURE 8-27 Using a fixture to assist with drilling.

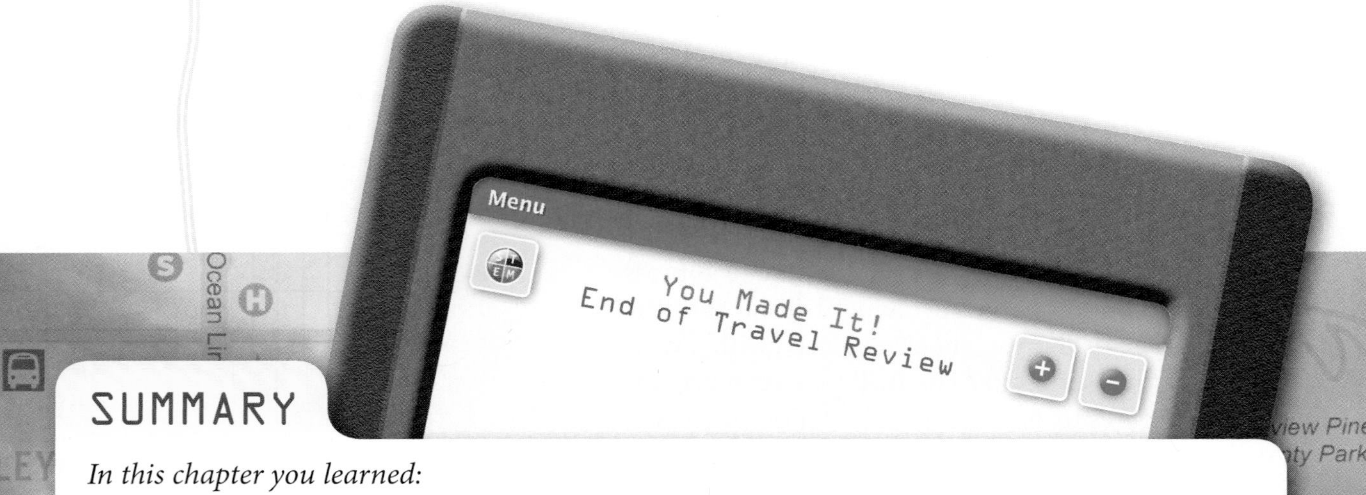

SUMMARY

In this chapter you learned:

- A prototype is a full-scale working model used to test a design concept.
- An alloy is a metal made by combining two or more metals into a single compound.
- Plastics are synthetic, human-made materials.
- Mechanical properties include hardness, toughness, elasticity, plasticity, brittleness, ductility, and strength.
- Strength describes the four forces that can be applied to a material: tension, compression, torsion, and shear.
- Hand tools are simple machines that help engineers fabricate prototypes.
- Power tools use electrical energy instead of human energy to fabricate prototypes.

VOCABULARY

Write a definition for each term in your own words. After you have finished, compare your answers with the definitions provided in this chapter.

Model
Prototype
Fabrication
Ferrous metals
Nonferrous metals
Alloy
Plastics
Mechanical properties

STRETCH YOUR KNOWLEDGE

Provide thoughtful, written responses to the following questions.

1. Explain why engineers must fabricate and test a prototype of a design concept.
2. Explain why engineers must understand the mechanical properties of materials.
3. How do engineers determine the mechanical properties of various materials?
4. Identify the four forces that can be applied to a material. Find and describe a real-world example of these forces in action.
5. Select a common hand tool. Analyze how the tool works and identify what simple machines are involved.
6. Research and find a metal alloy. Identify the metals that were combined to form the alloy. Hypothesize about why these metals were combined.

Onward to Next Destination

CHAPTER 9
Energy

Before You Begin

Think about these questions as you study the concepts in this chapter:

1. What is energy?
2. How do engineers measure work?
3. Is there a difference between energy and power?
4. Where does energy come from?
5. Why is sustainability important to our future?
6. What is the law of conservation of energy?
7. Are there different types of energy?
8. Does energy come in different forms?

Engineering in Action

Engineers and scientists are developing new ways to power cars and trucks without using petroleum. Many cars around the world instead use natural gas as a fuel source. Fuel cells, whose only by-product is water, show great promise as a clean power source. New types of batteries also are being developed to power cars that will drive longer distances before they need to be recharged.

Perhaps you are familiar with the 1980s science fiction movie *Back to the Future* in which an eccentric scientist develops a way to power his car from garbage. Well, science fiction is close to reality today. We can use garbage to produce energy. Although this technology has not yet been refined to the level necessary to power cars, it is reasonable to think that one day we will be able to pull into a landfill and say, "Fill 'er up!"

1

2

SECTION 1: WHAT IS ENERGY?

As we noted in Chapter 2, energy is simply the ability to do work. Your body uses energy when you jog, ride your skateboard or bicycle, or play video games (see Figure 9-1). A car uses energy when it takes us from one place to another. Energy provides much more than movement, though. It can also provide light, heat, and sound. In a large concert hall, all three of these **forms** of energy would be present. Energy is contained in all kinds of matter and can be converted from one form to another.

© ISTOCKPHOTO/BEANO5

FIGURE 9-1 Energy powers humans as well as electronic devices such as Nintendo Wii® motion wands.

Work and Power

In our daily lives, we usually think of work as a job where people go to earn a living or sometimes as a task to be completed. In engineering and physics, however, work has a very different meaning. We define **work** as the transfer of energy from one physical system to another. Work causes movement (including movement caused by motors). Work is done only if the "push" actually moves the object. If the object does not move, then no work has been done. Work is directly related to energy and is often the visible result you see. Usually you cannot see energy, but you can see the result of energy when it does work.

We measure work by how far an object moved and how much force it took to move it (see Figure 9-2). Mathematically, we calculate work by multiplying the force times the distance the object traveled (work = force × distance). The formula is:

$$W = F \times d$$

© CENGAGE LEARNING 2013

FIGURE 9-2 We calculate work by multiplying the force applied to an object by the distance the object moves.

work
Work is the transfer of energy from one physical system to another.

Math in Engineering

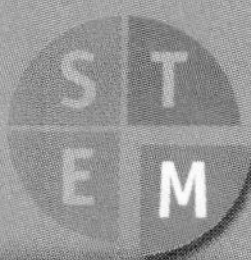

We use the International System (SI) of units, commonly referred to as the *metric system,* to measure in engineering and scientific fields. In the metric system, we measure force in newtons and distance in meters. Therefore, we would measure work in units of **newton-meters.** We call this unit a **joule.**

We use the term **power** to describe how much work has been or could be done in a certain amount of time. Power is the mathematical description of how much work is done. We calculate power by dividing work by time (power = work/time). The mathematical formula is written:

$$P = W/t$$

Horsepower is one common unit for measuring mechanical power in the U.S. customary system of measurement. This unit is based on the fact that an average horse could lift a 550-pound weight 1 foot in 1 second (see Figure 9-3). If you were to experiment with some weights (under adult supervision by a trained professional), you would find that an average teenager in good condition can produce only about 0.1–0.2 horsepower for several minutes.

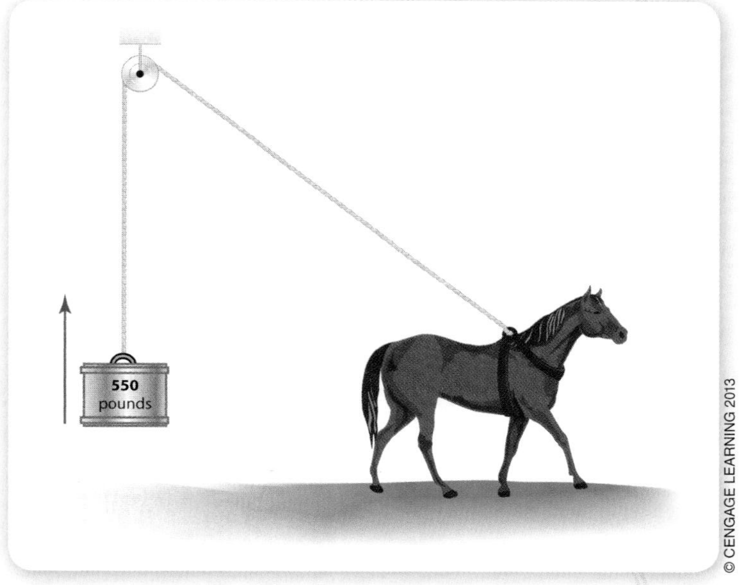

FIGURE 9-3 This is the operational definition of horsepower.

Why Study Energy?

Learning about energy is important because it is incorporated into the design of so many devices and products we use today. We would not be able to use any of the wonderful technological devices that we now enjoy without energy. Energy literally powers almost everything we depend on for both survival and recreation. Sources of energy are an important part of our economy and national security.

Some sources of energy are in limited supply and may be depleted in the near future if carelessly wasted. The conversion of energy into more usable forms of energy or power often produces negative consequences such as pollution. We should not waste energy; we should conserve it and

power
Power is how much work has been or could be done in a certain amount of time. Power is the mathematical description of how much work is done.

Science in Engineering

Fossil fuels, as the name implies, are produced from the remains of plants and animals that lived millions of years ago. The dying and dead plants and animals were covered by layers of sediment. The Earth's crust shifted and buried their remains. Heat and pressure transformed these massive deposits of prehistoric remains into fossil fuels (oil, coal, and natural gas).

use it wisely. Responsible citizens should understand not only the importance of conserving energy but also that using energy sometimes has negative impacts.

Sources of Energy

We can classify energy sources as exhaustible, renewable, or inexhaustible. We explain each source in the following sections. The sun is the greatest source of energy, and many forms of energy we use today can be traced indirectly to the sun. For example, **fossil fuels** (oil, coal, and natural gas) are the result of plants and animals such as dinosaurs that lived long ago. They depended on the sun for their survival, just as we do today.

fossil fuels
Fossil fuels are fuels produced by deposits of ancient plants and animals.

exhaustible energy
Exhaustible energy is any source of energy that is limited and cannot be replaced when it is used, such as oil, coal, and natural gas.

renewable energy
Renewable energy is any source of energy that can be replaced, such as ethanol and biomass.

Exhaustible Energy Fossil fuels are the primary sources of **exhaustible energy**. We call them *exhaustible* because we cannot replace them when they are used up—at least not for a million years. Thus, this source of energy is in limited supply. Our readily available supply of oil could potentially be used up in your lifetime (current estimates are about 40 years) if we do not practice energy conservation.

Exhaustible energy sources currently provide more than 85 percent of all the energy consumed in the United States. We depend on foreign countries to provide most of our oil, and these countries could reduce or eliminate our supply of oil at any time. This poses a potential threat to national security. We use oil not only as an energy source but also as a critical component in making plastics. An artificial reduction in the supply of oil could have a disastrous effect on our economy and on our ability to protect ourselves.

We convert fossil fuels into usable energy through **combustion**, or burning. Burning fossil fuels produce air pollution. The resultants of this air pollution include acid rain, which kills plants and forests; lung cancer in humans; and the destruction of the ozone layer, which protects our planet from the sun's rays.

Renewable Energy **Renewable energy** is energy that can be replaced. Examples of renewable energy sources include ethanol, hydrogen, and biomass.

Generally, these energy sources are less polluting than nonrenewable energy sources. Renewable energy sources will become increasingly important in the future. Renewable energy sources are vital to our economy and our national security. We can use plants such as corn, soy beans, seaweed, and sugarcane to make ethanol (a substitute for gasoline). We call this type of energy **biomass**. The term also refers to living or recently dead biological material that can be used as fuel.

Hydrogen fuel can be used to power cars as an alternative to fossil fuels. Hydrogen fuel is generated by separating water into hydrogen and oxygen through a process called electrolysis. Solar or wind energy can generate the electricity needed for electrolysis.

Inexhaustible Energy **Inexhaustible energy** is an energy source that cannot be used up, for example, solar, wind, water, and geothermal. These sources are absolutely critical to our future and the Earth's environment. The responsible use and management of natural resources is termed **sustainability**. Sustainability implies that we design engineering solutions with our future and the environment in mind.

inexhaustible energy
Inexhaustible energy is any energy source that cannot be used up, such as solar, wind, water, and geothermal.

sustainability
Sustainability implies that we design engineering solutions with future needs and the environment in mind.

The sun is the ultimate source of all renewable energy on Earth. We can capture **solar energy**, energy from the sun, in many ways. We use only a miniscule amount of the available solar energy that reaches Earth. Many houses take advantage of **passive solar heating**, in which the sun's energy passes through windows and is absorbed by stone, slate, or other materials in the house (see Figure 9-4). This heat is drawn out of the air during the day and then reflected back into the air at night when the temperatures are cooler. This is not a new process. Humans have used this principle for thousands of years.

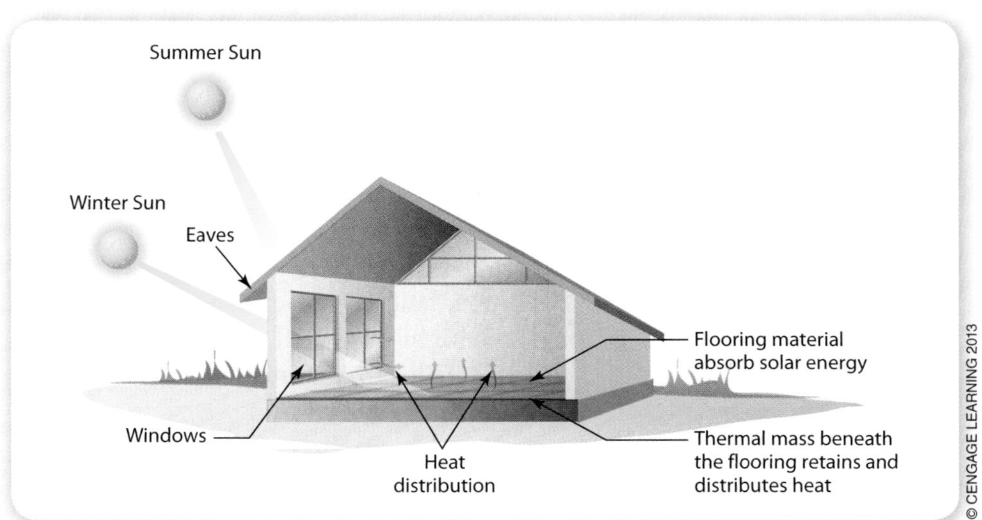

FIGURE 9-4 *Passive solar heating saves money and other resources.*

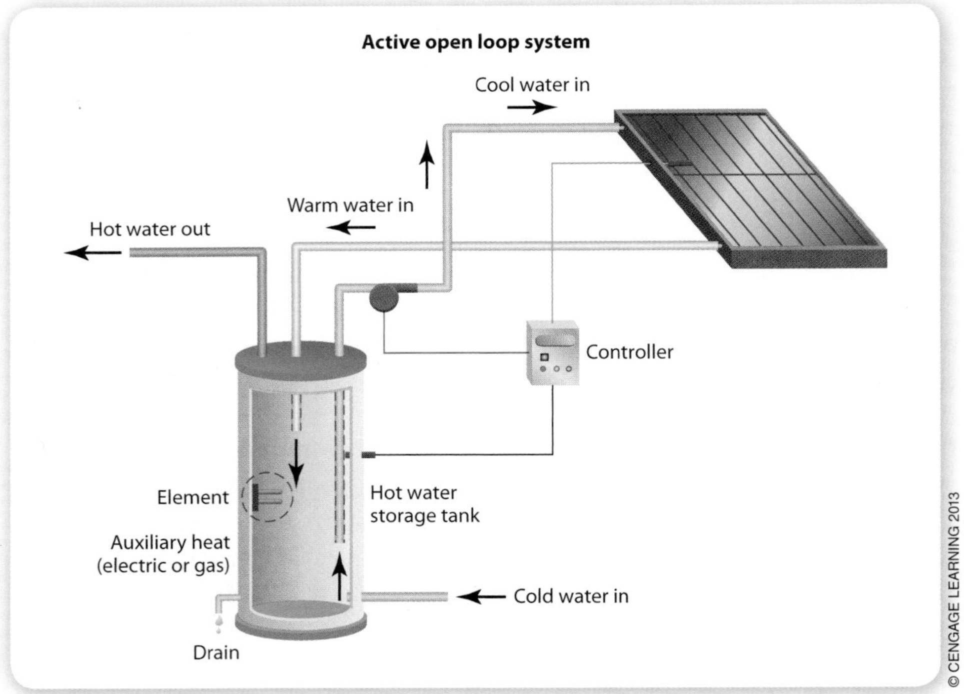

FIGURE 9-5 Active solar heating applied to a water heater.

Active solar heating is more efficient and involves a liquid circulating through tubes. The tubes are typically encased in a box under glass and mounted on a roof. As the liquid is heated, it circulates through the pipes to heat water or the entire house (see Figure 9-5).

FIGURE 9-6 Photovoltaic cells can provide electricity for houses and businesses.

A **photovoltaic cell** is a device that converts the sun's energy into electricity. We can group photovoltaic cells together in an **array** (see Figure 9-6). The cells are grouped together because individual cells do not produce enough electricity to be useful. We can combine one or more arrays with batteries to store the electricity generated by the photovoltaic cells. We use photovoltaic to power emergency phones along highways, satellites, the International Space Station, private houses, and even solar cars.

Wind is another inexhaustible energy source. Humans have used wind energy for thousands of years. You have probably

FIGURE 9-7 **Wind farms are becoming an increasingly important source of electricity.**

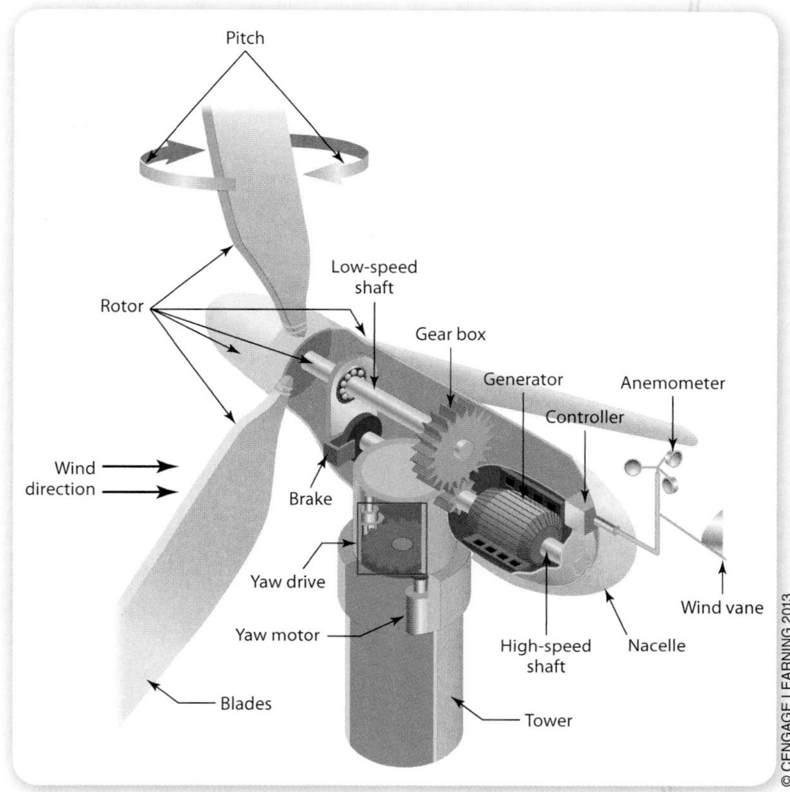

FIGURE 9-8 **Internal components of a wind turbine.**

seen windmills on farms that were used to draw water from the ground. We can also use wind to power large **wind turbine generators** that generate electricity. A turbine is a device that produces electricity from rotary motion (see Figure 9-7). The wind possesses mechanical energy by virtue of motion (discussed in more detail later in this chapter). Wind turns large blades that turn a shaft in the turbine (see Figure 9-8). The generator produces an electrical current (see Chapter 14). Often dozens or even hundreds of large wind turbines are placed together on wind farms, which are located in places with strong and frequent winds.

The United States receives only 1 percent of its total electricity from wind. Denmark, Spain, and Germany, on the other hand, are approaching 10 percent of their electrical consumption from wind. Using more wind energy would benefit the environment because wind does not produce any pollution or hazardous waste. It is also readily available without mining or drilling, is not radioactive, and does not require the transportation of dangerous substances.

Water is another energy source that people have been using for thousands of years. Water turned large waterwheels to power machine tools

Science in Engineering

A kilowatt is 1000 watts, a measure of the load on an electrical circuit. A kilowatt hour (kWh) is the work performed by 1 kilowatt of electric power in 1 hour. The United States has enough wind turbines to produce more than 12,000 megawatts of electricity (a megawatt is 1000 kilowatts). Electricity is sold to consumers by the kilowatt hour.

© ISTOCKPHOTO/LAWRENCE FREYTAG

FIGURE 9-9 The Grand Coulee Dam on the Columbia River, in Washington state, produces more hydroelectric power than any other dam in the United States and provides irrigation to farms in a semi-arid part of the country.

during the Industrial Revolution. Before that, waterwheels powered grist mills to crush corn and grains into meal. Today we use the force of falling water to turn giant turbines to create electricity, similar to how wind turns turbines. We call electricity generated from water **hydropower** or **hydroelectricity**.

Hydroelectric power plants are located inside dams (see Figures 9-9 and 9-10). Dams serve three valuable purposes: (1) they control flooding to protect people and farms, (2) they produce hydroelectricity, and (3) the reservoirs behind the dams provide recreational areas for boating and fishing.

Hydroelectricity is produced by using the force of falling water. The dam raises the level of the water and actually increases the water's potential energy (see the next section on types and forms of energy). The water falls with greater force through pipes in the

Did You Know?

Over the past several years, wind energy production has been increasing by more than 25 percent per year in the United States. The country now has enough wind turbines to power nearly 3 million average-sized American houses.

dam to turn turbines and power the giant generators (see Figure 9-11).

Geothermal energy is energy stored in the Earth in the form of heat. A layer of rock called the *mantle* lies deep within the Earth. The molten outer core superheats the mantle. Any water that seeps into the mantle turns to steam, which increases its volume and pressure. This eventually causes the steam to escape through cracks in the Earth. We find this steam and hot water in geysers and natural hot springs. We can use this steam to heat houses and buildings.

Iceland is a world leader in using geothermal energy to heat houses, businesses, and government buildings. We do not have to wait for water to naturally seep into the mantle to produce steam. We can pump water deep into the Earth through pipes, turning it into steam that can rise to the surface through another set of pipes. We can also use the superheated steam to power large steam turbine generators to produce electricity.

FIGURE 9-10 Many turbines are required in a hydropower plant to produce enough electricity to serve a city.

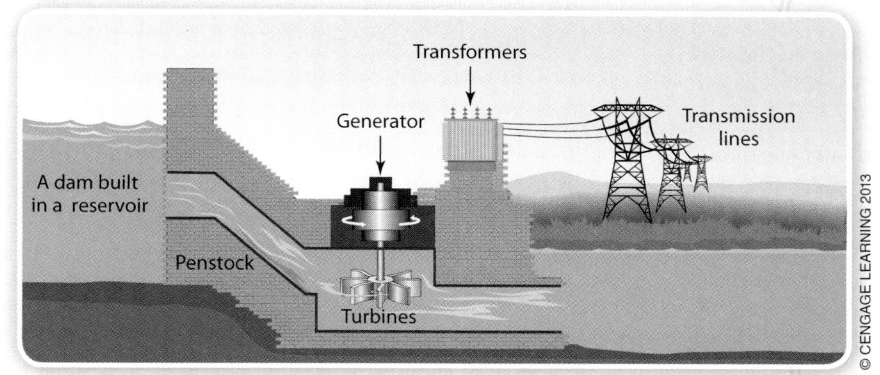

FIGURE 9-11 Hydroelectricity is produced by falling water passing through turbines inside a dam.

Did You Know?

According to the American Wind Energy Association, the electricity produced by the U.S. wind turbine fleet is approximately equal to burning 9 million tons of coal. This much coal would require a line of 10-ton trucks more than 3400 miles long, the distance from Miami to Seattle!

Expressed differently, our wind energy production is currently saving 50 million barrels of oil each year.

Research and present to the class your recommendation for an inexhaustible energy source to reduce our dependency on fossil fuels. Explain how the energy source you selected is better for the environment and humans. Describe how this energy is converted into a usable form. Compare and contrast production costs for producing this energy with fossil fuels. What actions would be required, either by individuals or by the government, for this inexhaustible energy source to be widely adopted?

SECTION 2: TYPES AND FORMS OF ENERGY

law of conservation of energy

The law of conservation of energy states that energy can be neither created nor destroyed. There is a fixed amount of energy present in different forms that can be converted from one form to another, but energy does not go away.

The **law of conservation of energy** states that energy can be neither created nor destroyed. This means that energy is present in different forms and can be converted from one form into another. We cannot "make" new energy—we can only convert it from one form into another. When energy is used, it does not disappear; the energy is simply transformed from one form into another form. Not all forms of energy are useful to us. For example, when an automobile consumes a gallon of gasoline, only about 30 percent of the energy actually is used to move the car. The remaining 70 percent is lost as heat energy into the atmosphere. This is a highly inefficient energy-power system.

We use the term **energy conversion** to describe changes in energy. All forms can be converted into another form. Energy can also be transferred from one object to another. As we will discuss, there are many different forms of energy. An automobile is a good example of several forms, energy conversion, and the transfer of energy from one object to another.

An automobile is powered by an engine that converts gasoline (a fossil fuel) into heat energy. As the gasoline vapors are ignited in the engine, they rapidly expand and force pistons down, turning a crankshaft. Thus, heat energy is converted into mechanical energy (see "Forms of Energy" later in the chapter). The pistons' mechanical energy is transferred to the crankshaft, still as mechanical energy, but now rotating rather than linear. The crankshaft transfers the mechanical energy through the transmission and drive shaft to the wheels. At the same time, electrical energy is being produced by a generator that is also powered by the rotating crankshaft. This electricity provides the spark that ignites the fuel that keeps the cycle going.

A car radio converts radio waves to electrical energy, which is amplified and played through speakers that convert the electrical energy to sound

Career Spotlight

Name:
Roger (Jui-Chen) Chang

Title:
Senior Mechanical Engineer, Arup

Job Description:
Chang works with architects to create mechanical systems for a variety of buildings, from airports to museums. One of his main concerns is energy efficiency, a hot topic in building design.

© CENGAGE LEARNING 2013

"Everyone is talking about helping to reduce global warming," Chang says. "This is the engineering side of that, and it's where a lot of the work is occurring. Buildings consume 40 percent of the energy in the United States, so for any sort of climate-change strategy to work, we must address building design."

When a building is being planned, Chang develops a computer model that includes key components such as walls, windows, lighting systems, and cooling and heating systems. The model allows him to understand the interior air flow, temperatures, and humidity before the building is actually constructed. The design process can take as long as two or three years for such complicated projects as airports and laboratories.

One of Chang's most interesting projects is the Syracuse Center of Excellence. This research center studies energy-efficient building practices, so the structure must demonstrate energy efficiency itself. Chang has designed automatic systems such as blinds that raise and lower in response to daylight. He has also developed a system that uses the ground to heat the building in the winter and draw away heat in the summer.

Chang has always had an interest in architecture, but he felt he was not a good enough artist to do architectural drawing. "Then I realized that there's an engineering side to architecture, where you don't have to do the drawings," he says. "You can still have a tremendous impact. As an engineer, it's very rewarding to help design a building that people are productive in."

Education:
Chang earned his bachelor's and master's degrees in mechanical engineering at the Massachusetts Institute of Technology (MIT).

"When I went to college, I thought I might do biotechnology work, but I decided it didn't interest me that much," he says. "I heard about a competition at MIT where you design a robot from scratch. It was fascinating, and that's how I got involved in mechanical engineering."

Advice to Students:
Chang says that students interested in engineering need to do well in such standard courses as math and science. But he urges them to also take humanities courses such as literature and art. "In the coming years, it's important to be well rounded," he says. "To address issues like climate change, engineers will have to pull inspiration and creativity from what they see around the world."

energy that you can hear. When the brakes are applied, they use mechanical energy to apply pressure to the wheels. The friction of the brake pad rubbing against the wheel transforms the mechanical energy into heat energy that is both transferred to the wheel and lost into the atmosphere. This ought to give you something to think about the next time you go for a car ride.

7 Types of Energy

Energy has two types: potential and kinetic. **Potential energy** is stored energy that is readily available under certain conditions. It is the energy of an object that is determined by its position or chemical structure. **Kinetic energy** is energy that is in motion, or what happens when potential energy is released.

The following example shows how potential energy changes into kinetic energy. Imagine you are on a skateboard at the top of a half-pipe. You have potential energy because of your position—in this case, your elevation. As soon as you start down the half-pipe, your potential energy is released. You now have kinetic energy because you are in motion. All moving objects have kinetic energy, whether or not they are moving under their own power. A thrown baseball has kinetic energy. So does a moving automobile.

An object's chemical structure also determines its potential energy. A tree log is an example of potential energy because of the chemical structure that makes up its wood. The log has potential energy as firewood. When the wood burns, its potential energy is converted to heat energy.

8 Forms of Energy

Energy can take six forms: mechanical, chemical, electrical, heat, nuclear, and light. The following sections describe each form.

Mechanical Moving objects have **mechanical energy**, one of the most common types of energy you see every day, often in things such as transportation devices (bicycles, cars, and even airplanes) (see Figure 9-12). People often think of mechanical energy when they talk about machines that do work for us. However, mechanical energy is not limited to just the machines in these examples—*you* have mechanical energy when you walk or run.

Other forms of mechanical energy include the motion of water or sound. It is easy to imagine the energy contained in the force of falling water or ocean waves crashing on the beach. Similarly, sound is nothing more than pressure waves moving through the air and striking against your eardrum. A microphone converts the mechanical energy of these sound waves into electrical energy. A speaker converts electrical energy back into mechanical energy (sound waves). So, a microphone and a speaker serve the same purpose but in reverse.

Chemical **Chemical energy** is stored energy. Growing plants store energy within their cells. Using the sun's radiant energy, they convert carbon dioxide, water, and minerals into pulp or fiber through a process called photosynthesis. Both burning and eating the plant material releases this stored energy. When a log burns, it converts the stored chemical energy into heat energy.

A battery is another example of stored chemical energy (see Figure 9-13). Essentially, a battery is a device that converts stored chemical energy into electrical energy (see Chapter 13). This is one of the most common but most important sources of chemical energy we use today. Millions of devices are made possible because of the stored chemical energy in a battery—devices ranging from your wristwatch to MP3 player.

FIGURE 9-12 Roller coasters have mechanical energy.

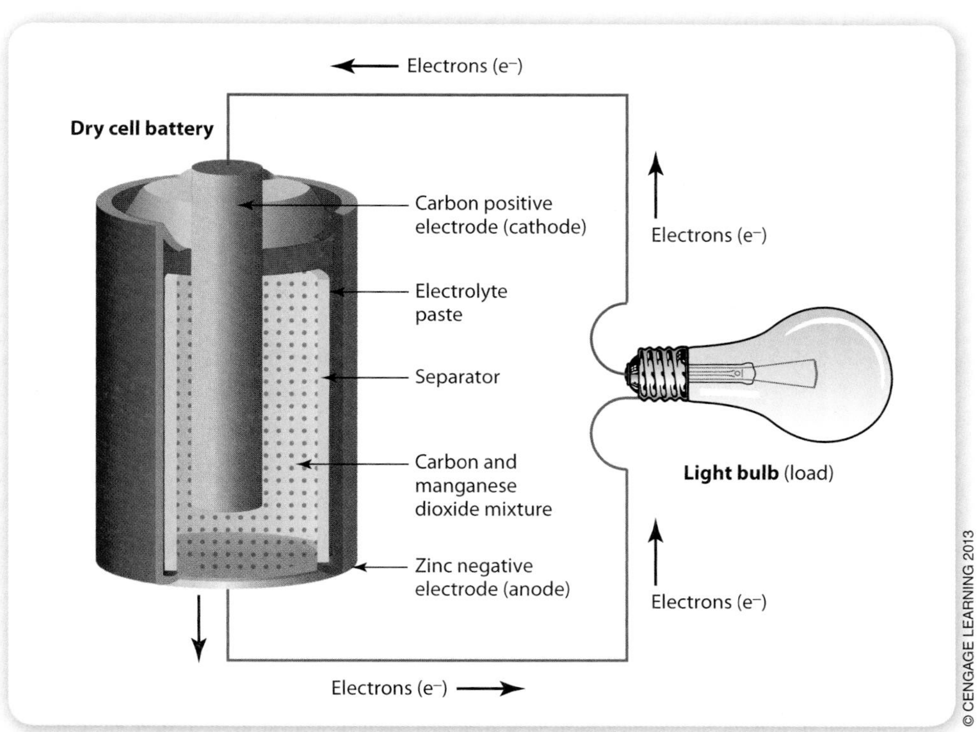

FIGURE 9-13 A dry cell battery uses chemical energy to produce electricity.

Science in Engineering

When you eat plants, your body converts the stored chemical energy into energy that you use to heat and power your body.

FIGURE 9-14 Solar arrays power communication satellites.

Many cars today have onboard global positioning system (GPS) navigation devices that can help drivers with street directions; and display restaurant and other business locations. Many smartphones also have GPS built into them. Satellites rely on batteries for their operation and use solar arrays to recharge the batteries (Figure 9-14). In addition to being helpful to drivers, GPS and communication satellites are crucial to our economy, safety, and national security.

Electrical **Electrical energy** is the energy that is produced by the flow of electrons. Electrons flow when negatively charged particles (electrons) are attracted to positively charged particles (protons). Electrical energy flows easily through certain materials known as **conductors**. One primary advantage of electrical energy is that it can move long distances before it is used. Electricity might be put to use hundreds of miles from where it was initially produced.

Electrical energy in your house is converted to heat energy and mechanical energy through motors that power many appliances. An example where these two forms of energy are combined (heat and mechanical) is a hair dryer. Electrical energy can also be converted into light energy through a light bulb. Electrical energy is one of the most personal forms of energy we use every day (Figure 9-15).

FIGURE 9-15 Electricity powers many everyday devices.

Heat **Heat** is a form of energy that can be defined as energy in transit. Heat always flows from the *warmer* to the *cooler* of two objects. Heat can move

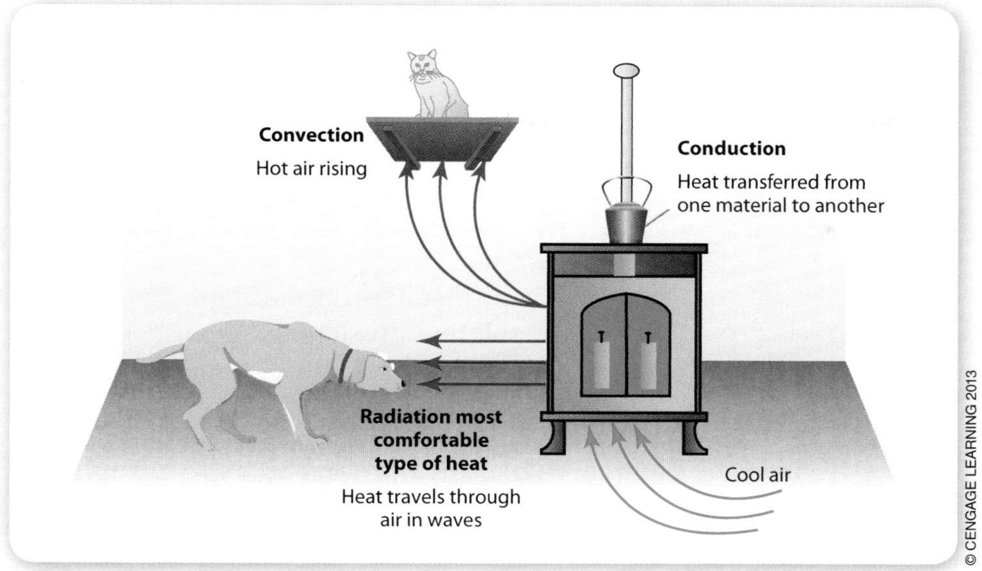

FIGURE 9-16 **Heat energy transferred by radiation, conduction, and convection.**

or transfer through three different methods: radiation, convection, and conduction (see Figure 9-16).

Radiation is the term for the way in which heat travels through air in the form of waves. Toasting a piece of bread in a toaster is an example of heat energy being transferred by radiation. Or perhaps you have observed a mirage as the heat rises off a paved road in the summer.

Convection is the term for the way in which heat travels by movement or circulation within fluids, either liquids or gases. This is how water heats on a stove. You have probably observed water bubbling as it boils. Watch how the bubbles rise through the water. The hot water rises to the top, and the cooler water settles to the bottom of the pan. Convection is also the principle behind the forced air furnace used in many houses. Hot air blows into a room and rises, while the cold air settles to the floor and is drawn into the furnace through the cold air return duct.

Conduction is the term for the way in which heat travels through a solid material. This is caused when the heated molecules become agitated and bump into each other. The molecules will give off some of their heat to the ones they bump into until all the molecules are at the same temperature. This principle can by used to either heat or cool materials. When a pan is placed on an electric stove's heating element, the heat is transferred from the heating element to the pan's bottom through conduction.

Light **Light energy** is also a form of radiant energy. Technically, it is called *electromagnetic radiation*. Light cannot be stored because it is always in motion. Therefore, light energy is a type of potential energy. Light travels through space

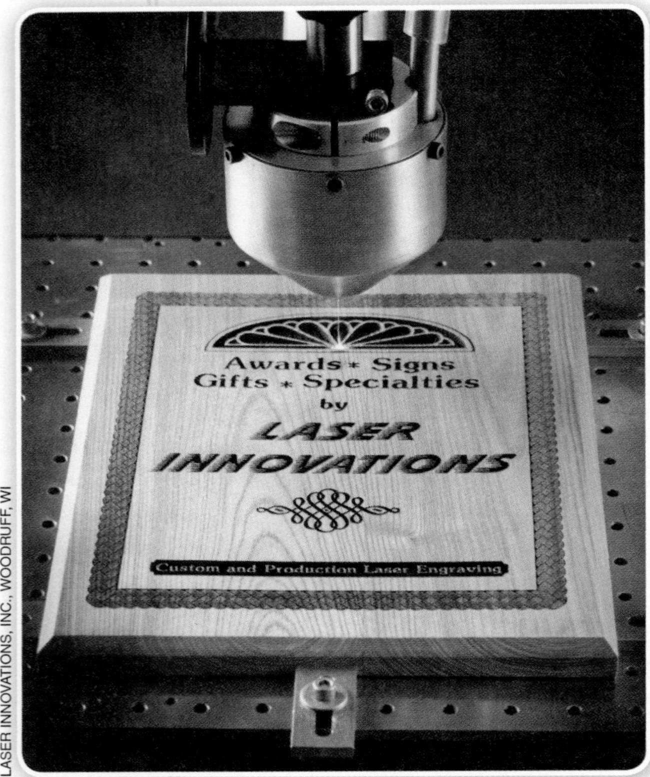

LASER INNOVATIONS, INC., WOODRUFF, WI

FIGURE 9-17 A laser engraver uses light energy to cut or etch material.

as a wave. The greater the wave frequency, the more energy it contains. Some forms, such as laser, are capable of highly focused uses (see Figure 9-17). We cannot see other forms of electromagnetic radiation such as x-rays, infrared waves, microwaves, and radio waves.

Nuclear Energy **Nuclear energy** is the energy that results from combining (fusing) or splitting (fissioning) atoms. The term **fusion** describes the combining of two or more atomic particles to form a heavier nucleus (the nucleus is the center of the atom). The light emitted by stars was produced by fusion.

Fission describes the process of an atom's nucleus splitting into its many smaller subparticles. Fission is the process that produces electricity in nuclear power. Splitting an atom through nuclear fission releases a great amount of heat. In a nuclear power plant, this process is controlled (see Figures 9-18 and 9-19), and

© A.L. SPANGLER/SHUTTERSTOCK.COM

FIGURE 9-18 Many nuclear power plants have huge cooling towers that allow the steam to escape after passing through the turbines.

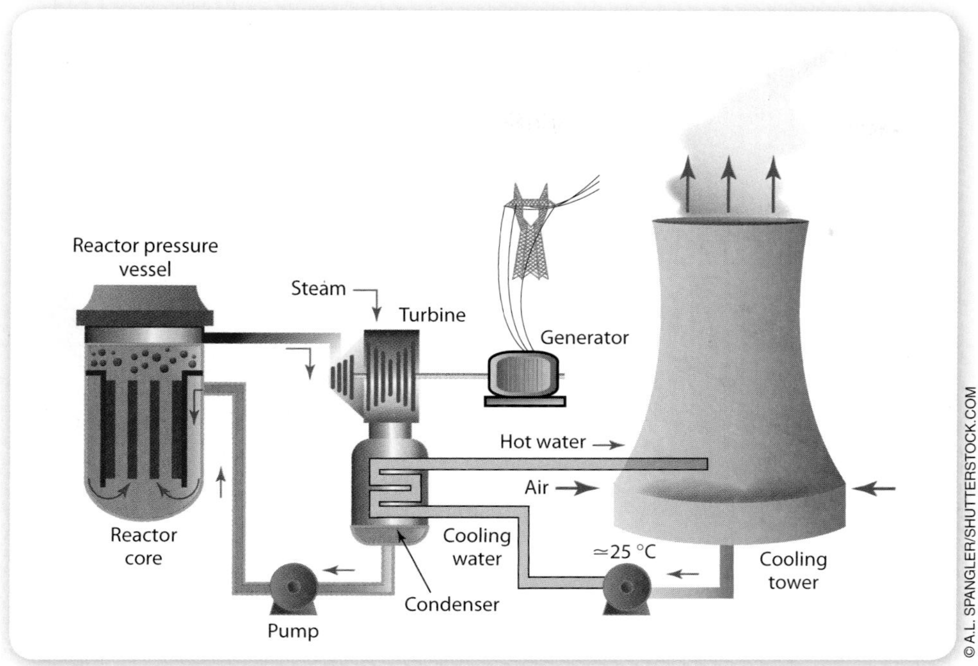

FIGURE 9-19 **Diagram of how a nuclear power plant produces electricity.**

the resulting heat is used to boil water. The boiling water produces steam, which is used to turn large steam turbines.

Did You Know?

Today, more than 400 nuclear power plants operate in 31 countries around the world. They generate electricity for some 1 billion people.

SUMMARY

In this chapter you learned:

- Energy is the ability to do work.
- Work is measured in terms of force and distance, which is mathematically expressed as $W = F \times d$.
- Power is the rate at which work is done (power = work/time).
- There are two types of energy: potential and kinetic.
- Potential energy is energy that is stored.
- Kinetic energy is energy in motion.
- Energy sources can be classified as exhaustible, renewable, or inexhaustible.
- The six main forms of energy are mechanical, heat, chemical, electrical, light, and nuclear.
- The law of conservation of energy states that energy can be neither created nor destroyed; it simply changes form.

VOCABULARY

Write a definition for these terms in your own words. After you have finished, compare your answers with the definitions provided in the chapter.

Work
Power
Fossil fuels
Exhaustible energy
Renewable energy
Inexhaustible energy
Sustainability
Law of conservation of energy

STRETCH YOUR KNOWLEDGE

Provide thoughtful, written responses to the following questions.

1. Why is it important for an engineer to know and understand the various forms of energy?
2. Explain what the law of conservation of energy is in your own words.
3. Why is it dangerous for our country to be dependent on fossil fuels?
4. What are some of the global consequences of burning fossil fuels?
5. Explain the difference among exhaustible, inexhaustible, and renewable energy sources.
6. Describe an advantage and a disadvantage of an exhaustible, an inexhaustible, and a renewable energy source.

Onward to Next Destination

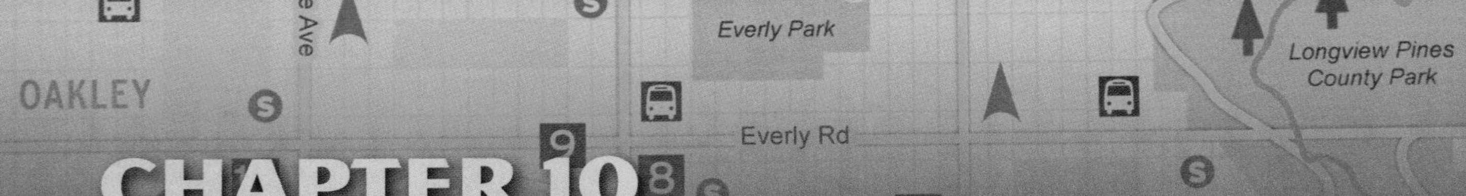

CHAPTER 10
Sustainable Architecture

Menu

Before You Begin

Think about these questions as you study the concepts in this chapter:

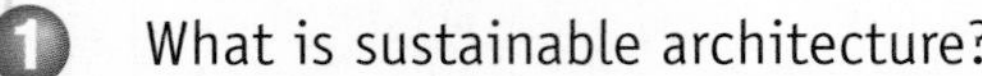

1. What is sustainable architecture?

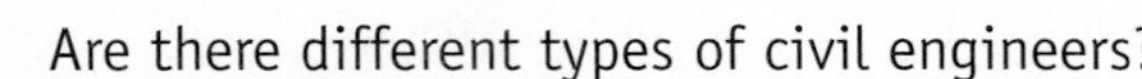

2. Are there different types of civil engineers?
3. Why is an environmental impact study important?
4. How is land legally described in the United States?
5. What are the elements and principles of design in architecture?
6. What are plot plans, floor plans, and elevation drawings?
7. What are the different styles of residential architecture?

FIGURE 10-1 The Burj Tower, in Dubai, is the tallest building in the world.

Engineering in Action

S T E M

How can a country claim to have "the world's tallest building"? To test that claim, we need a standard way of measuring the heights of buildings. According to the Chicago-based Council on Tall Buildings and Urban Habitat, "the height of a building is measured from the sidewalk level of the main entrance to the structural top of the building. This includes spires, but does not include television antennas, radio antennas, or flag poles." In addition, for the 10 tallest buildings in the world, there are four criteria used to describe the height: (1) the height of the structural top, (2) the highest occupied floor, (3) the top of the roof, and (4) the tip of the highest spire, pinnacle, antenna, mast, or flagpole on the building.

The new Burj Tower in Dubai, United Arab Emirates, is the tallest building in the world (Figure 10-1). This building is home to offices, a shopping mall, luxury apartments, and a hotel. Designing and building the Burj Tower posed many challenges that civil engineers and architects have worked together to solve.

SECTION 1: OVERVIEW OF CIVIL ENGINEERING AND SUSTAINABLE ARCHITECTURE

As you have learned throughout this book, there are many types of engineers and other professionals who design and develop the products we use and the buildings in which we live. You have also discovered the importance of sustainablity, or ensuring that our actions do not negatively impact our world. Civil engineers and architects have a big influence on our world because of the many spectacular structures they design. An architect and a civil engineer are both licensed professionals, but they have different jobs and responsibilities. Nevertheless, both must consider sustainablity in their jobs.

architect
An architect is a person who designs and supervises the construction of buildings. The architect is responsible for the building design, including how the space is laid out and how the building relates to its environment.

civil engineer
A civil engineer is a person who designs and supervises the construction of public works projects (such as highways, bridges, sanitation facilities, and water-treatment plants).

An **architect** is a person who designs and supervises the construction of buildings. The architect is responsible for the building design, including how the space is laid out and how the building relates to its environment. The architect is responsible for turning people's needs and wants into a pleasing building. Architects must understand construction techniques well enough to clearly communicate how the building is to be built. Similarly, an architect must know local and national building codes to ensure the public's safety and well-being. Architecture is a rigorous college program that combines math, science, and art. Good design is based on all three.

© SHAWN KASHOU/SHUTTERSTOCK.COM

FIGURE 10-2 Civil engineers often design and oversee the construction of very large projects.

A **civil engineer** is a person who is trained in the design and construction of public works projects (such as highways, bridges, sanitation facilities, and water-treatment plants) (Figure 10-2). A civil engineer is largely responsible for the site design, construction management, and the inner workings of a building. The structural design of a building is also part of the civil engineer's responsibilities, as are the mechanical, plumbing, electrical, and heating-and-cooling and ventilation systems. The civil engineer and architect work together for the success of a project.

Constructing a building using techniques that are environmentally friendly is called sustainable building or green building. Green building considers the structure's energy efficiency, such as electrical use, its water use, and its life cycle. A green building may cost more initially to build, but it will have lower operating costs over its life span. The U.S. Green Building Council has developed guidelines to assist with green building. Their system is called Leadership in Energy and Environmental Design (LEED). The LEED certification is recognized across the nation as the standard for sustainable buildings.

© JAMES M PHELPS, JR/SHUTTERSTOCK.COM

FIGURE 10-3 Structural engineers design a variety of structures, including dams, bridges, and stadiums.

Roles and Responsibilities of Civil Engineers

Within civil engineering are several different disciplines. These areas of specialization include structural, waste treatment, transportation, geotechnical, water management, and construction management. Regardless of the area of specialization, civil engineers and architects must be able to work well in teams.

Structural engineers design such things as sports stadiums, buildings, bridges, towers, and dams (Figure 10-3). **Waste-treatment engineers** design and analyze water-treatment facilities. This includes sanitary waste treatment for household water and waste, industrial waste, and potable water (purification of drinking water).

Transportation engineers are involved with the design and analysis of roads and highways, traffic-control signals, railways, and airports (Figure 10-4). **Geotechnical engineers** make sure the ground can support the weight of the structure, that is, whether the rock and soil support the load (Figure 10-5).

Water-management engineers design dams and structures to control the flow of water. This also includes drainage systems and waterways for transportation. **Construction-management engineers** oversee the actual building of the structure. The work of these engineers involves reviewing plans, ordering materials, and scheduling subcontractors. In general, they are responsible for quality control and ensuring that each structure is built as designed.

green building

Green building is both the process of building in an environmentally responsible way and the resulting structure.

structural engineer

A structural engineer is a person who designs structures such as sports stadiums, buildings, bridges, towers, and dams.

FIGURE 10-4 Transportation engineers design systems for moving people and cargo by road, railway, and air.

Projects

In civil engineering, the term **project** is used to refer to a very large undertaking that includes both planning and construction. Projects vary widely according to the type of civil engineering. As already noted, projects include roads and highways, bridges, dams, water-treatment facilities, freshwater-supply systems, the structural elements of buildings, and the mechanical systems in them.

Project Design The word *design* can be used as a verb or as a noun. As a verb, *to design* refers to the process you go through to develop a plan. This plan will have many different parts. It really does not matter if you are designing a bridge, dam, or building. Many of the design processes you use are the same.

When used as a noun, *the design* refers to the final plan or solution. The design will be represented through drawings, models (both computer-generated and physical models), and a proposal. Often the design is presented to the client in a formal presentation. When the client accepts the design proposal, the construction phase may begin. A civil engineering project includes both the design and construction phases.

3

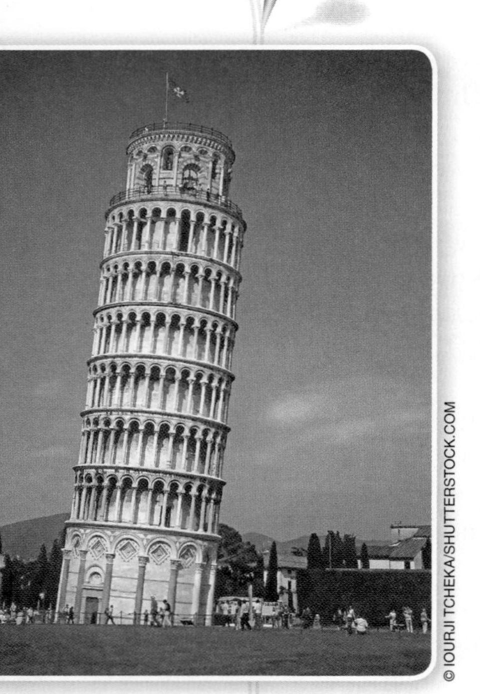

FIGURE 10-5 The famous Leaning Tower of Pisa in Italy shows the importance of geotechnical engineering.

Project Documentation Civil engineers use project documentation to record their ideas and activities, from their first idea for a project to its final development. Project documentation includes several things such as portfolios, journals, sketches and drawings, digital pictures, and computer files to record the design and construction process (Figure 10-6). Project documentation is very important to provide an accurate record of the project. Sometimes these documents become evidence in legal cases. Engineers and architects have a legal and ethical responsibility to maintain an accurate record of the design and construction of projects.

A project portfolio is a record of a project from its beginning to completion. Engineers use portfolios to show specific design aspects and their progress through a project. An engineer's portfolio, in the format of an engineer's notebook, is a legal document that can be used as evidence in a court of law.

The portfolio, like a diary, is a record of original ideas and research related to a project. The portfolio also records the progress being made on a project. Often a portfolio will include sketches, drawings, renderings, specifications, and charts and graphs.

A journal is another tool used by civil engineers. A journal is a daily record of writings, sketches, and research materials that describe the design process.

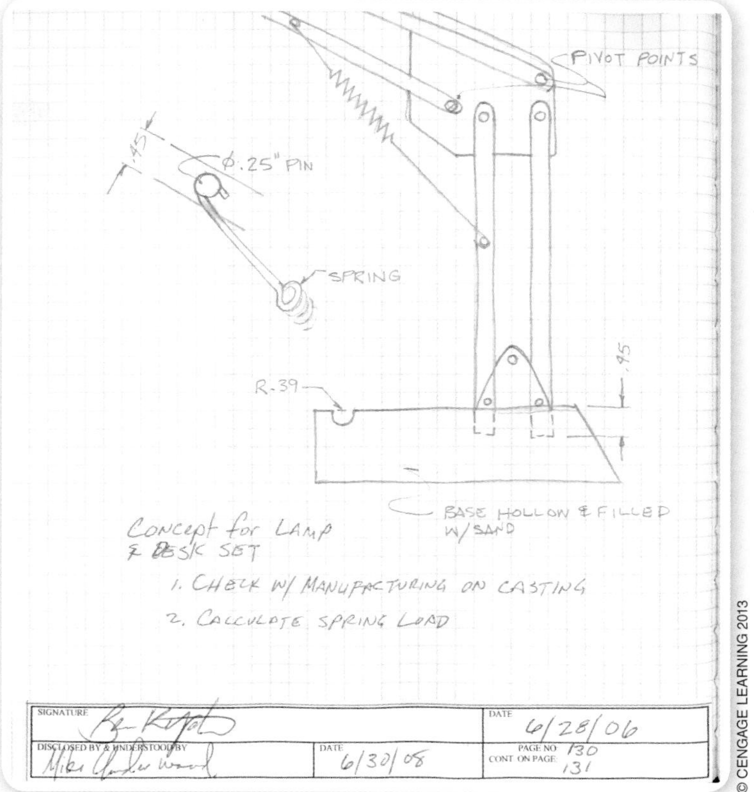

FIGURE 10-6 An engineer's notebook page might include concept sketches, notes, and dated signatures.

Project Planning

Project planning involves working in teams. You have probably heard the expression "Two heads are better than one." This is a true statement and the reason why different experts are brought together to form the project planning team. As described earlier in this chapter, there are many different specialties within civil engineering. Both the type of project and the project design will determine which specialists are needed on the project planning team.

An individual trained in structural stress analysis may not be the best person to design the electrical supply system for a building or test the soil to determine if it can bear the weight of the building. Similarly, a civil engineer trained to calculate heating and cooling needs and airflow through ducts may not be the best person to design the plumbing system.

Site Selection and Regulations Civil engineering projects are always located on a site (Figure 10-7). This is the primary difference between civil engineers and engineers who work in manufacturing and whose products are made in a manufacturing facility. Civil engineers may work in offices, but each project is located at a site.

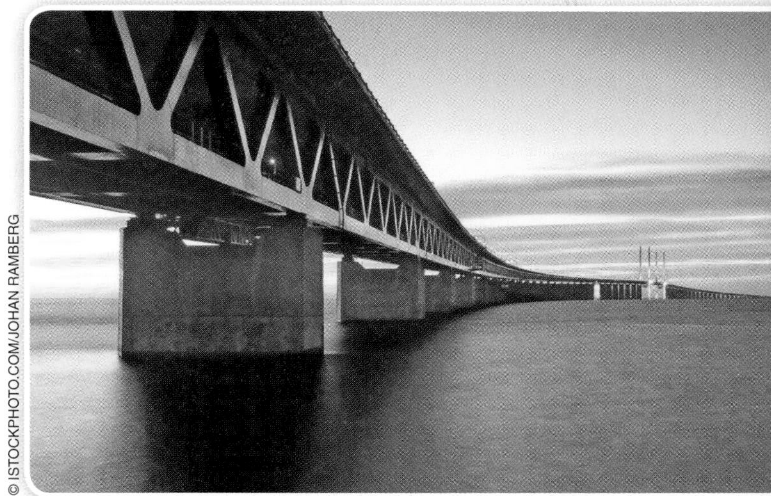

FIGURE 10-7 Civil engineers design projects to be built on a site, not in a factory. The Oresund Bridge between Denmark and Sweden is the world's longest single bridge carrying both road and railway traffic.

The selection of the site for a project is a critical decision. A good deal of research must be done in many different areas. For example, you need to know who the past owners of the site were and what they used the site for. Are there legal restrictions on what the land can be used for (zoning laws)? What is the adjoining property used for?

4

One of the most important considerations in site selection is the impact on the environment. How will human activity at this site affect others? Remember, pollution or contaminants from human activity seep into the ground and also run downhill or downstream. Are there wetlands in the area that will be impacted? How will the local wildlife be affected by the project? These questions and many more must be answered in an **environmental impact study** before the project can begin.

environmental impact study

An environmental impact study assesses the impact of a proposed project on the environment.

Soil type is an important consideration in all civil engineering projects. The Leaning Tower of Pisa is one of the most famous examples of a building constructed on soil that could not sustain the weight of the building. Why are very tall skyscrapers built in New York City and Chicago but not in Los Angeles? Soil type is important in all projects, not just buildings. Can you imagine the consequences of not involving geotechnical civil engineers in the planning for a large dam?

Site Planning

Civil engineers have legal and ethical responsibilities on any projects they undertake. A responsible design will make the best use of the property and minimize the impact on the environment. Project designs should be pleasing to the eye and blend with the natural surroundings. Space should be both attractive and well used for the specific purpose of the project.

Codes and building requirements (national, state, and local) will impact the use of the site and the design. The local codes may affect the type of structure that can be built and where it is located. The availability or lack of certain utilities will also affect the site planning. Landscaping is also an important part of site planning.

Did You Know?

Manufacturing and construction processes both produce a product. The difference between construction and manufacturing is whether the product is built on-site or in a facility.

Math in Engineering

S T E M

Mathematics is very important in surveying. Surveying uses principles of geometry and trigonometry. Calculus is used to accurately describe the area of lakes.

Description of Property Land is valuable. All land in the United States has a legal description. This legal description is considered a public record and is typically recorded at the county courthouse. The purpose of the description is to make sure that all landowners know exactly where their boundaries are. 5

One of the most common methods to establish legal boundaries is through the use of **surveys** (Figure 10-8). Surveys have been used in the United States since the 1790s. A rectangular survey is based on principal meridians and baselines. Meridians (lines of longitude) are the lines that run north and south; baselines run east to west. The Prime Meridian passes through Greenwich, England, and is the meridian defined as 0. It divides Earth into the eastern and western hemispheres. Today GPS (Global Positioning System) is also used to describe property.

survey
A survey is a method used to establish legal boundaries of land.

Space Allocation Planning Space allocations are made according to the needs of the user. Depending on the intended use of the space, having enough room for people to move about may be an important consideration. For example, businesses that have frequent short-visit customers, such as gas stations and banks, need easy access to roads or highways. Businesses that have customers who stay longer, such as shopping malls and grocery stores, need more parking spaces. Shopping malls need enough interior space between stores to allow hundreds of people to move about easily.

How do large numbers of people get quickly and safely into a sports arena or concert hall? Have you ever been stuck in a traffic jam on your way to a major event? Space allocation planning is critical to anticipate how traffic will flow. Civil engineers and architects need to plan for how people and goods will arrive and depart from the structures they are designing

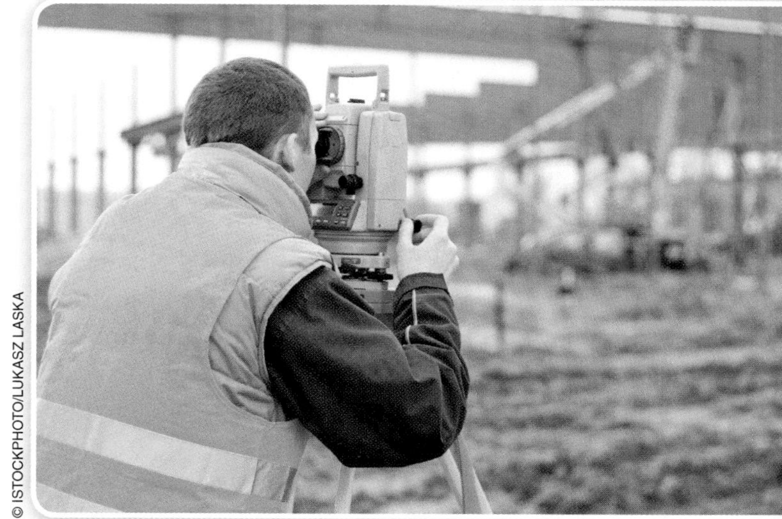

FIGURE 10-8 Surveyors provide accurate legal descriptions of land.

FIGURE 10-9 Site planning is important for traffic flow, safety, and getting utilities to the site.

FIGURE 10-10 Civil engineers must plan how to get the electricity produced at a dam to cities where it is needed.

(Figure 10-9). They also need to plan for how traffic will move *within* any structure.

Similarly, industrial or public works projects have specialized use requirements. Space allocations for an industrial site may include larger spaces for shipping and receiving, either by rail or by semi-truck. They may also need much larger supplies of resources such as electricity, water, freshwater, and natural gas. How these resources will arrive at the structure safely is an important part of the planning process. Will they be in underground pipelines and power lines? Will a road be laid over the pipelines?

Planning for getting re-sources into or out of a public works site is an important consideration in space allocation planning. Electricity produced at a dam is of little value if it cannot be shipped to customers over large power-transmission lines (Figure 10-10). Power lines for electricity or pipelines for liquids and gases are very expensive. Civil engineers plan the shortest and safest routes possible for these supply lines.

Site Plan Layout Site layout almost always involves significant amounts of grading. **Grading** is the term used to describe moving and leveling earth. Planning for grading is based on the space allocation and use previously described, as well as on the contours of the site. **Contours** are lines on a map that show the topographical outlines of the land's physical characteristics or lines that join points of similar elevation.

contours
Contours are lines on a map that show topographical outlines or elevations of land.

The contours of a site can influence the orientation of the structure. Because of the time and expense involved in moving and relocating large amounts of earth, the architect or civil engineer will pay close attention to how the structure might be oriented to take advantage of existing contours.

Top soil is a very valuable resource. For this reason, it is usually removed first and set aside to be redistributed after final grading. Grading must be done carefully to ensure that water will drain away from the structure. Civil engineers must plan to minimize erosion of the land.

The location of certain elements on the site such as driveways must be planned to minimize expense while still maintaining ease of use. For example, a short driveway is much cheaper to grade and build than a long driveway. But if that driveway is too steep as a result, a short driveway is a poor solution. Perhaps a longer driveway would better balance the steepness of the terrain. This is just one example of the kind of decisions that architects and civil engineers must make when designing the site plan layout. As mentioned earlier, utilities are a major part of the planning process as well.

architecture
Architecture is the art and science of designing and erecting buildings.

sustainable architecture
Sustainable architecture, sometimes termed green architecture, refers to designing a structure with the environment in mind.

SECTION 2: ARCHITECTURE

Architecture may be defined as the art and science of designing and erecting buildings. It truly is a combination of both art and science that can provide lasting beauty that will be valued for generations. The artistic part is the impression that first meets the eye. Do the lines flow naturally? Where do they lead your eyes? Does the design blend with its environment?

Sustainable architecture, sometimes termed green architecture, refers to designing a structure with the environment in mind. Architects consider the building's location. Is it close to mass transit? Architects decide which building materials will be used. Are the materials made from recycled products? Architects also consider the building's energy use. Can solar energy be used?

Frank Lloyd Wright is one of the most famous American architects. He was renowned for how his uniquely designed houses and buildings blended naturally with their surrounding environments, whether built in a desert, on a prairie, or in a woodland (Figure 10-11). Research Frank Lloyd Wright, and see if you can find the house he designed and built over a river.

The science, of course, is the part that ensures the building will stand strong for centuries and be safe for the occupants. Is the space used efficiently? As a visitor, are you drawn into the main spaces? Does air flow naturally throughout the building? Does the building take full advantage of solar heating? Many of today's buildings reach incredible heights, such as the proposed Freedom Tower to replace the World Trade Center in New York City

FIGURE 10-11 Frank Lloyd Wright's designs blended naturally with their environments.

Math in Engineering

Mathematics is critically important in the design of large skyscrapers. Systems such as plumbing become more complicated as building height increases. The water must be pumped to the top at various levels, and the wastewater in drainpipes develops incredible force as it falls. Mathematical models allow engineers to design systems to address these problems.

© CHRISTOPHER PENLER/SHUTTERSTOCK.COM

FIGURE 10-12 Buildings such as the new Freedom Tower reach incredible heights and require sound scientific and mathematical models.

(Figure 10-12). As the height of the building increases, so does the emphasis on sound scientific principles.

Buildings are designed for a purpose. Understanding a client's needs or the primary purpose of a building is critically important in a well-designed building. Traffic flow, discussed earlier for planning the site, is also a primary factor in the building design. Consider the difference in traffic flow between a fast-food restaurant and an elegant restaurant. Understanding how the building will be used is essential to a good design.

Weather can be very harsh on a building's roof and exterior. Wind, rain, hail, and the blistering heat from a summer's sun can take their toll on a building's exterior. It takes an understanding of science to ensure that appropriate materials are selected that will endure this punishment from nature for years to come. Science is also involved in designing how the water will run off the building's roof and sides and away from the foundation on the ground. It takes art to make these functional elements attractive. Buildings also need to be constructed to withstand weather extremes, such as hurricanes, tornadoes, and perhaps even earthquakes. Local building codes will dictate how a building should be built to withstand the extreme conditions of that particular region.

Architectural Styles

Architectural styles have changed dramatically over time. The European settlers who colonized the United States brought with them the building styles and techniques from their homelands. So housing styles across the country vary based on the area's initial settlers. Common architectural styles found in the United States are shown in Figure 10-13.

© HARRY HU/SHUTTERSTOCK.COM

(a)

© SUSAN LAW CAIN/SHUTTERSTOCK.COM

(b)

© IRIANA SHIYAN/SHUTTERSTOCK.COM

(c)

© KIM SEIDL/ SHUTTERSTOCK.COM

(d)

© DVANDE/SHUTTERSTOCK.COM

(e)

© NELSON FONTAINE/ SHUTTERSTOCK.COM

(f)

© MICHELLE MARSAN/ SHUTTERSTOCK.COM

(g)

FIGURE 10-13 Architectural styles vary widely around the nation. The images here show examples of some of these styles: (a) colonial, (b) Victorian, (c) Craftsman, (d) row house, (e) ranch, (f) Cape Cod, (g) Spanish. Can you identify the architectural elements in each of these buildings?

Science in Engineering

Engineers and architects must understand the transfer of heat energy. Large buildings have different heating and cooling needs on each side of the building as well as on different floors. Heating and cooling systems place huge energy demands on large buildings. Increasingly, architects and civil engineers must design green structures that take advantage of natural heating and cooling to save energy.

form
Form is the principle of design that is described by lines and geometric shapes.

texture
Texture refers to the roughness or smoothness, including reflective properties, of a material or building.

Regardless of the architectural style, all styles combine elements and principles of design. These elements and principles include form, texture, rhythm, balance, proportion, and unity. **Form** is the principle of design that is described by lines and geometric shapes. The form of a structure is often dictated by its function. **Texture** refers to the roughness or smoothness, including reflective properties, of a material or building.

Rhythm is the term used to describe the illusion of flow or movement created by having a regularly repeated pattern of lines, planes, or surface treatments. **Balance** is the principle of design dealing with the various areas of a structure as they relate to an imaginary centerline; sometimes this is referred to as *symmetry*. **Proportion** is a principle of design that deals with the size and shape of areas and their relationship to one another. **Unity** is the principle of design that ties the various elements of a structure's design together.

rhythm
Rhythm is the illusion of flow or movement created by having a regularly repeated pattern of lines, planes, or surface treatments.

balance
Balance is the principle of design dealing with the various areas of a structure as they relate to an imaginary centerline; sometimes referred to as *symmetry*.

Working Drawings

Working drawings are used to describe and communicate the construction requirements. A complete set of working drawings for a house typically includes a site plan; floor plan; elevations; sections and details; schedules; and mechanical, plumbing, and electrical plans. Specialized symbols are used to represent items such as cabinets, sinks, and electrical switches and lights. Residential plans are commonly drawn to a scale of ¼ inch = 1 foot.

Floor plans provide builders and families a quick look at many aspects of a house's design. They are especially helpful in showing the space and size relationships of rooms and how people will move through the house. Floor plans can answer questions such as "Is the laundry room near the bedrooms?" "Is there enough space in the family room for a pool table?" "If someone comes in from outside with muddy shoes will her or she have to walk on the carpet?" (Figures 10-14 and 10-15).

Elevations are drawings that show the exterior of the house. What will the front look like? Will it be finished in brick, stone, or wood? Where are the windows on the side of the house? Elevations give a very accurate picture to builders and owners about what their future house will really look like when finished (Figure 10-16).

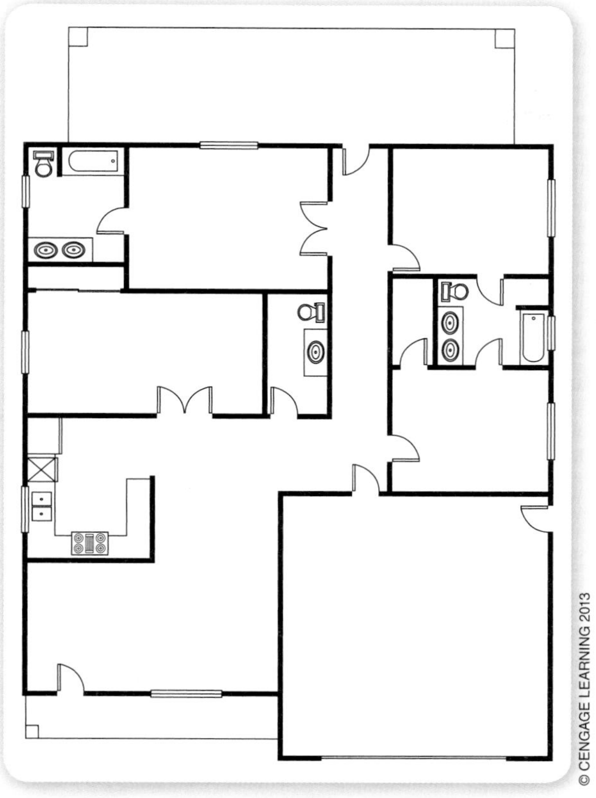

FIGURE 10-14 Floor plans are helpful for seeing room size, space usage, and traffic flow.

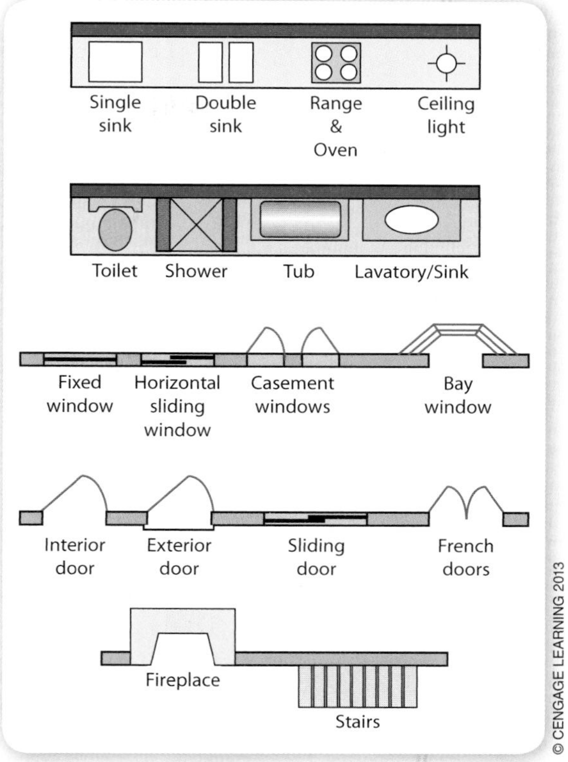

FIGURE 10-15 Architectural symbols are used in the floor plan to represent different components of the structure.

FIGURE 10-16 Elevations often give a client the best overall picture of a building.

proportion

A proportion is the relationship between one object and another, or between one size and another size.

unity

Unity is the principle of design that ties the various elements of a structure's design together.

Sections and details are drawings that describe interior details. **Sections** show the details through a cutting plane across the building. **Details** are drawings that enlarge a specific part to help a builder understand how the particular pieces fit together. The word *typical* may be used to mean that the same detail is repeated in other places in the building.

Schedules are tables that list specific details, such as information about windows and doors. Schedules can also be used to list the different light fixtures that will go in the house. The information in the schedules tells the builder what kinds of materials to order. The information is usually very specific about the sizes, models, and manufacturers of the parts.

Mechanical, electrical, plumbing, and protection systems are also all specified on drawings included in a complete set of working drawings. The mechanical plan will include information about the heating, ventilation, and air conditioning (HVAC) system of the building. The health and comfort of the people who live in the house are affected by the HVAC system. The heating and cooling units, along with the duct work for moving the air, must be carefully selected according to the size of the building.

The safety of the occupants is determined by the electrical, plumbing, and protection systems designed for the house. Strict codes regulate freshwater supply and wastewater removal from houses. Similarly, electrical codes specify the size of wire, number of lights and outlets on a circuit, and other information designed to protect the occupants. *Protection systems* is the general term for smoke and carbon monoxide detectors. These devices must be located in certain areas of a house such as a bedroom.

Engineering Challenge

Architectural styles vary from one part of the country to another. With your teacher's permission, work in teams to search the Web for famous houses and buildings that represent these different styles: arts and crafts, bungalow, Cape Cod, farmhouse, federal, Georgian, modern, Queen Anne, ranch, saltbox, Spanish, split-level, Tudor, and Victorian. For example, what style is the White House?

Working with your team, develop an electronic presentation of architectural styles. Include in your presentation small photos that represent each style. Provide a title page with the information your teacher requires for your class. Type the following information under each photo and architectural style:

1. the name of the architectural style
2. where the style originated
3. the geographical location where the style is most common
4. the decade or years of most common use
5. the typical number of stories for the style, and
6. the distinctive features of the style that make it identifiable and unique.

Science in Engineering

Engineers and architects must understand science to design large structures. Physics is the branch of science that helps structural engineers plan for the forces and loads that affect large structures.

SECTION 3: STRUCTURAL ENGINEERING

Structural engineering is the analysis of forces and loads and their effects on a structure. All structures have many different forces acting on them. A *force* is a push or pull on the structure. Force also can come from phenomena such as wind. The term **load** refers to the weight a structure must support (Figure 10-17). Load can be from people or furniture in a building, or it can come from cars on a bridge.

A structural engineer is a professional who studies the forces acting on structures and the design of such constructions as bridges, buildings, dams, and stadiums. There are three basic principles for any structure: (1) it must be stable, (2) it must be strong enough to withstand the forces and loads acting on it, and (3) it must be economical to build.

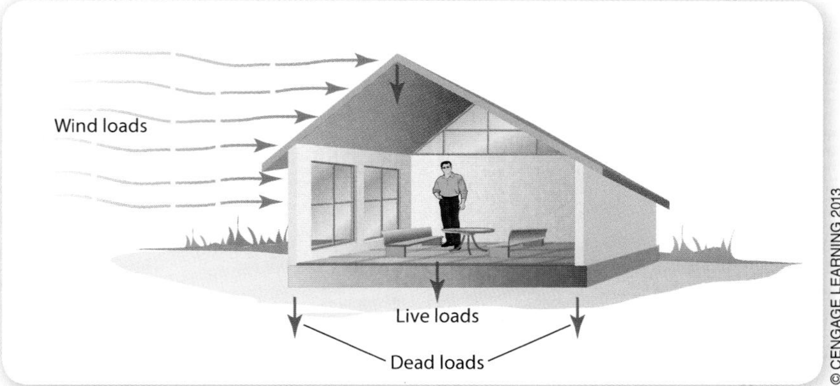

FIGURE 10-17 A dead load is the weight of the structural members, live loads are the people and furniture, and wind loads are the pressure against the exterior of the house.

Foundations

The **foundation** transfers a structure's loads to the ground. The foundation is the substructure of the building, and it is usually made from concrete. The entire structure rests on the foundation. Think about the incredible amount of weight a foundation must support for a tall building. The type of soil is critical in designing the foundation; how well the soil will be able to support weight becomes crucial to final design.

Columns and Beams

Columns and beams are important structural elements that support the loads placed on a commercial building. A **column** (or post) is a vertical structural

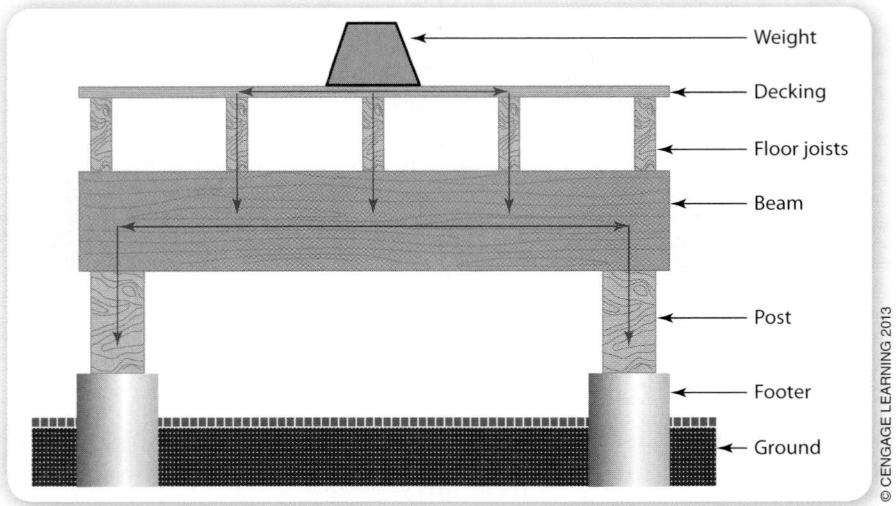

FIGURE 10-18 **Columns (posts) and beams transfer the load to the foundation.**

member that transfers loads to the structure's foundation. A **beam** is a horizontal structural member that supports roof or wall loads (Figure 10-18).

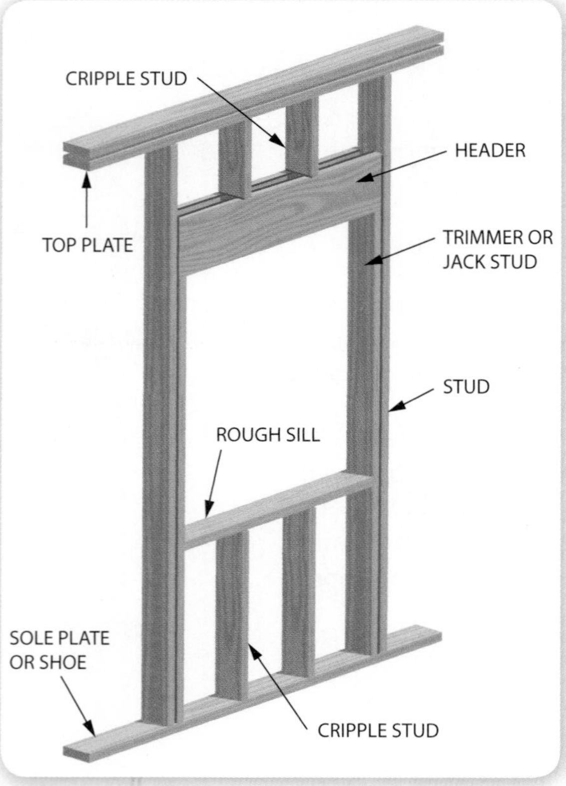

Columns and beams must be large enough to support the loads applied to them. Various materials can be used for beams and columns, including wood, engineered lumber, steel, and concrete. Engineered lumber is made from smaller pieces of wood that have been glued together with special glues. Engineered lumber is stronger than solid wood of the same size, and it will not warp or split the way solid lumber does.

Wall Sections

Loads in a house are carried by wall sections. The various components of a wall section are designed to distribute the house's weight. Vertical wall members that carry the load are called studs (see Figure 10-19). Loads over door openings and windows are called headers. Door and window openings are called rough openings because they are constructed larger than the actual door or window. This extra area allows the carpenter to level the doors and windows during installation.

FIGURE 10-19 **Members of a wall section.**

Roof Systems

The **roof system** is the top of the structure and the primary protection for a building's interior from weather. The roof system selected depends in part on regional weather conditions. There are numerous different styles of roofs. Common roof styles are shown in Figure 10-20. The finishing material on the roof is also important to how well the building will stand up to weather, sun, and fire. Common roof materials for houses include shingles made from asphalt or wood shakes, corrugated metal surfaces, and tile.

Pitch is the term used to describe the slope of the roof (Figure 10-21). Roofs on houses typically have a roof pitch of 3⁄12 or greater. A 3⁄12 pitch means that the roof rises 3 feet vertically for every 12 horizontal feet. This pitch is essential to help rain and snow run off the roof. Houses in areas where heavy snowfall is common have steeper roof pitches to prevent the snow from piling up too deeply. Snow is quite heavy and can cause a roof to collapse.

The structural members of a roof on a house are called **rafters** when they are erected on-site. This is the traditional method of building a roof. **Roof trusses** are premade at factories and then shipped to the site (Figure 10-22). Roof trusses are common in residential and commercial construction today. Trusses are stronger than traditional rafters built on-site. Roof trusses can span greater distances than traditional rafters and do not require intermediate supports.

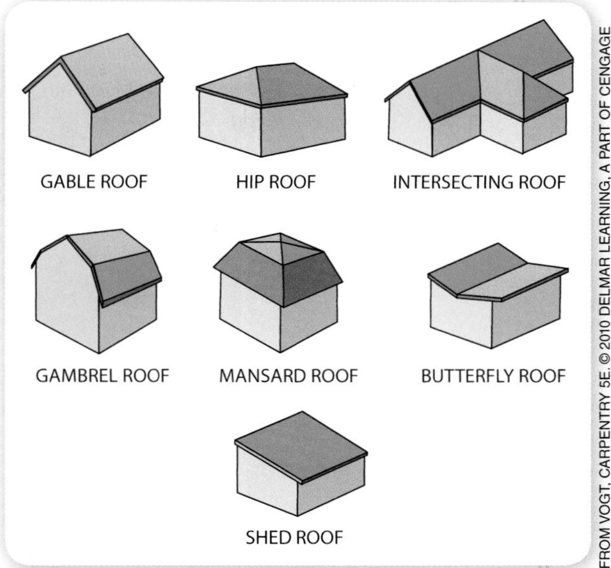

FIGURE 10-20 Different roof styles.

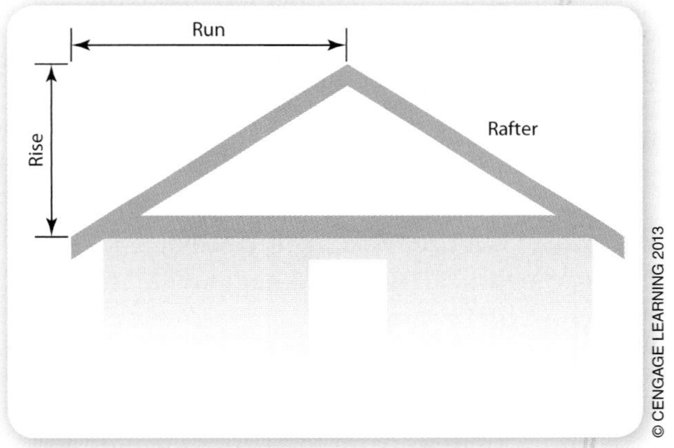

FIGURE 10-21 Roof pitch is measured in rise over run.

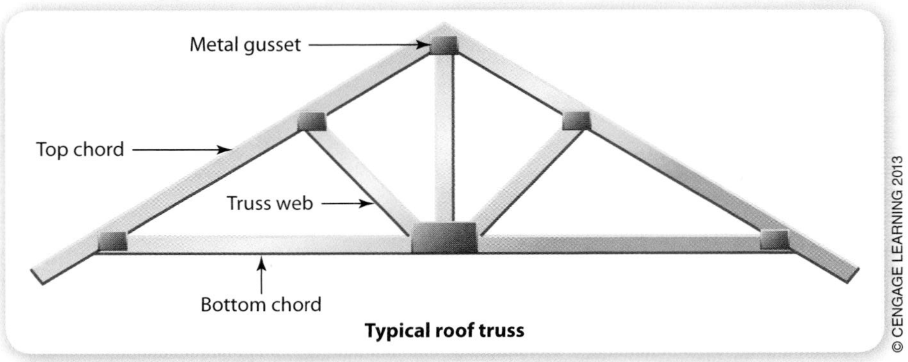

FIGURE 10-22 Roof trusses are stronger than traditional rafters.

SUMMARY

In this chapter you learned:

- A civil engineer is a person who is trained in the design and construction of public works projects (such as highways, bridges, sanitation facilities, and water-treatment plants) and who is largely responsible for the site design, construction management, and the inner workings of a building.
- Within civil engineering are several different specializations: structural, waste treatment, transportation, geotechnical, water management, and construction management.
- Civil engineers use project documentation to record their ideas and activities, from their first idea for a project to its final development.
- Civil engineering projects are always located on a site.
- All land in the United States has a legal description. This legal description is considered a public record.
- Sustainable architecture is designing structures that are environmentally friendly.
- Green building refers to both the process of building that is environmentally responsible and the resulting structure.
- Architectural styles combine elements and principles of design. These elements and principles include form, texture, rhythm, balance, proportion, and unity.
- Structural engineering is the analysis of forces and loads and their effects on a structure.

VOCABULARY

Write a definition for each term in your own words. After you have finished, compare your answers with the definitions provided in this chapter.

Architect
Civil engineer
Green building
Structural engineer
Environmental impact study
Survey
Contours
Architecture
Sustainable architecture
Form
Texture
Rhythm
Balance
Proportion
Unity

STRETCH YOUR KNOWLEDGE

Provide thoughtful, written responses to the following questions and assignments.

1. In what ways are civil engineers and architects similar, and in what ways are they different?
2. Why is teamwork important in civil engineering?
3. What are the different areas of specialization in civil engineering?
4. Why is it important for an engineer or architect to maintain accurate project documentation?
5. Why are elevation drawings often the most helpful to clients?
6. Why are sustainable architecture and green building important to our future?

Onward to Next Destination

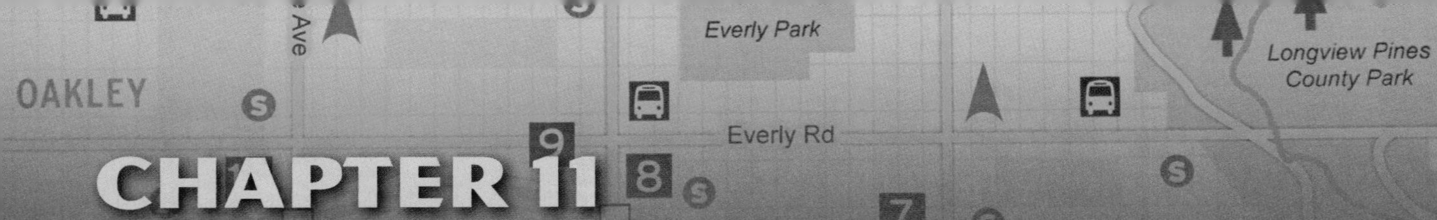

CHAPTER 11
Transfer of Energy

Menu

Before You Begin

Think about these questions as you study the concepts in this chapter:

What is a simple machine?

What is a mechanism?

How do engineers transfer energy so that the output movement is different than the input movement?

What types of movements can mechanisms change?

How do mechanisms make our everyday lives easier?

The Ferris wheel uses engineering principles to entertain its passengers.

Engineering in Action

As we know it today, the Ferris wheel came from the technical skills and engineering knowledge of a Pittsburgh, Pennsylvania, bridge builder named George Ferris. Ferris had graduated from Rensselaer Polytechnic Institute with a degree in civil engineering. He had worked for the railroads as a bridge builder and later founded a steel company in Pittsburgh. Ferris was approached at an engineering banquet in 1891 to design an American counterpart to the Eiffel Tower, the landmark of the 1889 Paris Exhibition. He sketched out a vertical merry-go-round on a napkin. A couple of years later, his Ferris wheel was unveiled at the 1893 Chicago World's Fair. The ride stood 264 feet high and, as you might expect, was constructed of steel. We need to thank the engineering education and technical skill of George Ferris for today's amusement ride.

SECTION 1: SIMPLE MACHINES

A **simple machine** is a basic device used to make work easier. Work is the transfer of energy from one physical system to another. During this process, energy is not created, only transferred. There are six simple machines: inclined plane, wedge, screw, lever, wheel and axle, and pulley. We discuss each simple machine separately in this chapter. If we combine two or more simple machines so that they function together to make work easier, the device is called a **compound machine.** A wheelbarrow is an example of a compound machine: it incorporates two simple machines (the lever and the wheel and axle).

Compound and simple machines use a concept termed **mechanical advantage**. Mechanical advantage is gaining increased force or extended motion by means of a machine. We typically describe mechanical advantage mathematically. Mechanical advantage is the ratio of force that performs the work to the force initially applied to the machine.

simple machine

A simple machine is a basic device that makes work easier.

compound machine

A compound machine is a combination of two or more simple machines.

mechanical advantage

Mechanical advantage is gaining increased force or motion by using a machine.

Inclined Plane

Have you ever helped someone move packages or furniture using a truck with a ramp? Sliding the packages up the ramp is an easier job than lifting the heavy packages. This process uses the mechanical advantage of the inclined plane. An **inclined plane** is a sloping surface that alters the force required to move the load in a perpendicular direction (see Figure 11-1). By applying force in one direction, horizontally, the packages move in a different direction, vertically.

If the slope or angle of the inclined plane is low (not very steep), a smaller amount of horizontal force is required. However, the length of the inclined plane must be longer. If the slope of the inclined plane is steep, then more force is required to raise an object to the same height. On the other hand, the length of the inclined plane is shorter. Civil engineers use mathematical calculations related to the inclined plane when they design the slope of roadways and railroad lines. Building codes mandate the slope of the inclined planes used for handicapped ramps on sidewalks and in buildings.

Did You Know?

Amusement rides, such as Ferris wheels and roller coasters, all use simple machines and mechanisms to change energy into motion. This motion provides excitement and entertainment for you and your friends. You could say that amusement rides are all work and all play.

FIGURE 11-1 The inclined plane transfers horizontal force to vertical force.

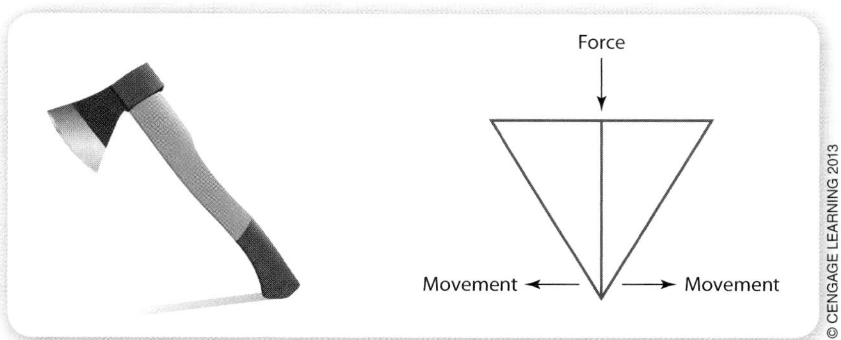

FIGURE 11-2 The ax is an example of how a wedge transfers energy.

Wedge

Have you ever used an ax to chop firewood for a campfire or used a knife to cut vegetables for dinner? If you have, then you have used the mechanical advantage of the wedge. The **wedge** converts motion in one direction into a splitting motion acting at right angles to the wedge, as illustrated in Figure 11-2. Sometimes we view a wedge as consisting of two inclined planes placed together. Force in one direction is transferred to two opposite directions, thus separating the object.

Screw

A **screw** is an inclined plane wrapped around a cylinder. A screw transfers circular motion to a straight-line movement, either in or out, depending on the direction the screw is turned (Figure 11-3). Just like the inclined plane, the slope of the screw threads affects both the force required to turn the screw and the amount the screw travels on each turn. Examine the two screws in Figure 11-4. We can install the screw in Figure 11-4a quickly because the slope of the inclined is very steep. The screw in Figure 11-4b takes longer to install because of the gradual slope of the inclined plane. However, we can install the screw in Figure 11-4b with very

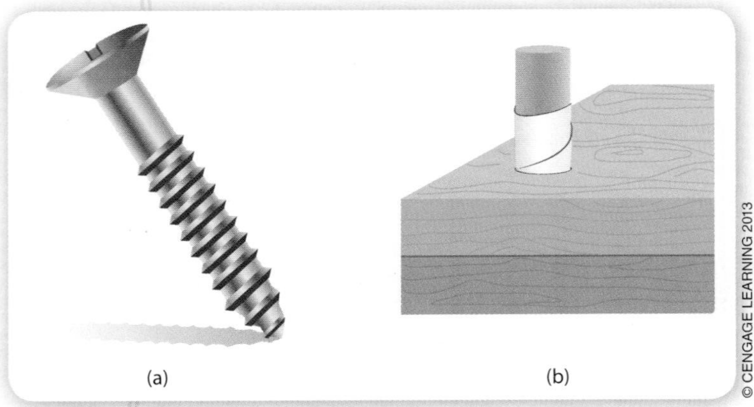

FIGURE 11-3 The screw is an inclined plane wrapped around a cylinder.

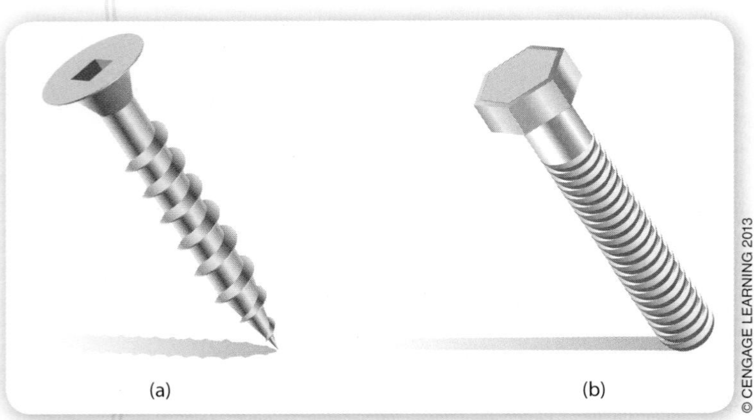

FIGURE 11-4 Screws can apply inclined planes with different slopes for different uses. (a) A screw with coarse threads can be driven quickly. (b) A screw with fine threads can be driven with greater accuracy.

accurate regulation of the force it applies. The screw can also be used as to install a lid on a jar, as a lifting device such as a car jack, or as a clamping tool such as a vise used on a workbench.

Lever

A **lever** is an arm that pivots on a point called the **fulcrum**. The lever converts movement from one side of the fulcrum to movement in the opposite direction on the other side of the fulcrum. The seesaw that children ride in a playground is an excellent example of a lever. On a seesaw, the distance from the fulcrum is equal on each side of the lever. Engineers, however, use levers with arms of different lengths to gain mechanical advantages. Depending on where the fulcrum is located, the lever can either increase the force transferred or change the distance over which the force is applied. Figure 11-5 shows the principles for levers that engineers use for their calculations.

Three types of levers are based on the location of the fulcrum. The

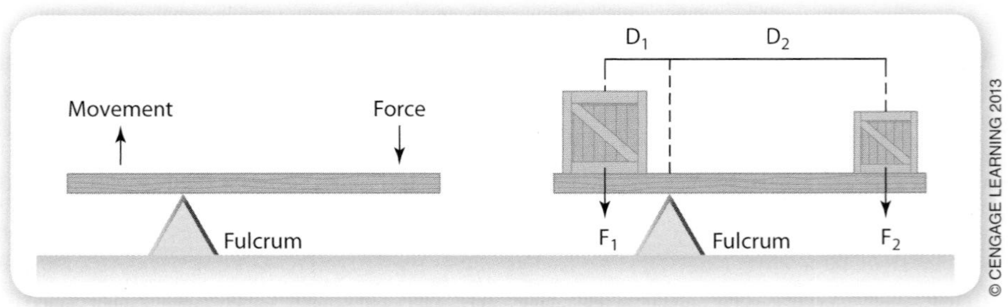

FIGURE 11-5 The lever is a very useful simple machine. It provides a mechanical advantage to help with work.

Math in Engineering

Engineers use mathematical calculations to determine the amount of load a lever can lift, the force required to lift a specific load, and the length of the lever arm needed to lift a specific load. The formula used is $F_1 \times D_1 = F_2 \times D_2$. The longer the lever's arm, the less force is needed to lift the same load.

seesaw in Figure 11-6 is an example of a first-class lever. Energy transfers from one side of the fulcrum to the other. The amount of mechanical advantage is based on the ratio of the length of the lever on each side of the fulcrum.

An example of a second-class lever is the wheelbarrow, where the wheel serves as the fulcrum (Figure 11-7). The length of the lever arms and the position of the load determine the mechanical advantage gained. A third-class

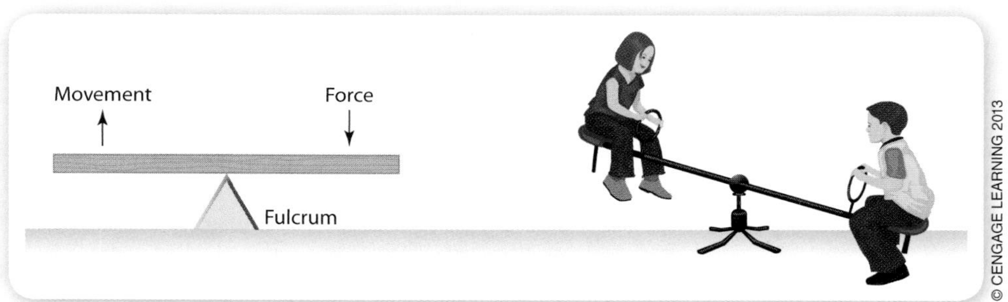

FIGURE 11-6 The first-class lever can be used to increase force and also to change the direction of the motion. A seesaw is an example of a first-class lever.

FIGURE 11-7 The second-class lever has its fulcrum at the opposite end of the arm from where the force is applied. A wheelbarrow is an example of a second-class lever.

Science in Engineering

It was Sir Isaac Newton's *Laws of Motion* (1687) that provided engineers with the scientific background to develop mathematical calculations related to levers. Newton's third law states, "For every action there is an equal and opposite reaction." When we force one side of the lever down, the opposite side moves up with an equal force. Engineers can, however, change the length of the lever arms in their design process to control the force that is required.

lever is the type of lever you use the most. Your arms serve as third-class levers. Other examples of third-class levers are when you use a softball or baseball bat, a mouse trap, and tweezers (Figure 11-8). Again, the length of the lever arm and the position of the force applied determine the mechanical advantage gained.

Wheel and Axle

A **wheel and axle** is a rotating lever (the wheel) that moves around the fulcrum (the axle). The wheel and axle uses the mechanical advantage of a second-class lever. The larger the wheel diameter, the greater the mechanical advantage that is gained (Figure 11-9). Examples of this second-class lever (wheel and axle) include a door knob and the pedal spindle on your bicycle.

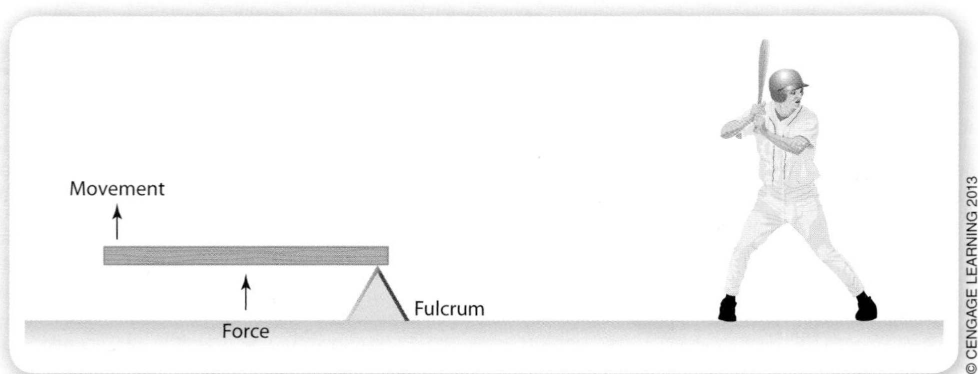

FIGURE 11-8 The third-class lever has the fulcrum at one end of the arm and the force is applied along the lever's arm. A baseball bat operates as a third-class lever.

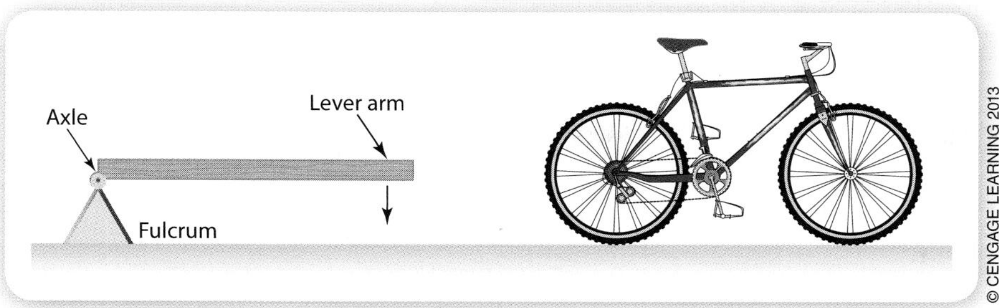

FIGURE 11-9 The wheel and axle allow for easier movement of objects.

Another way to use the wheel and axle is to reduce friction. Imagine pushing a box along a sidewalk. Now place the box on a wagon and try pulling the wagon with the box. Wheels on your bicycle, your skateboard, or a wagon use this concept. The larger the wheel, the less friction and the more easily the object will roll.

Pulley

When we wrap a rope or cable over a grooved wheel, we call the simple machine a **pulley**. As shown in Figure 11-10, the pulley works in two ways. First, it changes the direction of the force. Second, it can change the amount of force required to lift the load. When we group pulleys together, we create a **compound pulley** or a **block and tackle**. A compound pulley not only can reduce the force required to lift a load but also can increase the length of rope needed. We call this a **trade-off**. Figures 11-11a and 11-11b demonstrate how engineers calculate the mechanical advantages of pulleys.

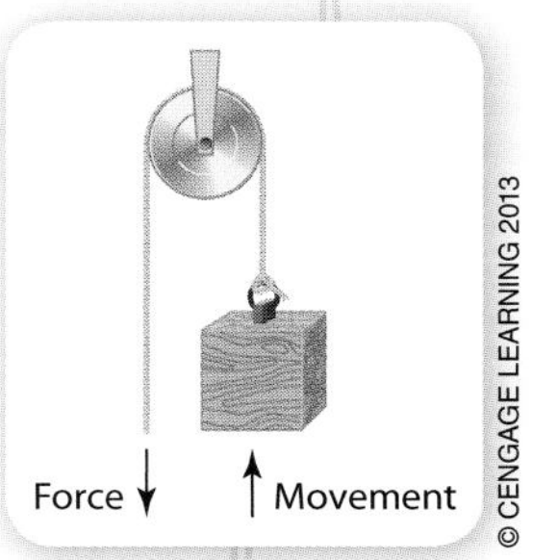

FIGURE 11-10 The pulley changes the direction of the force that is applied.

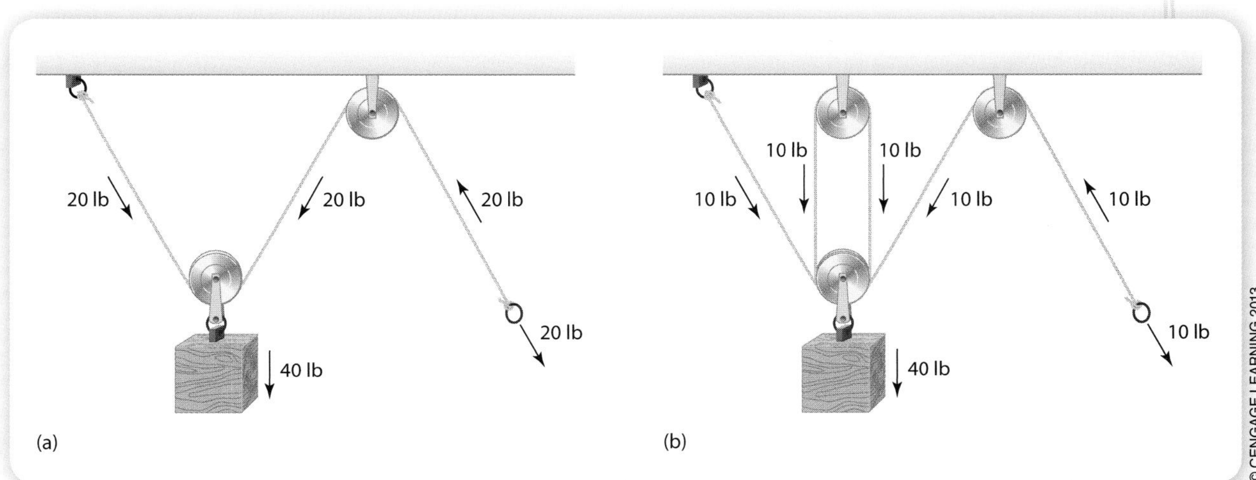

FIGURE 11-11 A compound pulley requires less force but more rope to lift the weight.

Career Spotlight

Name:
Oksana Wall

Title:
Consulting Structural Engineer, Celtic Engineering, Inc.

© CENGAGE LEARNING 2013

Job Description:
As a 13-year-old girl from Venezuela, Wall visited Disney World in Florida, and it immediately cast a spell on her. "It was such a happy and magical place," she says. Her father was an electrical engineer, and she was good at math and science, so she was already leaning toward an architecture or engineering career. But on that trip, she made up her mind to do engineering work for theme parks when she grew up.

After achieving her dream of working for Walt Disney World, Wall left for Celtic Engineering, Inc., which was started by her husband. Wall works with theme parks to create structures for rides and shows. For example, she helped design a multimedia show in Asia that includes a 360° movie experience. "The movie totally engulfs you, with several animated elements coming down from the ceiling," she says. "You feel like you're underwater."

Wall worked with other engineers on mechanisms that allow the show's elements to lower over the audience, perform their function, and then lift up again. The biggest element weighs about 25,000 pounds. "Behind the scenes there are at least 10 different mechanisms that work separately to bring the show together," she says.

Wall has a great time doing this kind of work. "It's very challenging doing structural engineering for entertainment because so much of what you do has never been done before in exactly that way," she says. "You have to adapt conventional engineering methods to something that is quite unique. It's fun never working on the same thing twice."

Education:
Wall received her bachelor's and master's degrees in civil engineering from the Florida Institute of Technology. She knew she wanted to work in theme parks, so she stayed focused on that goal in school. "I kept talking to all my teachers and to anybody else who would listen about what I wanted to do," she says. "Eventually, I spoke with other engineers to find out what I should be doing. Is this a good course to take? What is a good practical experience to have?"

Advice to Students:
Wall sees advantages in work that seems too difficult. "Don't get discouraged when something is hard," she says, "because, when things are really hard and you feel really dumb, that is when your brain is really stretching."

One difficulty Wall had to overcome is being a woman in engineering. "There are very few women in the field, but I never let that discourage me," she says. "When you have a hard time, think about your dream, and you will eventually get there if you keep on working at it."

Math in Engineering

Engineers can mathematically calculate the force required to raise an object based on the number of pulleys included in the block-and-tackle system. Because the load is distributed equally to all ropes in the pulley system, the more pulleys that are included, the less force is required to lift an object. Engineers use diagrams like those in Figures 11-11a and 11-11b to help with their mathematical calculations.

SECTION 2: MECHANISMS

Have you ever wondered how pedaling moves your bicycle? Have you ever thought about how the disc tray on your CD player opens and closes? These processes and many more are made possible by people designing and controlling devices called mechanisms. A **mechanism** is a device that transmits movements so that the output movement is different from the input movement. Mechanisms can change the movement's speed, its direction, or its type of motion. Motion can be in the form of linear motion, rotary motion, or reciprocating motion.

mechanism

A mechanism is a device that transmits movements so that the output movement is different than the input movement. It can be used to change the direction, speed, or type of movement.

linear motion

Linear motion is movement in a straight line.

Linear Motion

In **linear motion**, an object moves along a straight line such as that done by a grocery store checkout conveyor. The lever is a simple machine that provides linear motion.

Rack and Pinion A mechanism that converts rotary motion to linear motion is the **rack-and-pinion gear** (Figure 11-12). If you look at the mechanism that opens the CD drive on your computer, you will find a

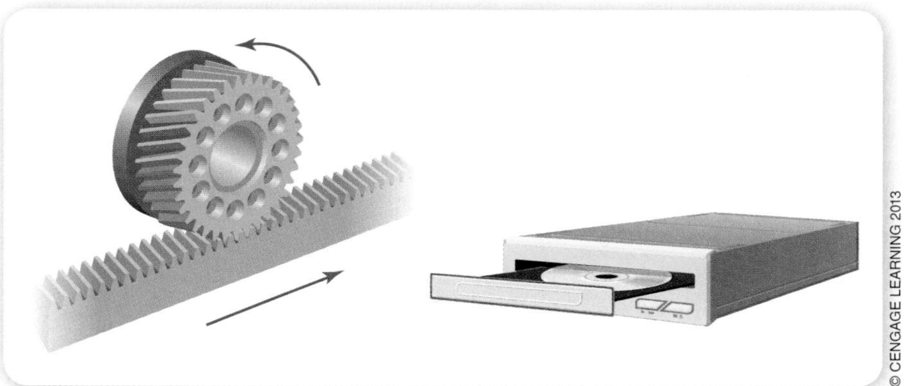

FIGURE 11-12 Rack-and-pinion gears convert rotary motion into linear motion. They are used to open CD drives.

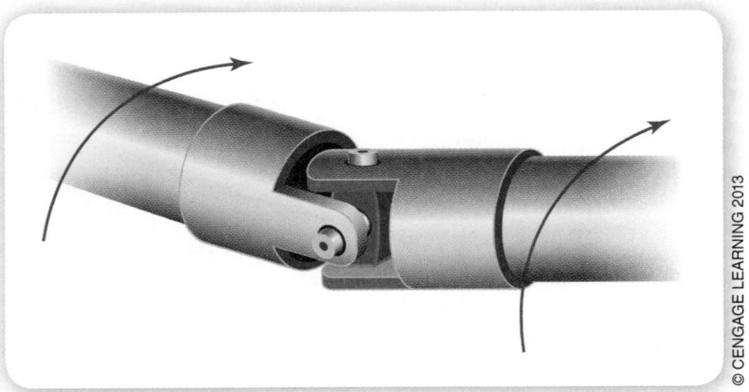

FIGURE 11-13 A universal joint is used to transfer rotary motion from an automobile's engine to its wheels.

rack-and-pinion gear. When you push the eject button, the computer uses a small motor to produce rotary motion. The rack-and-pinion gear converts this rotary motion into the linear movement of the CD tray opening and then closing.

Rotary Motion

rotary motion
Rotary motion is circular movement.

Revolving movement like a bicycle wheel or Ferris wheel is termed **rotary motion**. We measure rotary motion speed in revolutions per minute (rpm). We call the force of rotary motion **torque**.

Universal Joint The simplest means of transferring rotary motion is through a **universal joint** like that shown in Figure 11-13. The universal joint allows for the two axles to vary their alignment and also move slightly during the operation. Automobiles typically transfer rotary motion from the engine to the wheels through these universal joints.

Gears A **gear** is a circular device that transfers rotary motion using interlocking teeth. Engineers combine gears in a variety of ways to gain mechanical advantages for rotary motion. A **gear train** is a set of gears designed to increase or decrease speed or to increase or decrease torque. When speed increases, torque decreases; and when torque increases, speed decreases.

The number of teeth on the various gears in the gear train determines the **ratio** of the gear train. We call the first gear in a gear train the **drive gear** and the last gear in the gear train the **driven gear**. We call a gear used in the middle of the gear train an **idler gear** (Figure 11-14). A drive gear with more teeth than the driven gear will increase speed and decrease torque. A drive gear with fewer teeth than the driven gear will increase torque and decrease speed.

FIGURE 11-14 The gear train is used to transfer rotary motion. It can change the speed or torque of the motion. Each gear turns in the opposite direction of its drive gear. You can see gear trains at work on bicycles.

FIGURE 11-15 Bevel gears are used to change the direction of rotary motion in robotic arms.

Special Gears We use a **bevel gear** (Figure 11-15) to change the direction of rotary motion. Typically, bevel gears do not change the speed of rotation, but we can use them to transfer the movement of a robotic arm without changing speed or torque. We can also use a **crown-and-pinion gear** to change the direction of rotary motion, although this typically decreases the speed of rotation while increasing the torque. You can find

bevel gear
A bevel gear is a gear in which the axis of the drive gear forms an angle with the axis of the driven gear.

Engineering Challenge

ENGINEERING CHALLENGE 1

Your engineering team needs to design a set of gears or a gear train to transfer rotary motion. The input speed from a motor is 1560 rpm.

Design a gear train that will provide an output speed of 390 rpm rotating in the same direction as the input shaft.

Design a gear train that will provide an output speed of 4680 rpm rotating in the opposite direction from the input shaft.

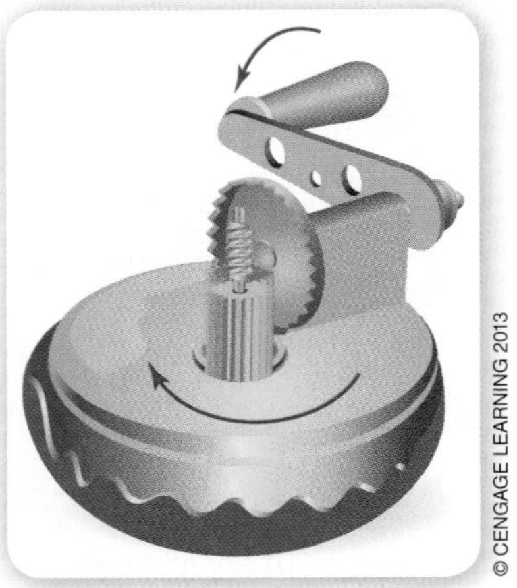

FIGURE 11-16 The crown-and-pinion gear changes the direction of the rotary motion and typically increases its torque. Fishing reels use crown-and-pinion gears.

FIGURE 11-17 The worm gear transfers rotary motion while changing direction and increasing torque.

5

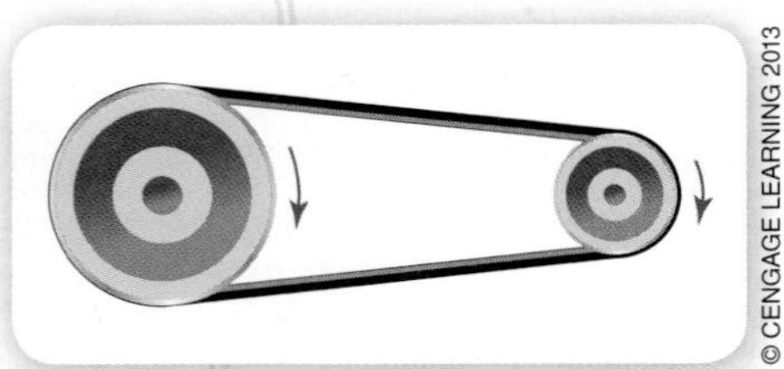

FIGURE 11-18 The V-belt and pulley are used to transfer rotary motion. This mechanism can change the speed or torque of the motion. Each pulley turns in the same direction.

a crown-and-pinion gear under the cover of a fishing reel (Figure 11-16). A **worm gear** (Figure 11-17) changes the direction of the rotation while increasing the torque of the movement.

V-Belts and Pulleys Engineers also can change rotary motion by using **V-belts and pulleys** as shown in Figure 11-18. Gears that mesh together change the direction of their rotation (e.g., clockwise to counterclockwise); pulleys using V-belts continue to turn in the same direction. For V-belts and pulleys, the determining factor for speed and torque is not the number of teeth per gear but the circumference of the pulleys.

Lead Screw We can convert rotary motion to linear motion by using a **lead screw**. This mechanism is not reversible. The metal-turning lathe in Figure 11-19 is an example of a lead screw.

reciprocating motion
Reciprocating motion is linear movement in a back-and-forth motion.

Reciprocating Motion

We call a movement that repeatedly goes back and forth a **reciprocating motion**. An example of a reciprocating motion is the needle of a sewing machine or the saw blade on a scroll saw.

Cam and Follower A common mechanism used to convert rotary motion into a reciprocating motion is a **cam and follower** (Figure 11-20). The cam is always the input device for the cam and follower. The follower, sometimes called a *lifter*, is always the driven component of this mechanism.

FIGURE 11-19 The metal-turning lathe is an excellent example of a typical use for the lead screw.

Crank and Slider Engineers can change rotary motion into reciprocating motion by using a mechanism termed a **crank**. Typically, the crank is connected to a rod that moves back and forth. Sometimes this rod is called the *slider* and at other times a *connecting rod*. The linear distance traveled by the rod is twice the offset of the crank. We can also reverse this motion so that the back-and-forth motion of the rod causes the rotary movement of the crank (Figure 11-21).

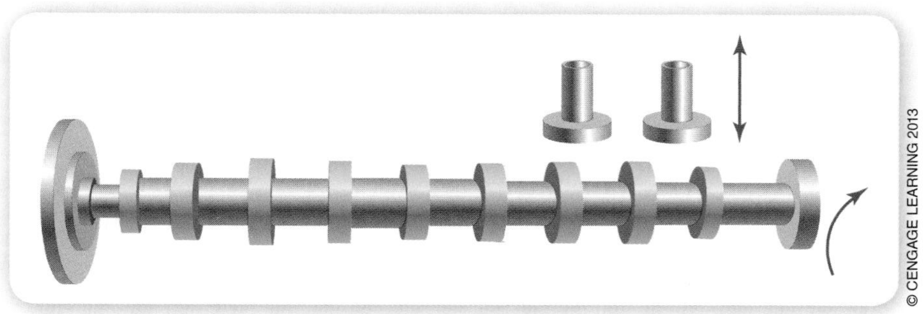

FIGURE 11-20 The cam and follower are used to transfer rotary motion into reciprocating motion. Camshafts and followers are used in automobile engines.

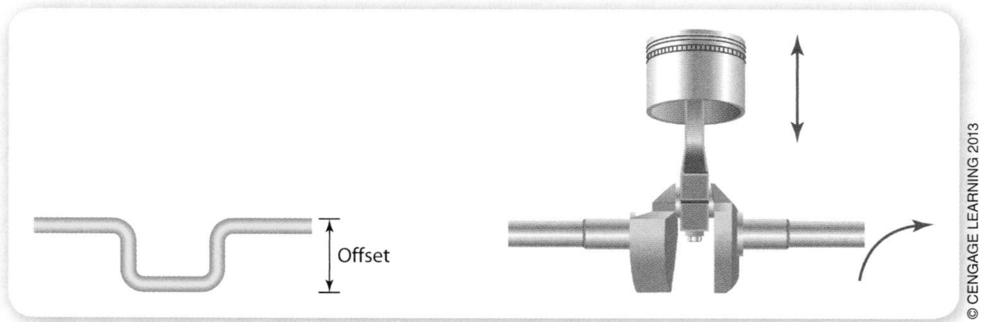

FIGURE 11-21 The crank is used to transfer rotary motion into a reciprocating motion. This transfer can also be reversed.

Engineering Challenge

ENGINEERING CHALLENGE 2

Design, build, and test an amusement ride that contains simple machines and mechanisms to move passengers (marbles) from a holding tank (waiting line) to a destination that is 24 inches higher and 24 inches to the side of the holding tank. Because this is an amusement ride, you should use as many simple machines and mechanisms as you can that will fit within the 24-inch-cube space.

SUMMARY

In this chapter you learned:

- Engineers use mechanical advantage to perform work. Mechanical advantage is a mathematical ratio of the force developed to the force applied.
- There are six simple machines: inclined plane, wedge, lever, pulley, screw, and wheel and axle.
- Two or more simple machines used together become a compound machine.
- Mechanisms change movement so that the output movement is different than the input movement.
- We can use mechanisms to change a movement's direction, speed, or type.
- Motion can be linear, rotary, or reciprocating.

VOCABULARY

Write a definition for each term in your own words. After you have finished, compare your answers with the definitions provided in this chapter.

Simple machine
Compound machine
Mechanical advantage
Mechanism
Linear motion
Rotary motion
Bevel gear
Reciprocating motion

STRETCH YOUR KNOWLEDGE

Provide thoughtful, written responses to the following questions.

1. How has the development of mechanisms affected people's daily lives over time (both in work and play)?
2. How have our economy and standard of living been affected by the harnessing of energy and the development of mechanisms?
3. Describe how your life would be different if we did not have mechanisms.
4. What types of engineers use mechanisms in their design work?

Onward to Next Destination

CHAPTER 12
Fluid Power

Menu

Before You Begin

Think about these questions as you study the concepts in this chapter:

 What is fluid power?

 What is pneumatics?

 What is hydraulics?

 How do engineers use fluid power?

 How does fluid power make our daily lives easier and safer?

 How can engineers determine how a fluid power system will operate?

Engineering in Action

S T E M

How did you get to school today? Did you ride in a car? Did you ride in a school bus? If you live in large city, did you take the train? All these types of transportation rely on **fluid power** to take you safely to school. Fluid power is the use of a fluid under pressure to transmit power. We classify fluids as either liquids or gases.

The car and the school bus use a liquid fluid power. When the driver steps on the brake pedal, liquid brake fluid pushes through a tube to the wheel. At the wheel, the fluid power applies the brakes. Just like water spraying out of a garden hose, liquid fluid power has a lot of force.

The train uses air fluid power to stop. The train is much heavier than a car or school bus, so it is much harder to stop. Air can be compressed more than a liquid can. Try squeezing a full plastic bottle of water. It is hard to compress. Now try squeezing an empty plastic bottle with its lid on tight. You can still easily compress the bottle. Therefore air does not have the force of a liquid. So how does air stop a train?

The first train brakes were very simple. Each train car had a crank located on its roof. A brakeman turned the crank to apply the brakes. The train engineer blew the train's whistle to signal the brakeman. The brakeman would then run from car to car turning the cranks. Brakemen had a very dangerous job (see Figure 12-1).

FIGURE 12-1 Working as a brakeman was dangerous.

fluid power
Fluid power is the use of a fluid under pressure to transmit power.

A man named George Westinghouse invented a device called the *fail-safe* braking system. Air did not have enough force to stop a large train, but it could be used to signal each train car's brakes. So for a train, each wheel's brakes are always on. An air signal releases the brakes, thus the term *fail-safe* is used. So trains use fluid power to transfer a control signal rather than to apply the stopping force.

Did You Know?

- George Westinghouse patented his fail-safe brake on March 5, 1872 (patent no. re.5,504).
- The U.S. Congress passed the Safety Appliance Act of 1893; the act required fail-safe brakes on all trains.
- George Westinghouse was awarded 361 patents.
- George Westinghouse and Thomas Edison were rivals in the early days of the electrical utility industry.

Science in Engineering

Ctesibius was a Greek scientist and inventor who lived in Alexandria, Egypt, from 285 to 222 B.C. Ctesibius began developing his theories at a young age. As a youth, he dropped a lead ball into a tube of air. The ball compressed the air, and he realized that compressed air was also a substance. Ctesibius wrote the first theories about the science of compressed air and how air could be controlled by pumps. He is known as the "father of pneumatics." Ctesibius is also credited for discovering the principles of the siphon.

SECTION 1: FLUID POWER SCIENCE

For centuries, fluid power science has helped society advance. Ancient Greeks, Chinese, and Egyptians all used fluid power to help with their work. All of these civilizations had scientists and engineers who studied the effects of fluid power. They commonly focused their studies on novel uses for fluid power rather than on applications that might benefit society as a whole.

hydraulics
Hydraulics is the transfer of power through a liquid.

pneumatics
Pneumatics is the transfer of power through a gas, usually air.

Types of Fluid Power

We divide fluid power into two types: hydraulics and pneumatics. **Hydraulics** is the transfer of power through a liquid. **Pneumatics** is the transfer of power through a gas, usually air.

Materials are made of matter. Matter has three different states: solid, liquid, and gas. The molecules of solid materials are strongly linked together. Liquids have molecules that are loosely bonded together. The molecules of gases are free to move. The term *fluid* refers to any material that flows, so we can call both liquids and gases *fluids*. Liquids and gases perform differently under pressure.

Look at Figure 12-2. Notice that when pressure is applied to both a liquid and a gas, the gas compresses more. This is because the molecules in the gas are less restricted and can move around to fill any voids. Hydraulic pressure is like pushing on a solid object. Pneumatic pressure is like pushing on a spring (see Figure 12-3).

FIGURE 12-2 How liquids and gases react to pressure.

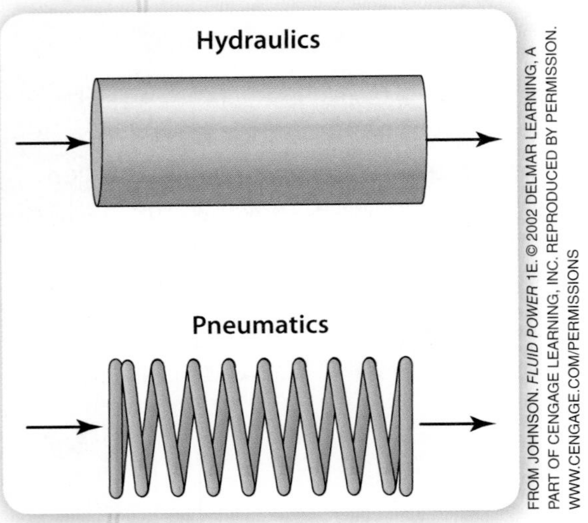

FIGURE 12-3 Hydraulics provides a firm transfer of power. Pneumatics transfers power like a spring.

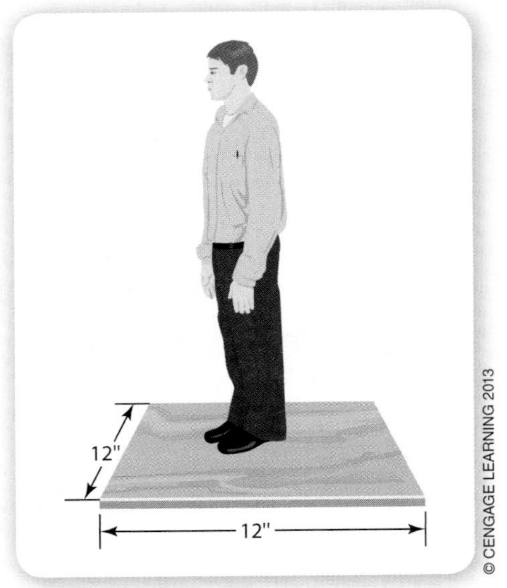

FIGURE 12-4 Pounds per square foot.

Pressure

The pressure in a fluid power system is the essential component that allows the system to operate. **Pressure** is the force on a surface area. It can be mathematically described as force divided by area (pressure = force/area). The student in Figure 12-4 weighs 120 pounds and is standing on a piece of plywood that is 1 foot by 1 foot. The pressure ($P = F/A$) in this example is 120 pounds per square foot.

Early scientists developed an understanding of the principles of hydraulics and pneumatics. Using these principles, engineers today can design fluid power systems to achieve their goals.

Fluid Power Formulas

By using mathematical formulas, engineers can design, calculate, and control fluid power systems. Many of these formulas were developed centuries ago.

Boyle's law
Boyle's law states that, for a fixed amount of gas kept at a fixed temperature, pressure and volume are inversely proportional.

Boyle's Law Boyle's law states: "For a fixed amount of gas kept at a fixed temperature, pressure and volume are inversely proportional." Boyle's law applies only to gases or pneumatics. This pneumatic principle was published by Robert Boyle in 1662 (Figure 12-5). Mathematically, we express Boyle's law as: $V_1 \times P_1 = V_2 \times P_2$.

FIGURE 12-5 **Robert Boyle.**

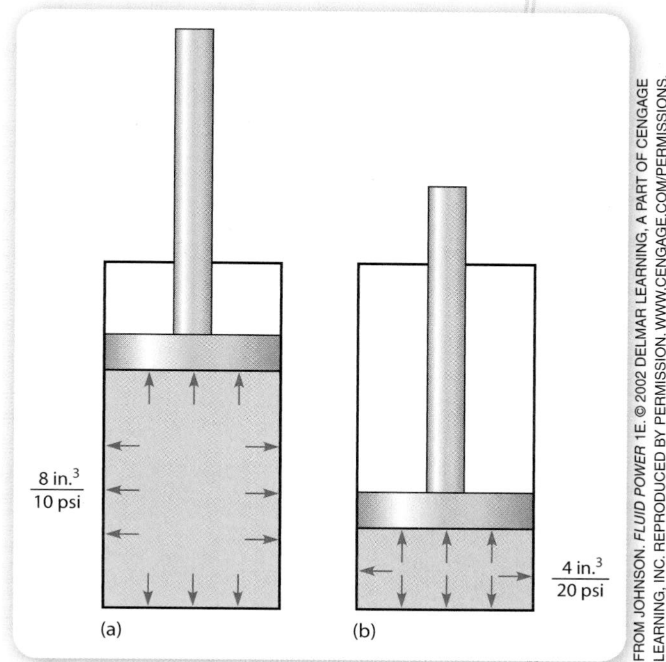

FIGURE 12-6 **Boyle's law in action.**

Take a look at Figure 12-6. The cylinder has a volume of 8 cubic inches and is under a pressure of 10 pounds per square inch (psi) (Figure 12-6a). When a pressure of 20 psi is applied (Figure 12-6b), what happens to the volume? Answer: It decreases; it is inversely proportional to the pressure applied. Mathematically, using Boyle's law for Figure 12-6 we find:

$$V_1 \times P_1 = V_2 \times P_2$$

$$8 \text{ in.}^3 \times 10 \text{ psi} = V_2 \times 20 \text{ psi}$$

$$V_2 = 4 \text{ in.}^3$$

Charles's Law **Charles's law** states: "The volume of a confined gas is proportional to its temperature, provided its pressure remains constant." Charles's law applies only to gases or pneumatics. The law was published in 1802 by Joseph Louis Gay-Lussac. In his publication, Gay-Lussac referred to the unpublished work of Jacques Charles dated 1787, so the principle was termed *Charles's law.*

To use Charles's law, you must use **absolute temperature**. Absolute temperature is a temperature scale that calls the point at which all molecular movement stops *zero* (0). Scientists call this scale the Kelvin (K) scale. Zero on the Kelvin scale is equal to −463° Fahrenheit. The mathematical equation for Charles's law is:

$$\frac{V_1}{T_1} = \frac{V_2}{T_2}$$

Charles's law

Charles's law states that the volume of a confined gas is proportional to its temperature, provided its pressure remains constant.

absolute temperature

Absolute temperature is the measurement system in which zero (0) is the point at which all molecular movement stops. It is also known as the Kelvin (K) scale; 0 degree K = −463 degrees Fahrenheit.

Math in Engineering

S T E M

To calculate fluid power, engineers use mathematical equations. An equation indicates that the mathematical calculation on the left side equals the mathematical calculations on the right side. By using this mathematical principle, engineers can design fluid power systems.

Pascal's law

Pascal's law states that when there is an increase in pressure at any point in a confined fluid, there is an equal increase at every other point in the container.

The cylinder in Figure 12-7 currently has a volume of 120 cubic inches at 60 K. If we raise the temperature to 90 K, what will happen to the volume? Answer: It would increase proportionally. Using Charles's law, the solution for this problem is:

$$\frac{120 \text{ in.}^3}{60 \text{ K}} = \frac{V}{90 \text{ K}}$$

$$V = 180 \text{ in.}^3$$

Pascal's Law Pascal's law states: "When there is an increase in pressure at any point in a confined fluid, there is an equal increase at every other point in the container." We can apply Pascal's law to both liquids and gases (both hydraulics and pneumatics). The principle was developed by Blaise Pascal, a French mathematician who lived in the seventeenth century (see Figure 12-8).

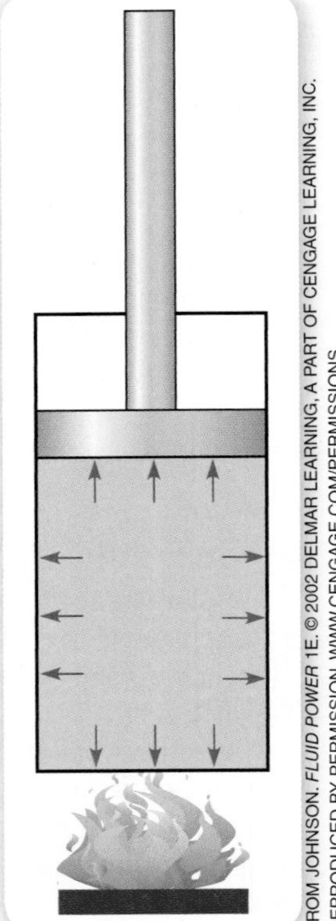

FROM JOHNSON. *FLUID POWER* 1E. © 2002 DELMAR LEARNING, A PART OF CENGAGE LEARNING, INC. REPRODUCED BY PERMISSION. WWW.CENGAGE.COM/PERMISSIONS.

FIGURE 12-7 Charles's law in action.

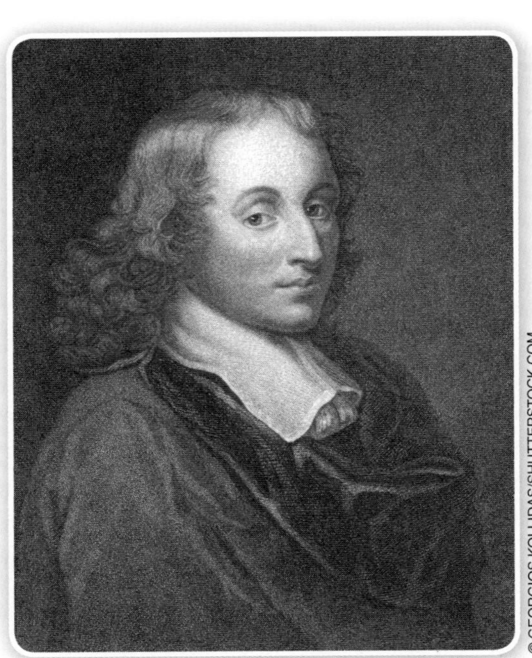

© GEORGIOS KOLLIDAS/SHUTTERSTOCK.COM

FIGURE 12-8 Blaise Pascal.

As shown in Figure 12-9, the pressure of a confined fluid is equal throughout a cylinder. The pressure on the wall is equal to the pressure on the piston. For fluid power devices, the area to which the pressure is applied is typically a round piston (see Figure 12-10). To determine the area the force is applied to, you must calculate the surface area of the round piston. The area of a circle is $A = \pi \times r \times r$.

Here is an example of an application of Pascal's law. In Figure 12-11a, 250 pounds are being applied to the piston. The resulting pressure in the

Science in Engineering

We can measure temperature in different scales: Kelvin, Celsius, and Fahrenheit. Scientists and engineers use formulas to convert temperatures from one scale to another. Two common elements of these measurement scales are (1) each has a zero (0) reading and (2) the distance between each degree reading is equal within each scale.

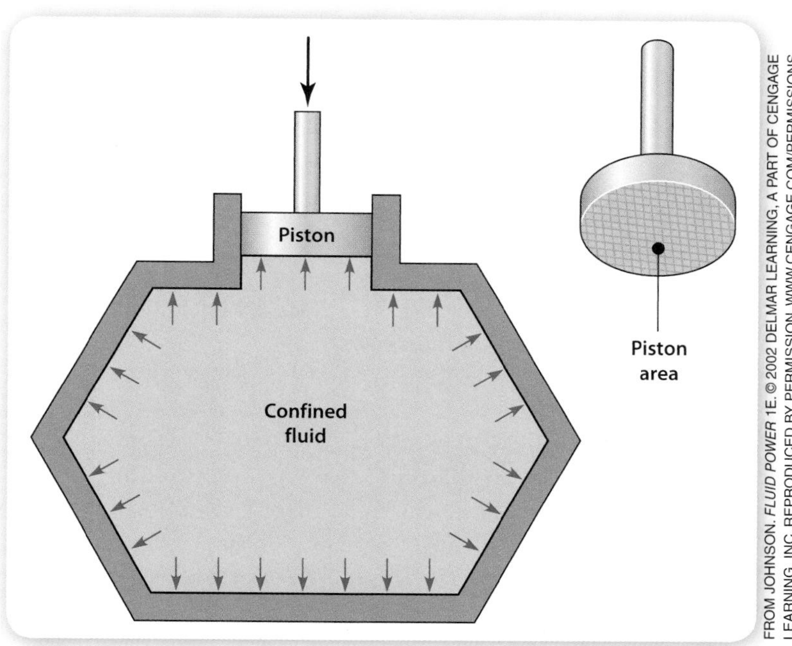

FIGURE 12-9 **Equal pressure throughout the cylinder.**

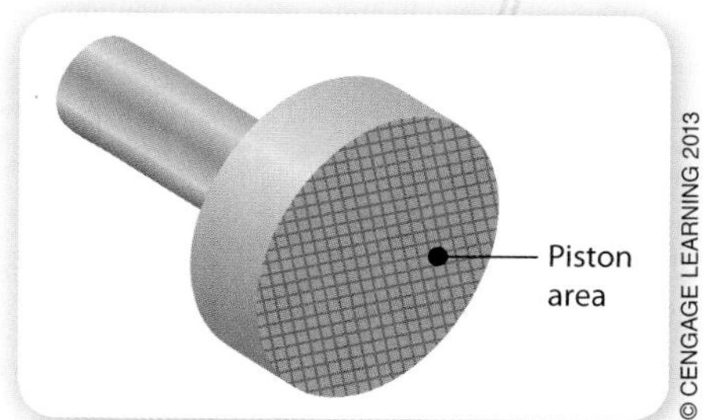

FIGURE 12-10 **The surface area of a piston.**

fluid is $P = F/A$ or $P = 250$ pounds divided by the piston's area. The piston's area is $A = \pi \times r \times r$ or $A = 3.14 \times ½ \times ½$ or $A = 0.785$ square inches. Therefore, $P = 250$ pounds/0.785 square inches or $P = 196$ psi.

The piston in Figure 12-11b has a much larger area. Its area $A = 3.14 \times 1.25 \times 1.25$ or 4.9 square inches. Therefore, the force exerted upward by this piston is:

$$P = \frac{F}{A}$$

$$196 \text{ psi} = \frac{F}{4.9 \text{ in.}^2}$$

$$F = 960 \text{ lb}$$

This is a lot more force and is an example of how fluid power can be used to increase mechanical advantage. This mechanical advantage gained by fluid power explains how the heavy earth mover in Figure 12-12 is able to accomplish its job.

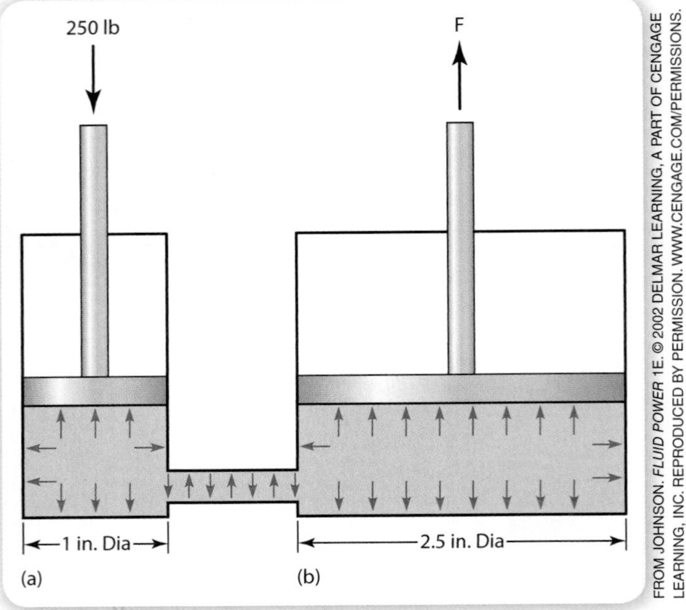

FIGURE 12-11 *Pascal's law in action.*

FIGURE 12-12 *Heavy equipment relies on fluid power.*

Engineering Challenge

ENGINEERING CHALLENGE 1

A. Engineers use Boyle's law to assist with the design process. Before fabricating prototypes, engineers use math to determine a solution for their design problems.

- In diagram 1, you desire to change the cylinder's current volume of 40 in.^{3} to a new volume of 20 in.^{3} The cylinder is currently under a pressure of 100 psi. What pressure do we need to reach the desired new volume?

B. Engineers use Charles's law to assist with the design process. Before fabricating prototypes, engineers use math to determine a solution for their design problems.

- In diagram 2, you want to change the cylinder's current volume of 100 in.^{3} to a new volume of 50 in.^{3} The cylinder is currently at a temperature of 48 K. What temperature do we need to reach the desired new volume?

C. Engineers use Pascal's law to assist with the design process. Before fabricating prototypes, engineers use math to determine a solution for their design problems.

- Determine what size piston B should be to lift the automobile in diagram 3. The automobile weighs 4000 pounds. The fluid power system has a pressure of 100 psi. What is the minimum diameter that piston B could be to raise this automobile?

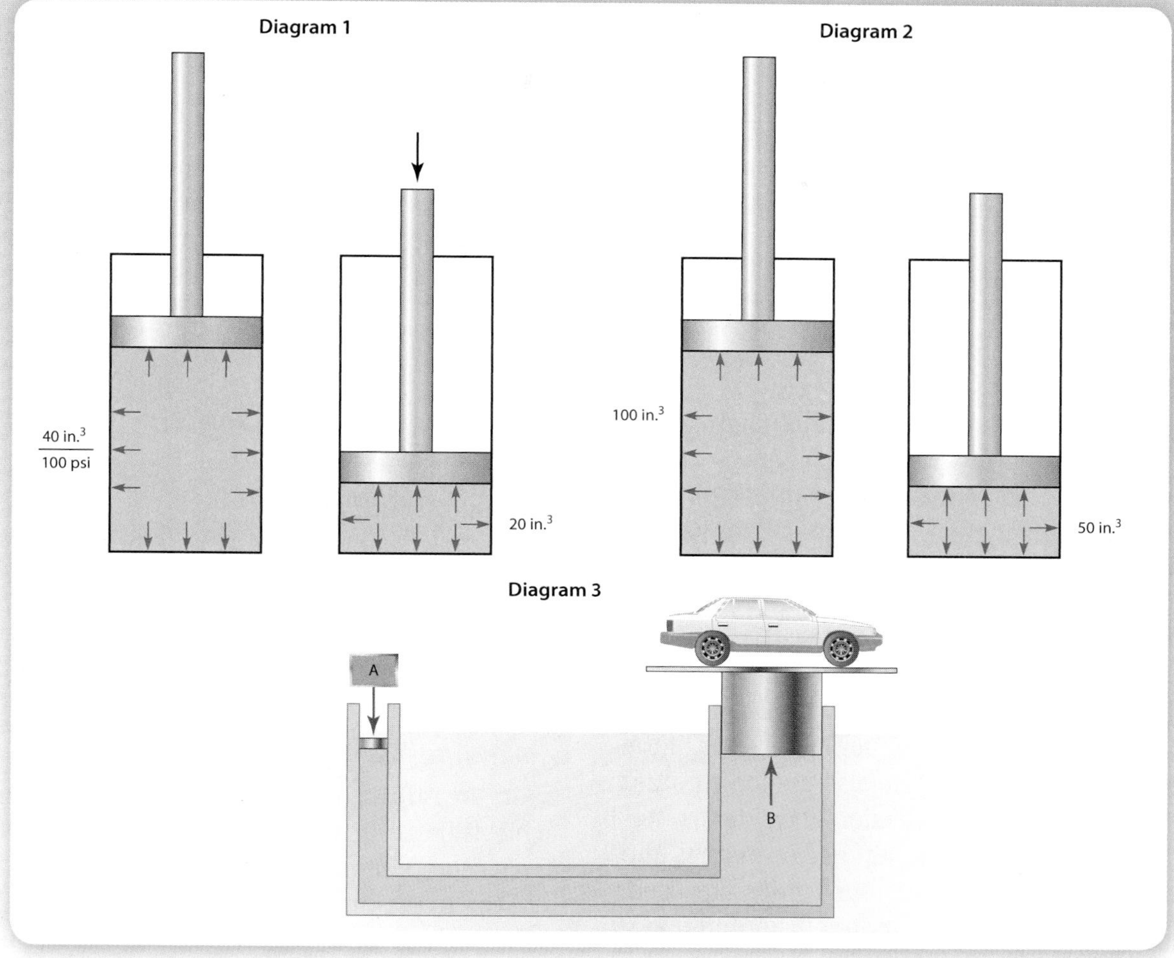

Diagram 1

40 in.^{3}

100 psi

20 in.^{3}

Diagram 2

100 in.^{3}

50 in.^{3}

Diagram 3

A

B

ENGINEERING CHALLENGE 1 Applying Boyle's, Charles's, and Pascal's laws.

SECTION 2: PNEUMATICS

Pneumatic Principles

The most common gas used for pneumatic power is air. Air does take up space, and it can be compressed. Because of the compression quality of air, however, pneumatics cannot be used for heavy loads. Pneumatics is well suited for operations that require a quick response and a low level of precision and have a light load. Pneumatics is also excellent for transferring a signal. Remember, that is how the fail-safe train brake system works.

Pneumatics is used in applications in which a hydraulic fluid leak is undesirable, such as in food-processing plants. Pneumatics is clean, and the fluid required for the system—air—is readily available. Pneumatics is typically used in control systems.

Career Spotlight

Name:
Ivan R. Diaz

Title:
Team Leader, Sustaining Engineering, Walt Disney World

Job Description:
When Ivan Diaz was a teenager, he watched a documentary on the engineers who designed Walt Disney World's attraction known as The Twilight Zone Tower of Terror. Their work inspired him to become an engineer himself. Diaz's career with Disney began during his junior year of college when he worked as an intern at the Disney World Resort as part of his coursework.

As an engineer with the Disney World Resort, Diaz served as a key member of a team that improved the popular Pirates of the Caribbean attraction. He also earned a patent for his automated attraction and ride maintenance system. This system uses antennas and receivers to track how much rides are used, allowing Disney to be more efficient in maintaining them.

Currently, Diaz works on Disney's transportation systems: the monorails, steam trains, buses, and watercraft. He and his team make sure the systems are reliable and up to date.

"I really have fun at work," Diaz says. "Fun and the great people I work with are what keep me at Disney. It's part of the magic."

Education:
Diaz graduated from Florida State University with a degree in electrical engineering. He furthered his education by earning a master's degree in business administration from the Roy E. Crummer School of Business at Rollins College. He has also earned his professional engineering license for the state of Florida.

As an undergraduate, Diaz got involved in a solar-car team. "That really helped me transition from what I learned in the classroom to real life," he says.

Advice to Students:
Diaz works to help young people become interested in math and science. He has served as a mentor for FIRST Robotics and FIRST LEGO League. The FIRST Robotics competition is a kind of game for high school students. At the beginning of the year, students across the country receive the same set of goals. Using a kit, they must build robots that help them achieve those goals. FIRST LEGO League is a similar competition for middle school students.

Diaz urges prospective engineers to take both physics and calculus in high school. "Math and science can sometimes seem boring," he says, "but here at Disney, the fun and excitement that you experience are all math- and science-driven."

At college, Diaz says, students should seek out internships during their freshman year. "You should start working in your area of focus as soon as possible so you can build professional skills, along with professional relationships," he says. "Your transition to the world of work begins when you start college."

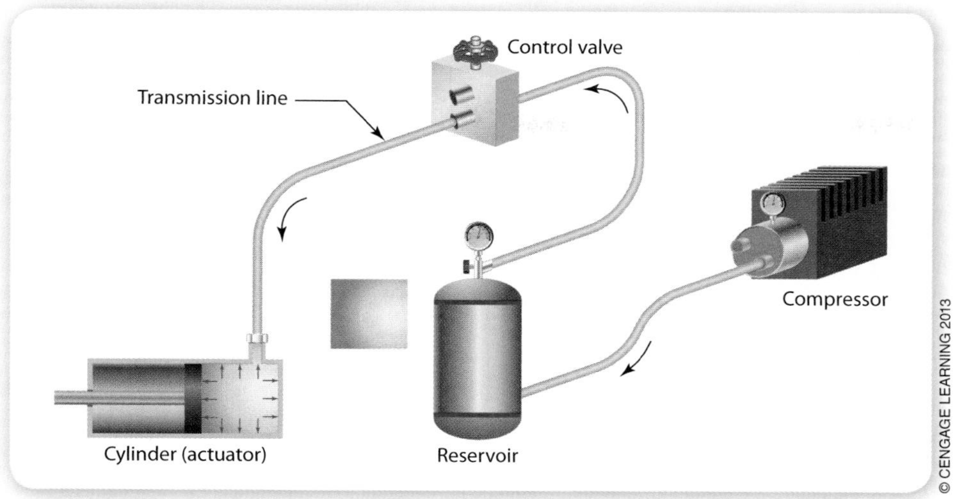

FIGURE 12-13 A pneumatic system.

Pneumatic System Components

For engineers to use pneumatics, they must design a pneumatic system. As we noted in Chapter 2, a system is a group of related components that work together to achieve a process. A typical pneumatic system consists of a compressor, a reservoir, transmission lines, control valves, and **actuators**. These components are shown in Figure 12-13.

The **compressor** is a device that increases the pressure of air and pumps the compressed air into a tank. The compressed air tank is called a **reservoir**. The compressed air is then routed to the desired location through **transmission lines**. Transmission lines are tubing through which the compressed air flows. **Control valves** regulate the flow of gases. The pneumatic system uses a **cylinder** with a movable piston to convert its fluid power into mechanical power in the form of linear motion.

SECTION 3: HYDRAULICS

Hydraulic Principles

A hydraulic system uses an oil-type liquid called **hydraulic fluid**. This fluid is typically made from petroleum products. The advantages of hydraulics include the fact that liquids are more difficult to compress than gases, so the hydraulic system can transfer greater amounts of power. The hydraulic fluid also helps lubricate the system's components. However, if the hydraulic fluid leaks, it may present a problem. Elevators, dentists' chairs, heavy equipment operations, and automobile braking systems use hydraulics. Hydraulic fluid pressure can easily be controlled.

Science in Engineering

Scientists are continually experimenting with different types of hydraulic fluids. Petroleum-based fluids are the most common. Because of environment concerns, however, vegetable oil-based fluids are now being tested. Scientific inquiry is beneficial to both engineering design and the environment.

Hydraulic System Components

The components of a hydraulic system are similar to those of the pneumatic system. Figure 12-14 shows the components of a hydraulic system. The pressure of the hydraulic fluid is provided by the system's pump. The hydraulic **pump** is a device that creates flow in the system. The reservoir stores the hydraulic fluid before it is pressurized by the pump. Transmission lines carry the pressurized hydraulic fluid to the control valve and actuators. The control valves regulate the flow and pressure of the hydraulic fluid. The actuators transfer the fluid power to mechanical power.

As with pneumatics, hydraulics uses an actuator called a *cylinder* to convert its fluid power into mechanical power in the form of linear motion. The advantages gained by fluid power cylinders apply to both hydraulic and pneumatics.

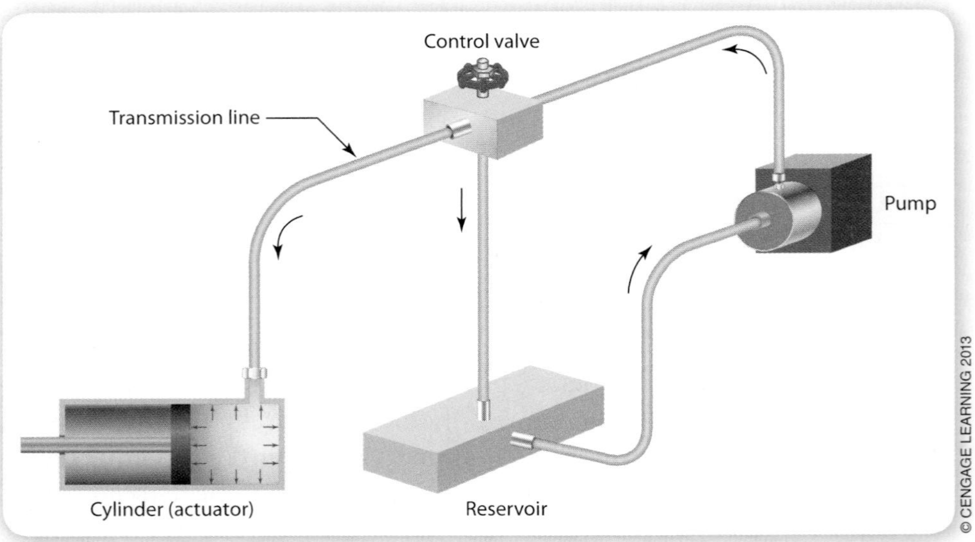

FIGURE 12-14 A hydraulic system.

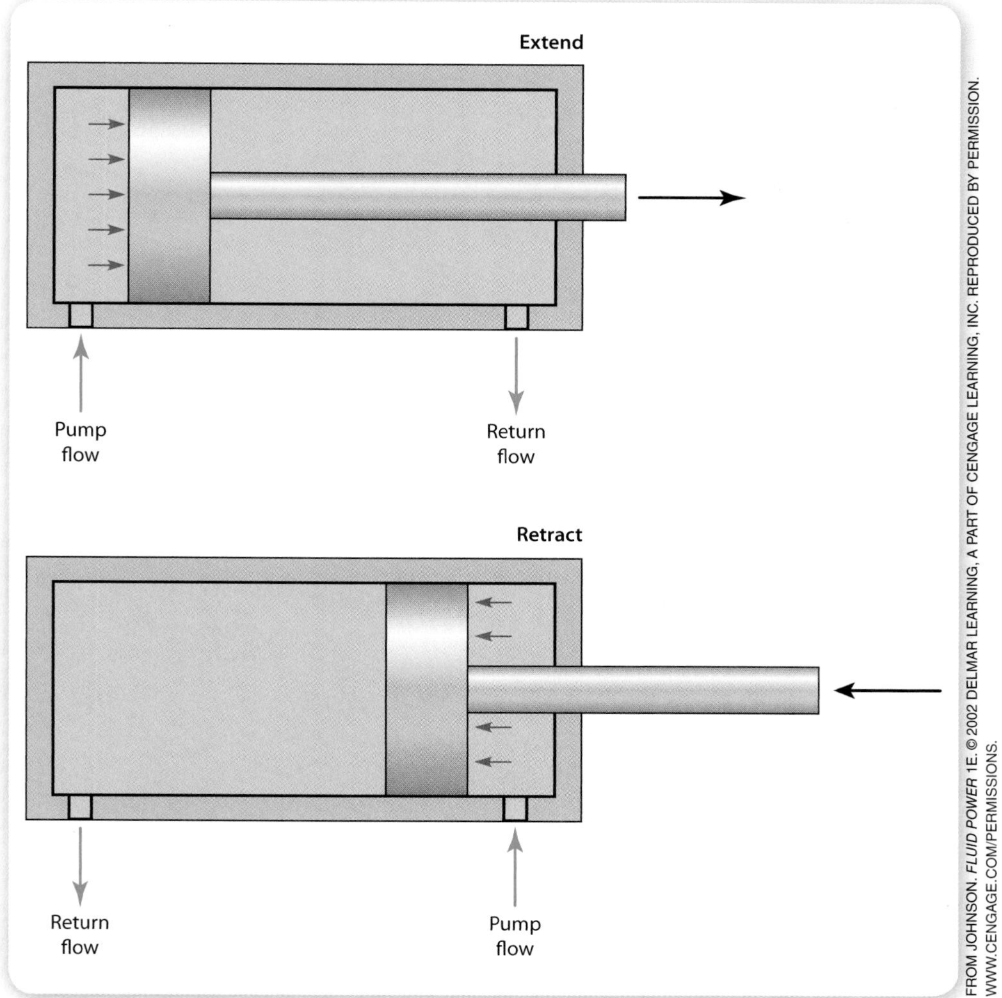

FIGURE 12-15 *Cylinders can extend and retract.*

Cylinders are devices that engineers can use in their design. The cylinders in Figure 12-15 can both extend and retract. Engineers use a two-way control valve to direct the pressurized fluid to one or the other side of the cylinder. Other types of cylinders, such as those shown in Figure 12-16, are spring loaded so that they will automatically return once the fluid power is released.

6

Based on Pascal's law, engineers can use two or more cylinders in a fluid power system. With the same pressure applied to both cylinders pictured in Figure 12-17, engineers can change the output force of each cylinder. How? By changing the piston surface area of each cylinder.

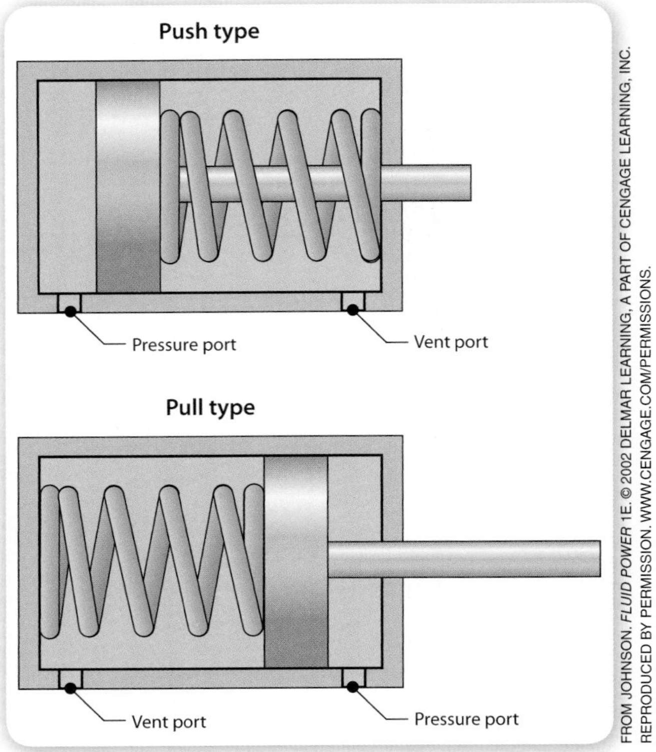

FIGURE 12-16 Spring-loaded cylinders.

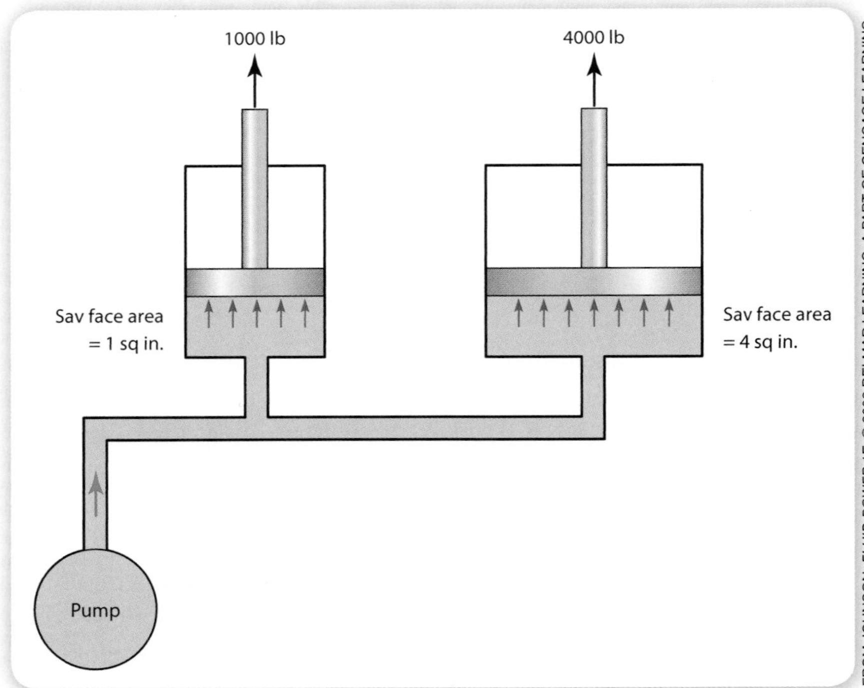

FIGURE 12-17 Understanding Pascal's law.

Math in Engineering

S T E M

Adjusting the distance that a piston travels in a fluid power cylinder can help engineers with their design. Look at the two cylinders in Figure 12-18. The diameter of piston A is 1 inch, and the diameter of piston B is 4 inches. If piston A moves 2 inches down the cylinder, how far will piston B move up the cylinder? Engineers use a mathematical equation to solve this problem. The answer is *x* inches.

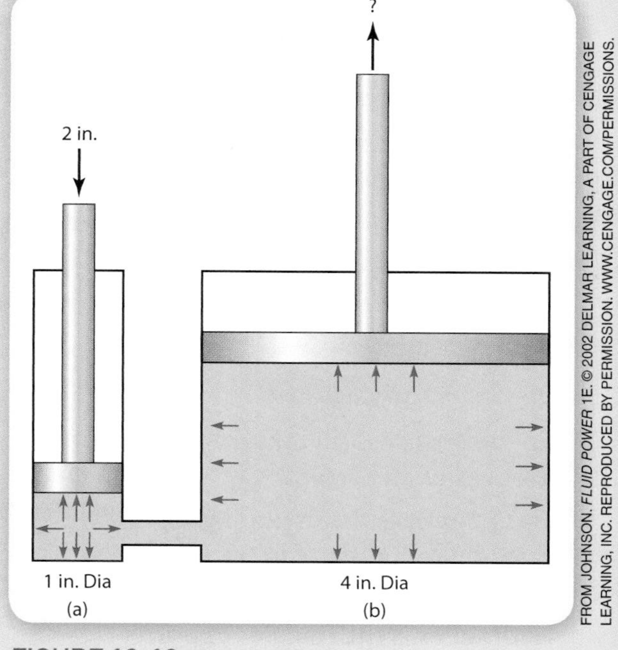

FIGURE 12-18 Using Pascal's law.

Engineering Challenge

ENGINEERING CHALLENGE 2

Your team will design, fabricate, and test a fluid power device that will move a tennis ball. Your device must lift the tennis ball 2 inches and then move it 2 inches to the left. At that point, the device must drop the ball. This operation is designed to resemble heavy equipment that uses fluid power to move earth.

Your team will be provided with basic fabrication materials such as lumber, string, glue, and fasteners. You can also use as many as six syringes of various sizes plus plastic tubing.

ENGINEERING CHALLENGE 2 Designing a lifting device.

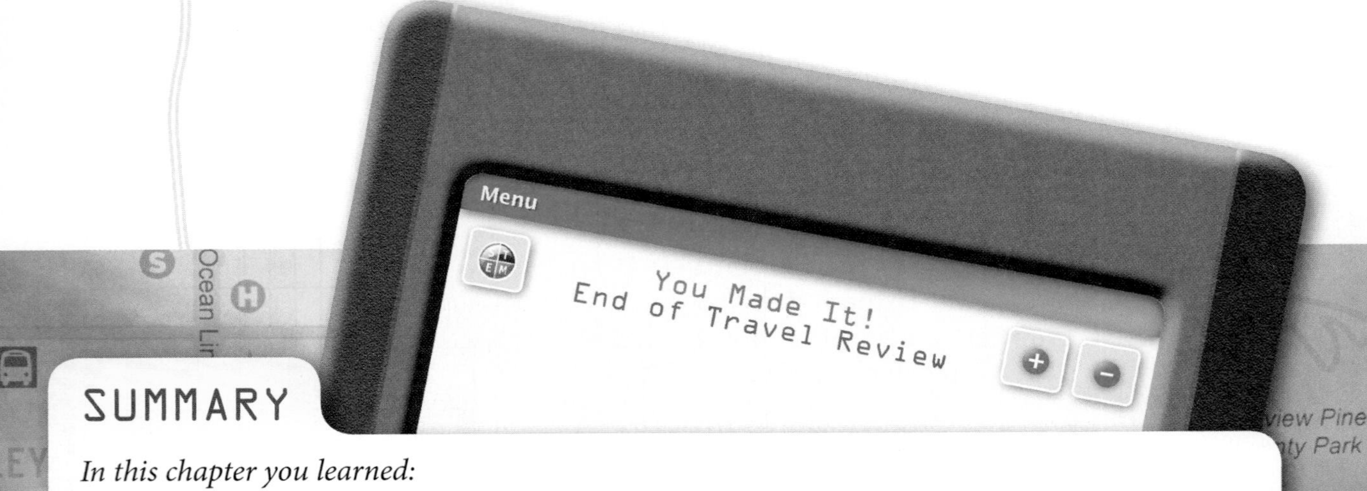

SUMMARY

In this chapter you learned:

- Engineers use fluid power to make our lives easier and safer.
- Fluid power is the use of liquids and gases under pressure to transmit power.
- Hydraulics is the transfer of power through a liquid.
- Pneumatics is the transfer of power through a gas, usually air.
- Boyle's law states that for pneumatics the volume and pressure are inversely proportional.
- Charles's law states that the volume of a gas is proportional to its absolute temperature.
- Pascal's law states that pressure in a container is equal throughout for both liquids and gases.

VOCABULARY

Write definitions for these terms in your own words. After you have finished, compare your answers with the definitions provided in this chapter.

Fluid power
Hydraulics
Pneumatics
Boyle's law
Charles's law
Absolute temperature
Pascal's law

STRETCH YOUR KNOWLEDGE

Provide thoughtful, written responses to the following questions.

1. Discuss why scientific research in fluid power is important to engineering design.
2. Describe why engineers use their math skills when designing a fluid power system.
3. In what ways has fluid power affected your life?
4. In what ways has fluid power affected society?
5. Where do you see fluid power being used in the future?

Onward to Next Destination

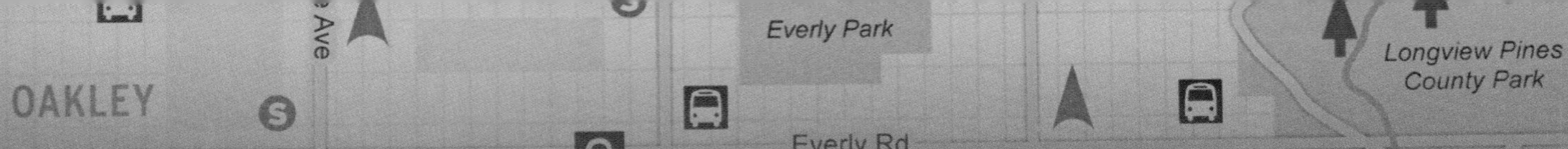

CHAPTER 13
Flight and Space

Menu

Before You Begin

Think about these questions as you study the concepts in this chapter:

1. How do airplanes fly?
2. What are the forces that affect flight?
3. What is aerodynamics?
4. What variables do engineers use when designing aircraft?
5. What is propulsion, and how is it created?
6. How do pilots control the flight of their aircraft?
7. What is the difference between typical airplane flight and spaceflight?

COURTESY OF MAARTEN VISSER-HOLLAND

Engineering in Action

Each and every day, thousands of people use air transportation in the United States. Commercial flying has become a way of life for Americans. However, flight would not be possible without the engineers and technologists who design and build the aircraft. The technological and engineering advances that propelled the aviation industry have occurred in a relatively very short period of time. Before 1900, travel by aircraft was only a dream. Today it is common. How has this happened? What does the future hold for flight and space transportation?

SECTION 1: FLIGHT TECHNOLOGY

History

lift

Lift is (1) a component of aerodynamic forces acting on an object in flight, (2) a force produced by an airfoil shape that works against gravity, (3) the force that acts at a right angle to the direction of motion through the air (created by differences in air pressure), and (4) the force that directly opposes the weight of an airplane and holds the airplane in the air.

Humans have always been fascinated by flight. Many Greek myths were stories about flight such as the myth of the winged horse Pegasus (see Figure 13-1). The Greeks also told the story of Icarus and Daedalus. Daedalus was an engineer who had been imprisoned with his son, Icarus. Daedalus designed and built wings out of feathers and wax so the pair could escape. But when Icarus felt the power of his wings in flight, he boldly flew too close to the sun. The sun's heat melted the wax, the wings fell apart, and Icarus fell into the sea below.

As early as 400 B.C., the Chinese had designed and constructed kites. These kites resembled birds. In the late 1400s, Leonardo da Vinci created more than 100 drawings of different flying machines. Like most early designs for flying machines, Leonardo's ornithopter design resembled a bird (see Figure 13-2). Even today, people are dreaming up new ways to fly faster and farther.

Forces on Flight

To leave the ground, an object such as an aircraft must overcome the pull of Earth's gravity. Gravity is one of the four forces that affect flight. A **force** is the transferring of energy to an object, typically by pushing or pulling on that object. We call the force that moves the aircraft away from Earth's gravitational pull **lift**. The more an aircraft weighs, the more lift that is required

FIGURE 13-1 *Pegasus, the flying horse of Greek mythology.*

to break Earth's gravitational pull. Early balloon flights used hot air to lift the balloons and overcome Earth's gravitation.

To move an aircraft or any other type of object forward, the engineer must overcome friction and wind resistance.

FIGURE 13-2 Leonardo da Vinci's ornithopter.

Did You Know?

Before the 1903 flight of the Wright brothers' *Flyer* (see Figure 13-3), the brothers extensively researched the best shape for the wings of their aircraft. Orville and Wilber Wright actually built a **wind tunnel** to test their designs. A wind tunnel is a device used by engineers to measure the effect of airflow across a device. Figure 13-4 shows the Wright brothers' wind tunnel and the sample airfoils used to test lift. The Wright brothers used the data they obtained to design their *Flyer*.

FIGURE 13-3 The Wright brothers' *Flyer.*

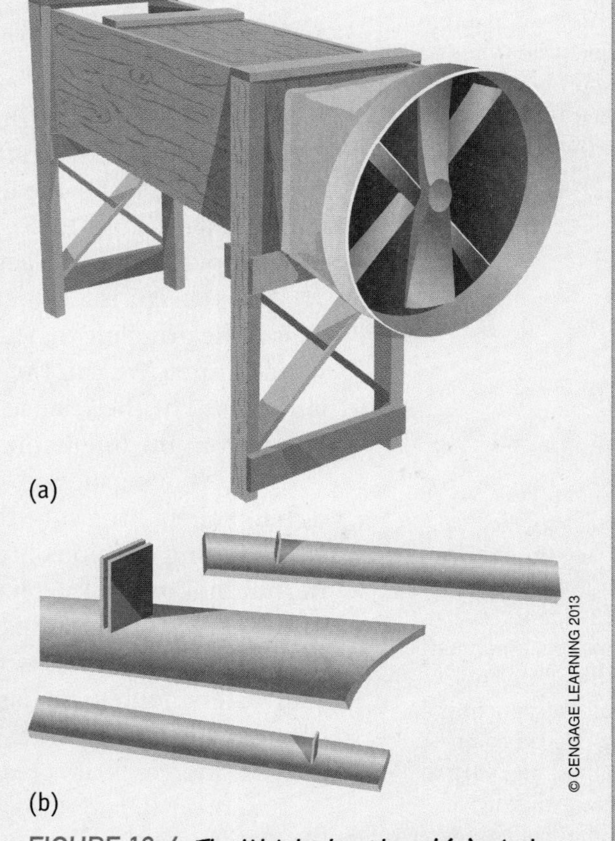

FIGURE 13-4 The Wright brothers' (a) wind tunnel and (b) sample airfoils.

drag
Drag is (1) a force that causes resistance to moving through the air, (2) resistance of the air (technically a fluid) against the forward movement of an airplane, (3) the force that acts opposite to the direction of motion (caused by friction and differences in air pressure), and (4) the resistance of the motion of an object through a fluid.

thrust
Thrust is (1) a force applied to a body to propel it in a desired direction, (2) the force that propels an aircraft in the direction of motion and is produced by engines, and (3) the force that moves an aircraft through the air. Thrust is generated by an aircraft's engines.

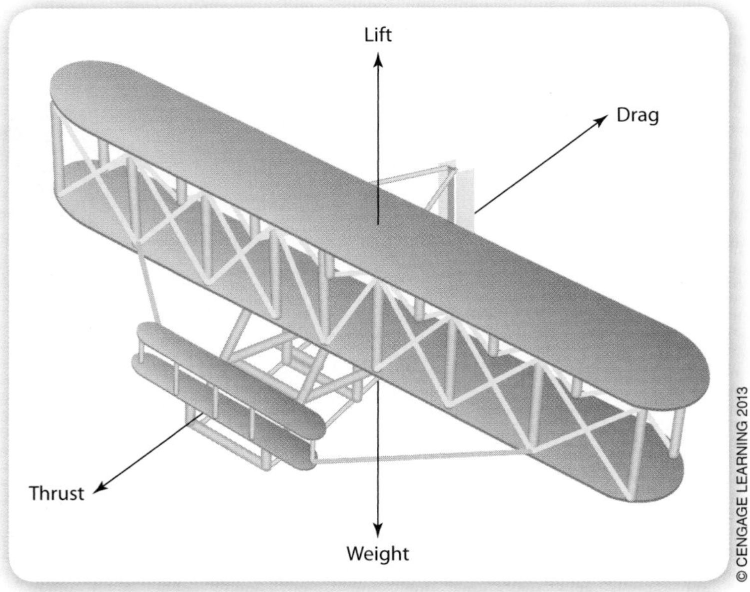

FIGURE 13-5 *The four forces on an aircraft.*

We call the combined friction and wind resistance on an aircraft **drag**. We call the energy an aircraft uses to overcome drag **thrust**. The more thrust produced, the faster the aircraft moves forward. However, with fixed-wing aircraft (see Figure 13-5), thrust also helps the aircraft develop lift because of the movement of air across the wings.

Wing design is a critical component of the lift developed by an aircraft. This is the principle that Orville and Wilbur Wright were studying with their wind tunnel. We call the principle **aerodynamics**. The term *aerodynamics* is also used by the science that deals with the motion of air and the resulting forces from the interaction of air and objects.

Aircraft wing design is critical to flight technology. Modern aircraft develop their lift using the shape of their wings. To understand how an aircraft wing develops lift, we must look at the scientific studies of Daniel Bernoulli. Look at the picture of the electric fan in Figure 13-6. The airflow in the center of the fan is very fast, whereas the airflow on the outer edge of the fan is slower. As a result, these two areas have different air pressures. The faster-moving air has a lower air pressure, and the slower-moving air has a greater air pressure. This is **Bernoulli's principle**: as the speed of a fluid (or gas) increases, its pressure decreases.

Based on Bernoulli's principle, engineers have developed wing designs like the one in Figure 13-7. The air flowing across the top of the wing moves faster than the air flowing along the lower wing surface. As a result, the air pressure under the wing is greater than the air pressure on the top of the wing. The higher air pressure (bottom of the wing) pushes the lower air

Science in Engineering

Daniel Bernoulli (1700–1782) was a Dutch-born scientist who is widely recognized for his work exploring the basic properties of fluid flow, pressure, and velocity. Today we know his work as Bernoulli's principle. He developed his theories while working in St. Petersburg, Russia. His theory was published in 1738 in a work entitled *Hydrodynamica*. Bernoulli's scientific investigations still provide today's engineers valuable theories about the flow of both fluids and gases.

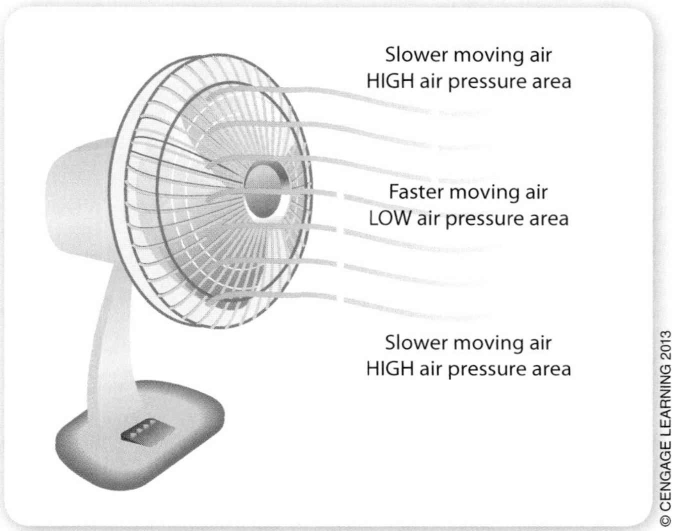

FIGURE 13-6 Air pressure differences produced by a fan.

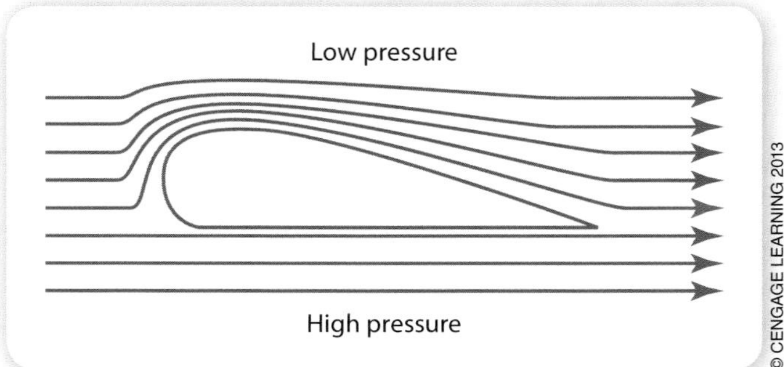

FIGURE 13-7 Bernoulli's principle as air flows along an airfoil.

aerodynamics

Aerodynamics is (1) the science that deals with the motion of air and the forces acting on objects as a result of the motion between the air and the object, and (2) the study of forces and the resulting motion of objects through the air.

Bernoulli's principle

Bernoulli's principle states that as the speed of a fluid increases, its pressure decreases.

FIGURE 13-8 Race cars use airfoils.

pressure (top of the wing). This difference in air pressure produces lift on the wing, and the wing rises.

Airfoil is another name for a wing. Look at the picture of the race car in Figure 13-8. Why are airfoils (wings) mounted on both the front and the rear of this race car? Are the airfoils producing lift or are they producing a downward force?

Figure 13-9 shows the main parts of a wing. The **leading edge** is the front surface of the wing, and the **trailing edge** is the rear edge of the wing. The distance between these two edges is the **chord**. The **mean chord** (*mean* is another term for average) is an imaginary line that is halfway between the wing's top and lower surfaces.

aspect ratio
Aspect ratio is the wingspan length divided by the chord.

The **wingspan** of an airfoil is the distance between its two ends. We use the wingspan to determine the wing's **aspect ratio**. The aspect ratio is a mathematical ratio that divides the wingspan by the chord. A high aspect ratio represents a long and narrow wing, whereas a low aspect ratio indicates a wing that is shaped more like a square (see Figure 13-10). Aircraft with high aspect ratios are gliders or long-range aircraft. Aircraft that need extra maneuverability have lower aspect ratios.

Engineering Challenge

ENGINEERING CHALLENGE 1

Following the Wright brothers' lead, you will design, fabricate, and test a set of wing designs. First, sketch two different wing side profiles using the Wright brothers' designs as samples. Your design can be very simple or you can try something new. We call this *experimentation*.

Take each design sketch to a copy machine and double its size. Then, with the two different shapes and two different sizes of wing side profile designs, fabricate each design using high-density Styrofoam. The size of your school's wind tunnel will determine the length for each wing. All four wings should have the same length. If your school does not have a wind tunnel, use a household fan for your testing.

Once you have fabricated the wings, test each wing design using the wind tunnel.

- Which wing design produced the most lift?
- Why did this design produce the most lift?
- Which wind design had the least amount of lift?
- Describe why you think these wing designs produced different amounts of lift.

Math in Engineering

A ratio is a mathematical comparison between two numbers. For example, your younger sister is one-half your height. This indicates a comparison between her height of 25 inches and your height of 50 inches. Ratios allow engineers to compare different designs. For aircraft, the engineer needs to know the aircraft design criteria in order to determine the aspect ratio for its wings.

The surface area of the wing is also a determining factor in the amount of lift the wing produces. The larger the wing surface area, the more lift produced. Along with the wing's surface area, aircraft speed also affects lift. The faster an aircraft flies, the more air moves across its wings. Greater air movement produces greater lift. That is why during takeoffs, aircraft travel down the runway gaining speed. When the aircraft reaches the speed that provides enough lift, it is able to leave the ground. Each aircraft has a minimum takeoff speed. One factor in determining takeoff speed is wing size.

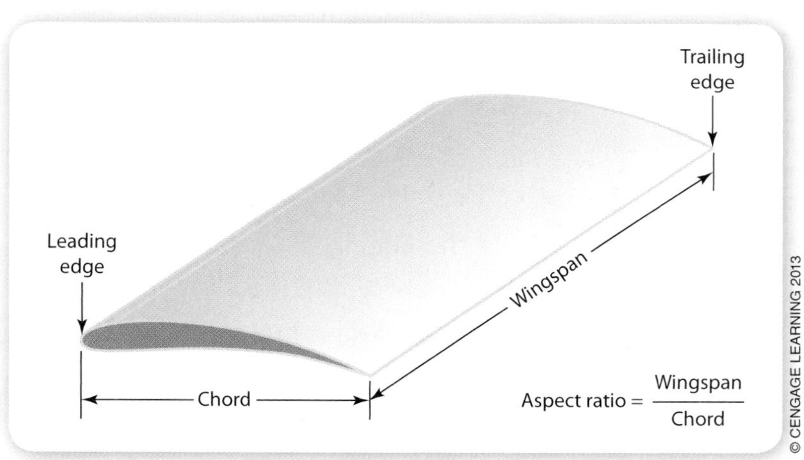

FIGURE 13-9 The components of a wing.

Engineers and aircraft pilots also need to understand a wing's **angle of attack**, or the measurement in degrees between the wing's chord and the movement of the wind against the leading edge. Most aircraft have a maximum angle of attack of 16°. At that point, the wing does not produce lift. If an aircraft attempts to fly at an angle of attack greater than 16°, the aircraft may stall.

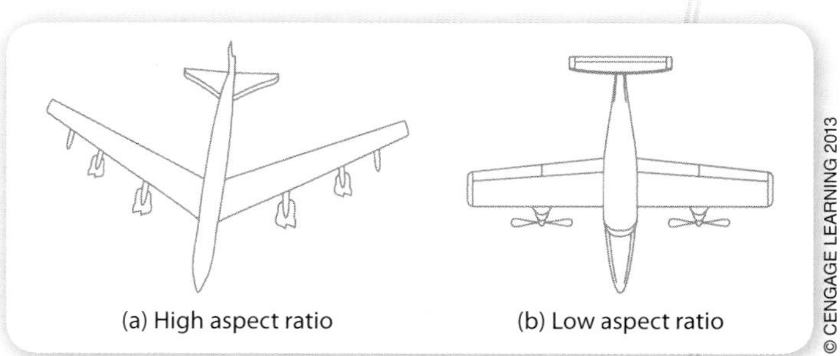

FIGURE 13-10 A comparison of aspect ratios.

Math in Engineering

Engineers must be able to calculate angles in degrees. Sometimes, though, even degrees do not provide engineers with the accuracy they need. We can further divide the measurement of degrees into minutes. Just as each hour of the day has 60 minutes, so does each degree. So an accurate angle reading may be "14 degrees, 23 minutes" (written 14° 23′).

Engineering Challenge

ENGINEERING CHALLENGE 2

Engineers need to calculate the amount of lift that a wing will produce. One item that they need to determine is the wing's surface area.

- Examine the three sets of wings pictured.
- Determine the surface area of each set of wings.
- Assuming all other factors are equal (mean chord and aircraft speed), which aircraft will provide the greatest lift?
- Determine the aspect ratio for each set of wings.
- Would the Boeing 747 or the F-18 be used for long-range flights? Why?
- Would the Boeing 747 or F-18 be used by stunt pilots? Why?
- Why did the Wright brothers' *Flyer* have two wings like the ones pictured?

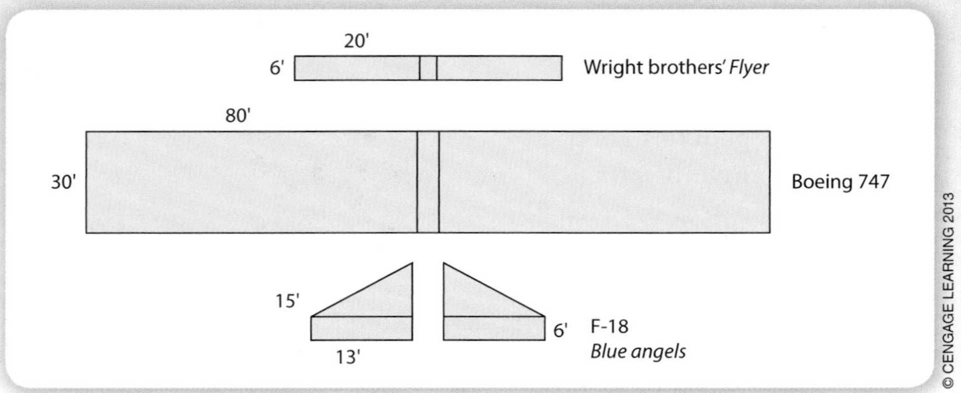

Calculating wing surface areas.

SECTION 2: AIRCRAFT PROPULSION AND CONTROL SYSTEMS

Internal-Combustion Engines

To raise *Flyer* off the North Carolina beach, Orville and Wilbur Wright needed it to move fast enough that the air moving along the wings would develop the required lift. It means that an aircraft requires propulsion to develop its forward motion.

propulsion
Propulsion is the means by which aircraft and spacecraft are moved forward. It is a combination of factors such as thrust (forward push), lift (upward push), drag (backward pull), and weight (downward pull).

The Wright Brothers designed and built a four-cylinder **internal-combustion engine** to develop this required propulsion (see Figure 13-11). The internal-combustion engine is a mechanical device that converts chemical energy (gasoline) into heat energy and then into mechanical energy (turning the propeller). Automobiles, tractors, lawn mowers, motorcycles, and boats also use internal-combustion engines.

A **propeller** is an airfoil mounted on a revolving shaft; its wing-shaped design moves air as it moves. The propeller creates a low pressure area in front of the propeller, thereby moving the aircraft forward because of the high pressure area behind the propeller.

The internal-combustion engine had been developed in the late 1800s and was the prime force helping to advance the automobile industry just as the Wright brothers were developing their aircraft. The internal-combustion engine is a combination of various simple machines and mechanisms (see Chapter 11). A piston moves up and down in a cylinder. One end of the piston is attached to the crank, which converts this reciprocating motion into rotational motion. This rotational motion is then used to turn the propeller. The Wright brothers' engine was an inline four-cylinder like the one in Figure 13-12.

Most internal-combustion engines operate on a four-stroke cycle (see Figure 13-13). The air–fuel mixture enters the combustion

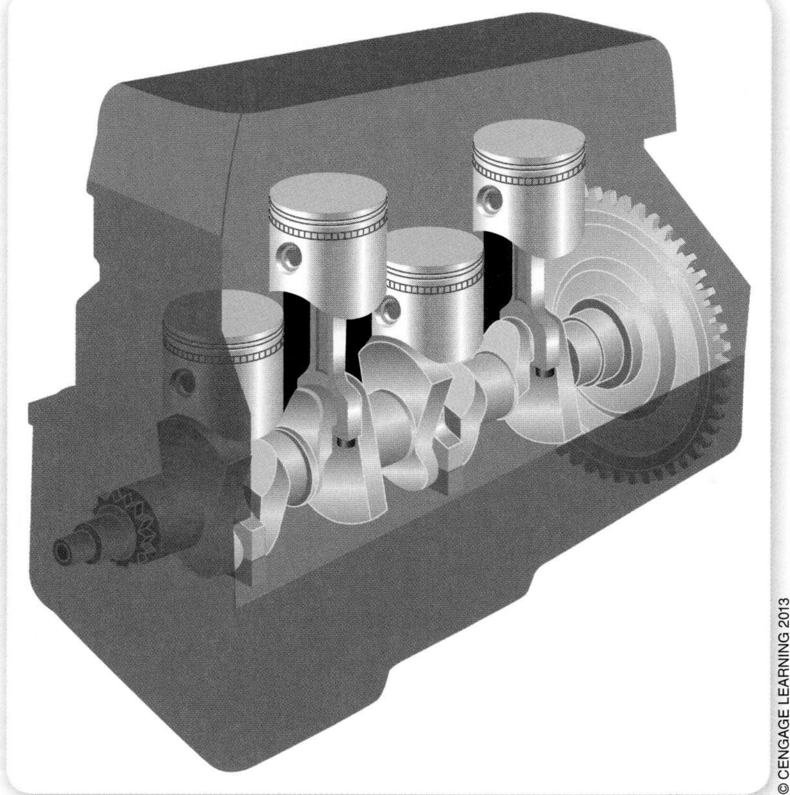

FIGURE 13-11 The Wright brothers' design used an internal-combustion engine similar to the engine shown in this drawing.

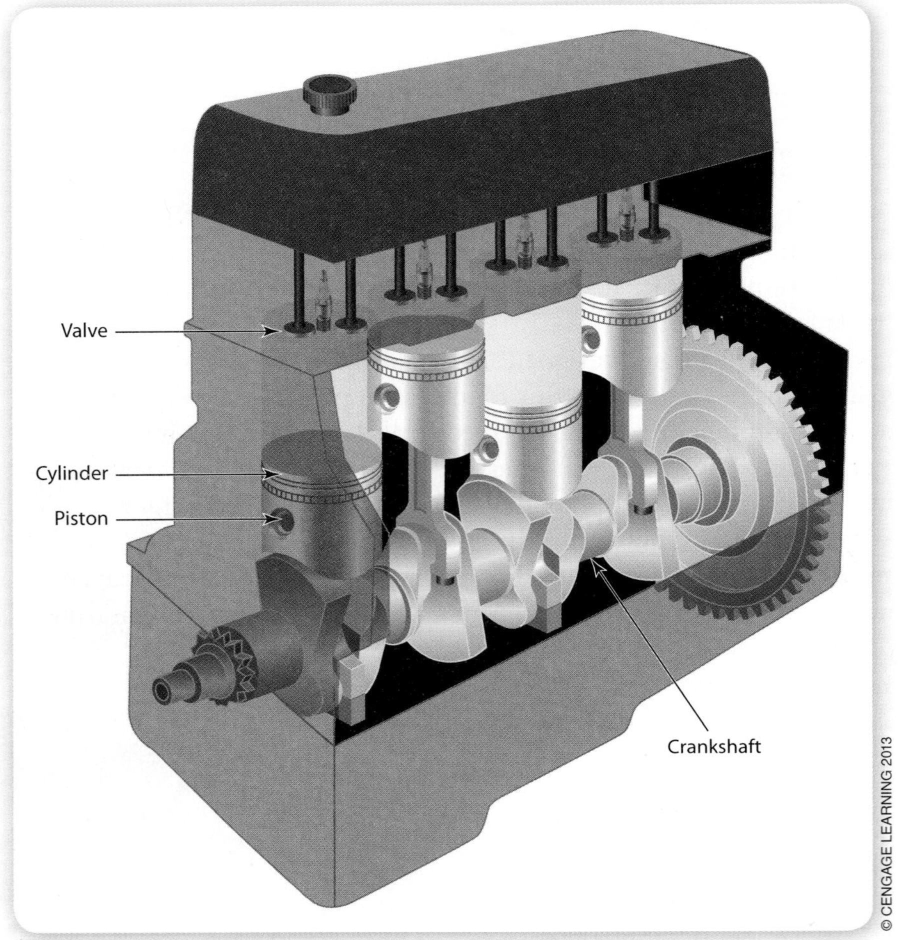

FIGURE 13-12 Components of an internal-combustion engine.

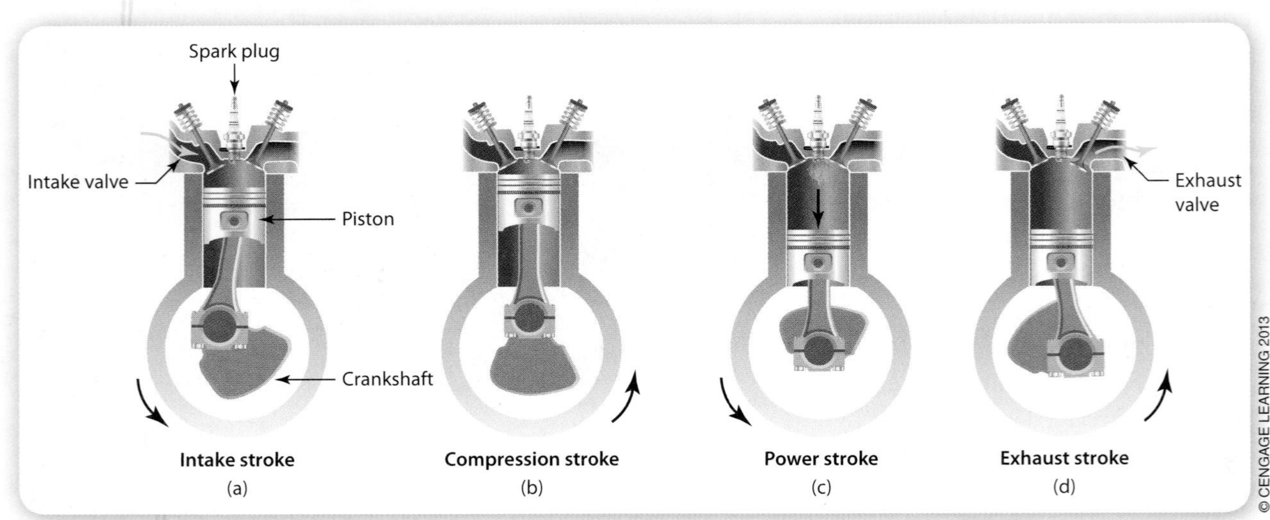

FIGURE 13-13 The four strokes of an internal-combustion engine.

chamber during the intake stroke. This explosive mixture is then pressurized during the compression stroke. A spark plug ignites the air–fuel mixture to start the power stroke; as its name implies, the power stroke develops the engine's force. During the exhaust stroke, the burnt air–fuel mixture is pushed out of the cylinder.

Jet Engines

Some modern aircraft use **jet engines** to produce their thrust. Jet engines develop thrust by moving air through a series of convergent and divergent ducts. When air moves through a straight pipe, its pressure and speed remain constant. But when air moves through pipes that either increase or decrease in area, the air's pressure and speed change. A convergent duct is like a funnel. As air passes through a convergent duct, its speed increases and its pressure decreases. A divergent duct is like the bell of a trumpet. The divergent duct increases the air's pressure but decreases its speed (see Figure 13-14). Just like the lift developed by aircraft wings, airflow through convergent and divergent ducts is also based on Bernoulli's principle.

Engineers use these changes in air speed and air pressure when designing jet engines. These actions are based on **Newton's third law of motion**: for every action, there is an equal and opposite reaction. In aerospace engineering, this principle of action and an equal reaction is very important. As shown in Figure 13-15, when air leaves a balloon (action), the balloon moves in the opposite direction (equal and opposite reaction).

Newton's third law of motion
Newton's third law of motion states that for every action, there is an equal and opposite reaction.

Air enters the jet engine through the air intake. It is then compressed in the compressor section. The compressed air next enters the combustion section, where fuel is added and ignited. The burning air–fuel mixture expands and forces the turbine wheels to rotate before it exits through the engine exhaust ducts. The turbine wheels are connected to the compressor and provide the energy to turn the compressor. The hot exhaust gas leaving the jet

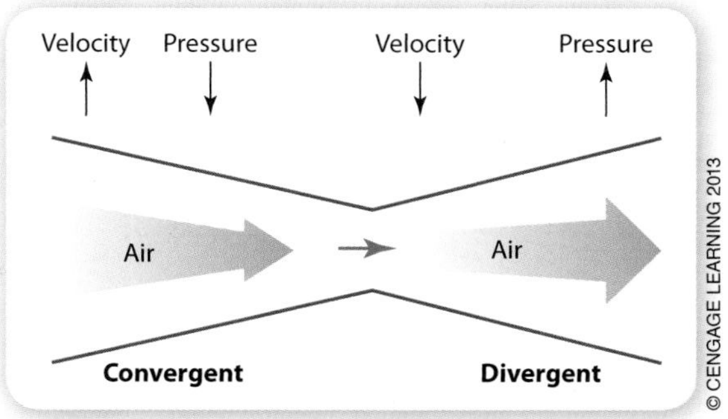

FIGURE 13-14 Bernoulli's principle with convergent and divergent ducts.

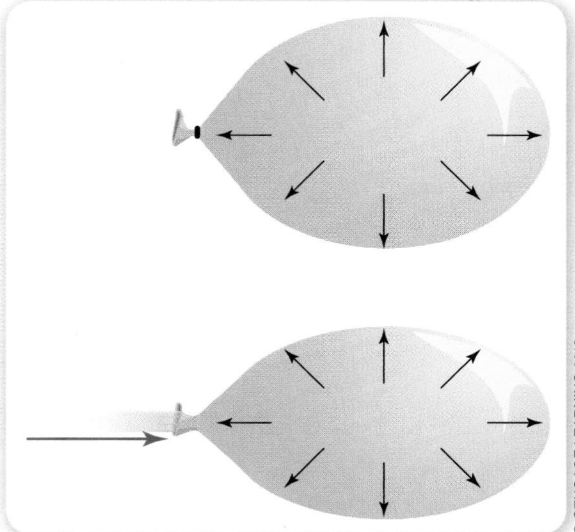

FIGURE 13-15 A balloon can demonstrate Newton's third law.

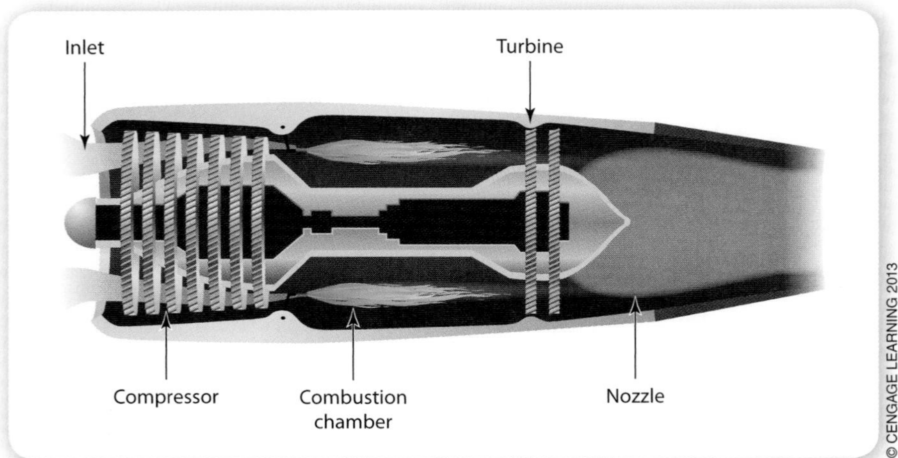

FIGURE 13-16 A turbojet engine.

engine (action) causes the aircraft to move forward (equal and opposite reaction). The more the jet engine turns and produces exhaust pressure, the more thrust the aircraft produces and the faster the aircraft flies.

Jet engines have many variations. The jet engine described and shown in Figure 13-16 is correctly termed a **turbojet engine**.

Aircraft Control Systems

6 Because aircraft operate in the sky, pilots need to be able to control the three possible directions of travel the aircraft could take. Each direction has a specific term that describes its movement. **Pitch** is the up or down movement of an aircraft. **Roll** is the clockwise or counterclockwise rotating motion of an aircraft. **Yaw** is an aircraft's left or right turning motion. Figure 13-17 shows these three movements.

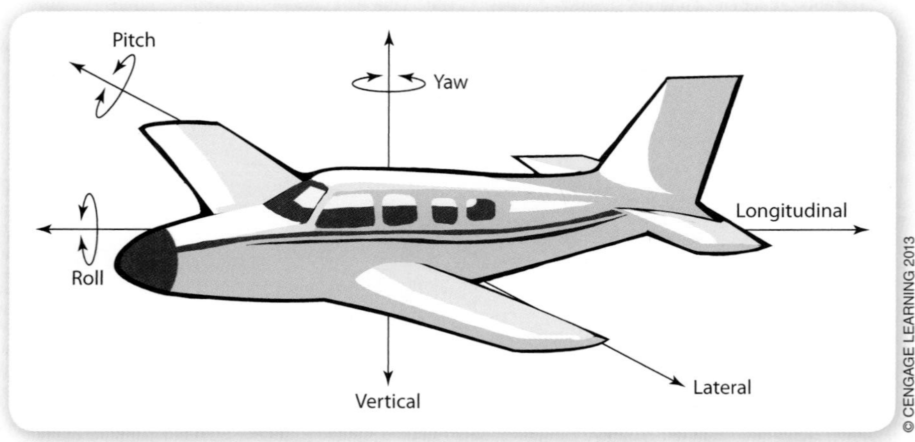

FIGURE 13-17 The movement of an aircraft.

An aircraft has three controlling surfaces to regulate these three movements (see Figure 13-18). The **elevators** control the pitch of the aircraft. They are located in the horizontal tail section. The **ailerons** control the roll of the aircraft, whereas the **rudder** controls the aircraft's yaw from the vertical tail section. The other parts of a typical aircraft are the cockpit, wings, fuselage, and tail section.

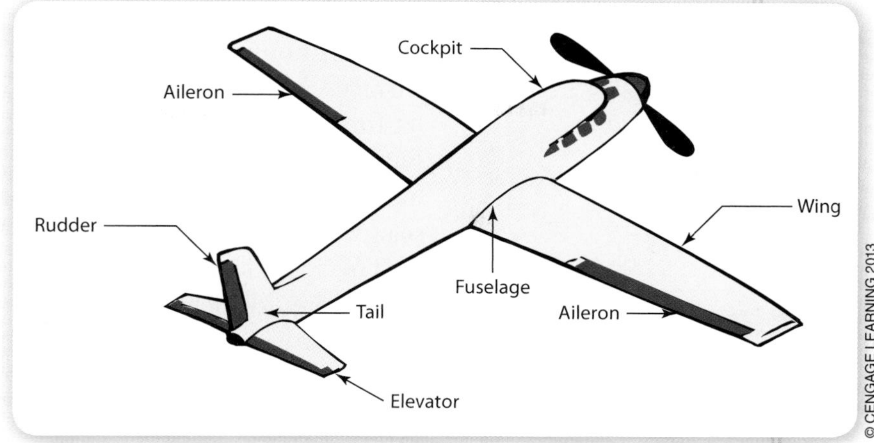

FIGURE 13-18 **The components and controlling surfaces of an aircraft.**

SECTION 3: ROCKETRY AND SPACEFLIGHT

7

The aerodynamics of aircraft flight is based on the effects of Earth's gravity and the availability of oxygen in the atmosphere. Flight in space, however, must account for different factors.

Rocketry

Solid-fueled and liquid-fueled **rockets** were developed because of the characteristics of travel beyond Earth's atmosphere. First, a spacecraft must develop a speed of more than 25,000 mile per hour to break the gravitational pull. Second, in space oxygen is absent, so turbojet engines and internal-combustion engines will not operate. Therefore, alternative types of propulsion were needed, which is why solid-fueled and liquid-fueled rockets were developed. Rockets develop thrust without having oxygen available to burn. Engineers term the types of chemical mixtures used in rockets to produce thrust **propellants**.

Solid-fueled rockets (Figure 13-19) develop thrust by igniting a solid fuel. The solid fuel has an opening through its center, where burning occurs. This combustion develops a powerful stream of hot gases that causes the rocket to move upward. Most solid-fueled rockets cannot be reused and must have a source of ignition.

Liquid-fueled rockets also produce a hot stream of gases that move the rocket upward. These rockets, however, provide a combustion chamber in which two different types of liquids are ignited (see Figure 13-20). The liquid-fuel storage tanks on these rockets can be reused. A plumbing system is required on a liquid-fueled rocket to direct the correct amount of fuel to the combustion chamber.

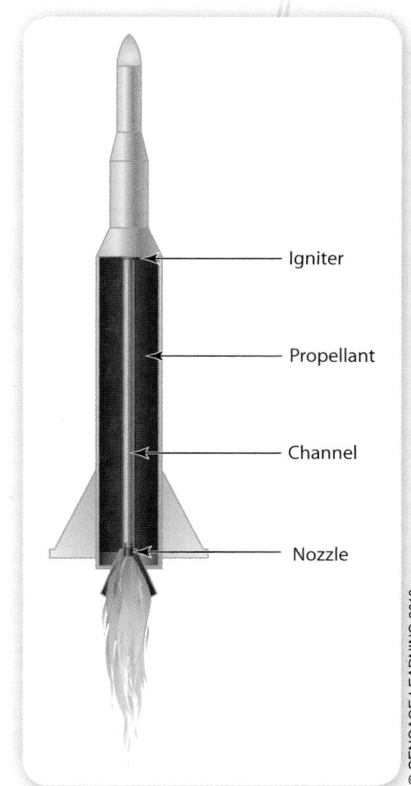

FIGURE 13-19 **A solid-fueled rocket.**

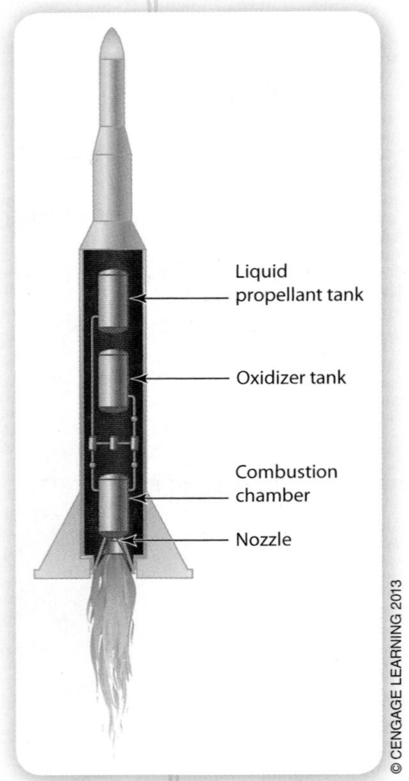

FIGURE 13-20 A liquid-fueled rocket.

Rockets have the same forces—drag, lift, thrust, and weight—as Earth-bound aircraft. They do not have the same controlling surfaces, however. The only controlling surfaces on a rocket are its fins. The path a rocket takes during its flight, as shown in Figure 13-21, is termed its **trajectory.** A rocket's **altitude** is the highest point of its flight. Altitude is measured as either above sea level or above Earth's surface.

The space shuttle (Figure 13-22) uses both solid-fueled rockets and liquid-fueled rockets during its liftoff. Two solid-fueled rockets provide additional thrust for the space shuttle during liftoff and then separate from the shuttle. The space shuttle's liquid-fueled rocket is fueled by the large external fuel tank during liftoff. The external fuel tank also separates before the shuttle leaves Earth's atmosphere. The space shuttle uses its liquid-fueled rocket during space travel. Without the drag produced by Earth's gravity and atmosphere, only a small amount of liquid fuel is needed to propel the shuttle in space. Internal fuel tanks carry this fuel.

Spaceflight

Spaceflight is the movement of a spacecraft into and through outer space. Rockets are the source of propulsion that provide the thrust that moves the spacecraft away from Earth's gravity. A **spacecraft** is a vehicle designed for spaceflight. Spacecraft can be manned or unmanned. A spacecraft can travel away from Earth or enter an orbit around Earth. Orbits are also classified by their altitude. *Low Earth orbits* are below

trajectory
Trajectory is the path an object takes during flight.

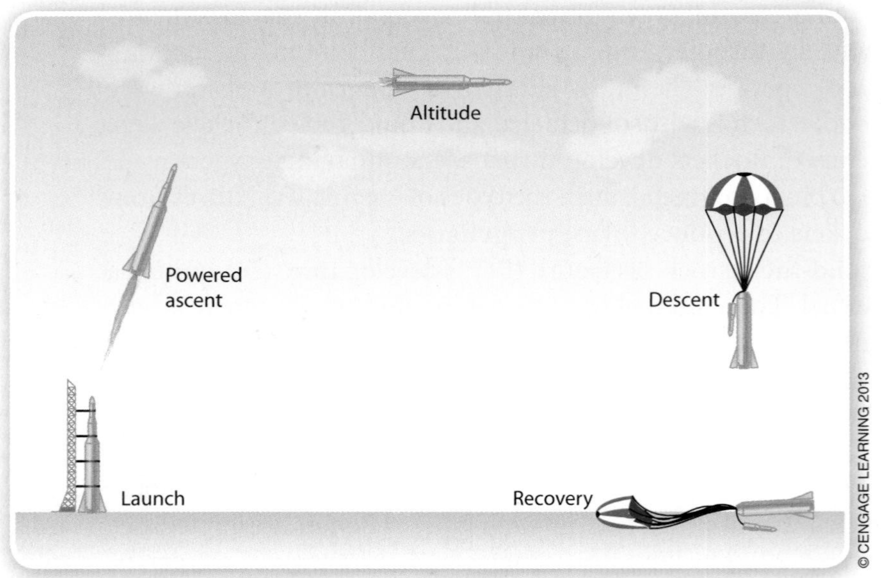

FIGURE 13-21 The trajectory of a rocket flight.

1240 miles. Orbits between 1240 and 22,240 miles are called *medium Earth orbits*. *High Earth orbits* are above 22,240 miles.

FIGURE 13-22 The space shuttle lifting off with its two external solid-fueled rockets.

Satellites are objects placed in Earth's orbit by humans. The first satellite was Sputnik 1. It was launched by the Soviet Union on October 4, 1957. Since that time, thousands of satellites have been sent into orbit for a variety of applications. These uses include communications, reconnaissance, navigation, and weather tracking and forecasting, some of which is military in nature and interest. Figure 13-23 shows a U.S. Air Force weather satellite.

Space stations are structures designed to allow humans to live in space. There are two basic types: monolithic and modular. *Monolithic space stations* are one-piece structures. They are launched into space as single units. The U.S. Skylab was a monolithic space station. *Modular space stations* are launched into space as separate sections and then assembled in outer space. The International Space Station (Figure 13-24) is a modular unit.

FIGURE 13-23 A U.S. Air Force weather satellite.

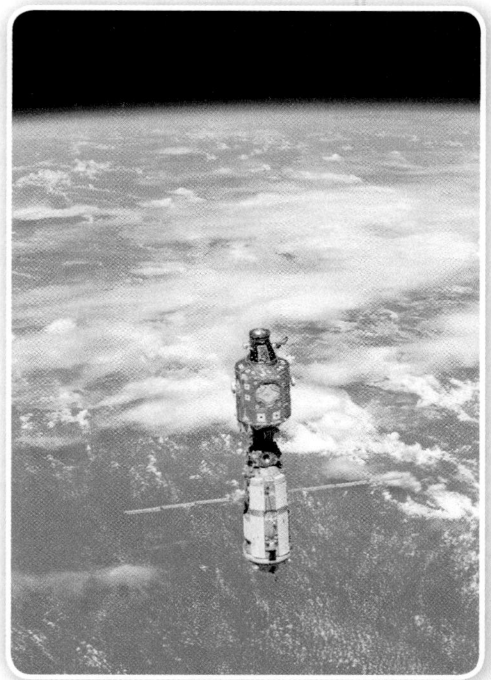

FIGURE 13-24 The International Space Station.

SUMMARY

In this chapter you learned:

- Aerodynamics is the study of flight.
- Airfoil design, wing surface area, wing angle of attack, and aircraft speed all affect the lift produced.
- Aerodynamic forces include thrust, drag, lift, and weight.
- Bernoulli's principle states that if the speed of a liquid increases, its pressure decreases.
- The turbojet engine is based on Newton's third law of motion.
- Pitch, the upward or downward movement of an aircraft, is controlled by the elevators.
- Roll, the rotating movement of an aircraft, is controlled by the ailerons.
- Yaw, the turning movement of an aircraft, is controlled by the rudder.
- There are two types of rockets: solid-fueled and liquid-fueled.

VOCABULARY

Write a definition for each term in your own words. After you have finished, compare your answers with the definitions provided in the chapter.

Lift
Drag
Thrust
Aerodynamics
Bernoulli's principle
Aspect ratio
Propulsion
Newton's third law of motion
Trajectory

STRETCH YOUR KNOWLEDGE

Provide thoughtful, written responses to the following questions:

1. Describe how Orville and Wilbur Wright's engineering design knowledge and skills changed society.
2. Describe why aeronautical engineers must understand both Bernoulli's principle and Newton's third law of motion.
3. Describe why an aircraft with a high aspect ratio is less maneuverable than an aircraft with a low aspect ratio.
4. What would be the advantage (if any) of aircraft that have movable wings?
5. Aircraft land and take off while facing into the wind. Why?
6. Compare and contrast the advantages of a monolithic space station and a modular space station.

Onward to Next Destination ▶

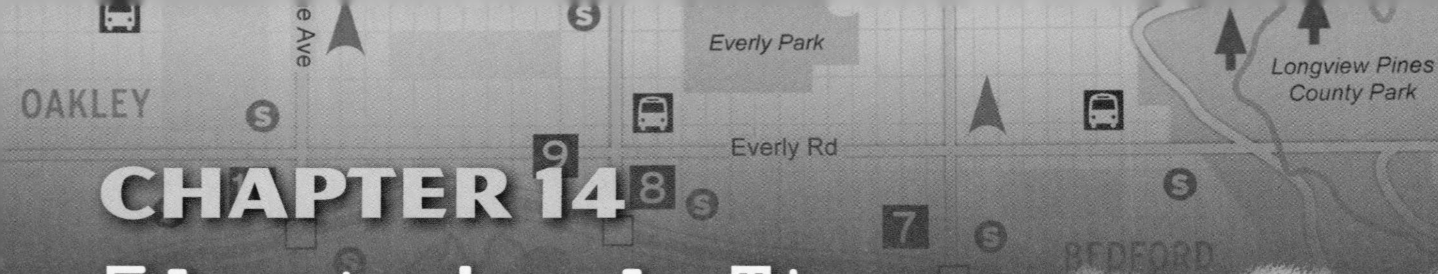

CHAPTER 14
Electrical Theory

Menu

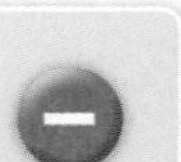

Before You Begin

Think about these questions as you study the concepts in this chapter:

 What is electricity?

 How does a magnetic field produce electricity?

 How can electricity produce a magnetic field?

 How does electricity make our life easier?

 How do engineers measure electricity?

 How do engineers mathematically calculate the flow of electrons using Ohm's law?

© MICHAEL SHAKE/SHUTTERSTOCK.COM

Engineering in Action

Conventional gasoline-powered vehicles are giving way to innovative hybrid vehicles. A hybrid vehicle uses both a gasoline engine and an electric motor to produce its power. The engineering required to design and build a hybrid vehicle, however, requires more than just adding an electric motor to a conventional vehicle. Engineers have designed a very sophisticated electric motor that also doubles as an electrical generator. In addition, the batteries, energy-storage devices, for hybrid vehicles are very advanced. Without the electrical theory concepts discussed in this chapter, engineers would not have been able to design today's hybrid vehicles.

SECTION 1: ELECTRON FLOW

Atomic Structure

All matter is made of small particles called **atoms**. Atoms are the building blocks of our universe. They are formed by varying numbers of **neutrons, protons**, and **electrons**. These parts of the atom are shown in Figure 14-1. Different materials have different numbers of neutrons, protons, and electrons. Scientists have developed the **periodic table of the elements** to display the different materials and their corresponding number of neutrons, protons, and electrons (Figure 14-2). The number of protons in an atom is termed its *atomic number*. The electrons orbit around the center of the atom. The center of the atom is called the *nucleus*, which contains protons and neutrons.

The electrons orbit at various levels based on the number of electrons the atom contains. Each level is called an *electron orbit*. The inner orbit contains two electrons. There are as many as eight electrons in the second orbit and in each orbit that follows.

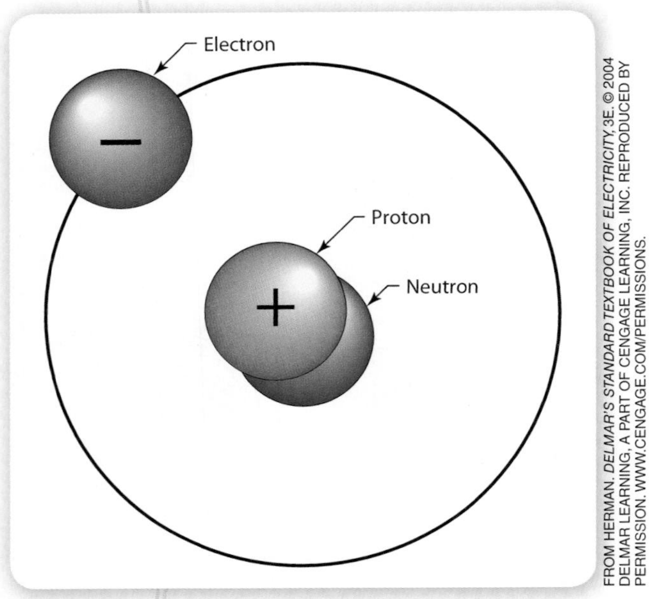

FIGURE 14-1 The parts of an atom.

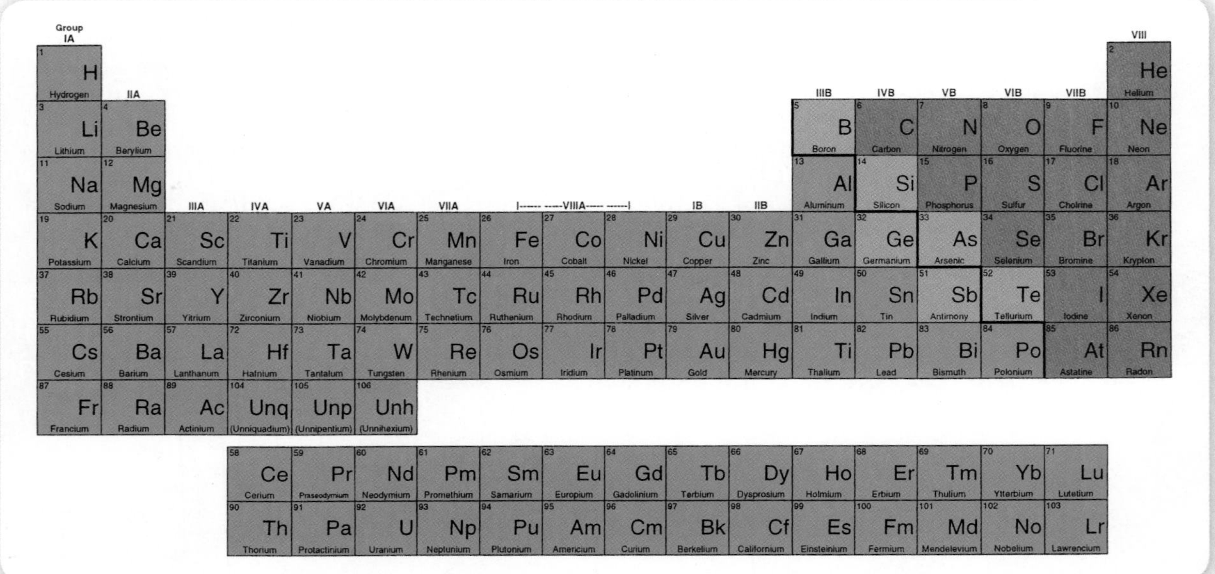

FIGURE 14-2 The periodic table of the elements.

Science in Engineering

Through scientific investigation, scientists have determined the number of electrons contained in the orbital levels of each element's atom. Investigation has determined that these orbits are not perfectly circular but vary. Scientists have named each of these orbitals based on the characteristic of the electrons' paths. The field of science that studies orbits is called *quantum mechanics*.

The electrons in the outermost orbit are called **valence electrons**. If only one, two, or three electrons are in the atom's outer orbit, they are termed *free electrons*. These electrons can easily move from one atom to another atom (Figure 14-3). The flow of these electrons from one atom to another along a pathway is called **electricity**.

valence electron
A valence electron is an electron on an atom's outermost orbit.

electricity
Electricity is the flow of electrons through a pathway.

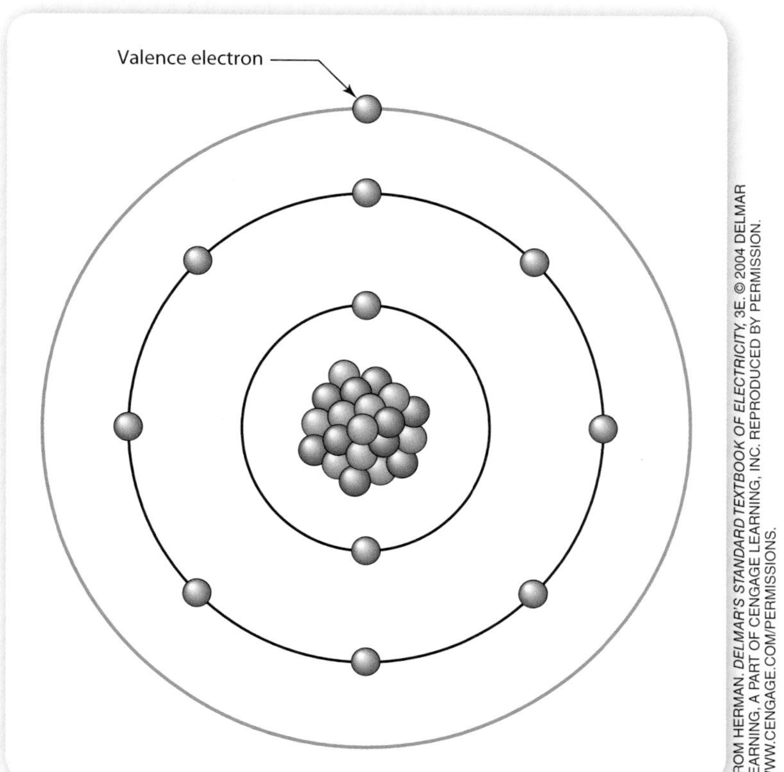

FIGURE 14-3 *Valence electrons are located on the outermost orbit of an atom.*

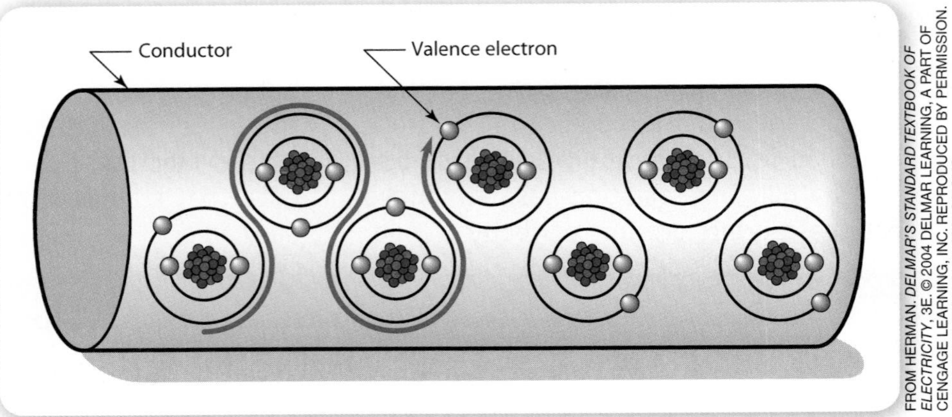

FIGURE 14-4 **Electrons flowing between atoms.**

Conductors

For the electrons to flow, there must be a pathway of material that contains atoms with valence electrons. Materials that are composed of atoms containing one, two, or three valence electrons are called **conductors**. Conductors allow electrons to flow through them easily. The flow of electrons through a conductor is shown in Figure 14-4.

electronegativity
Electronegativity is the measure of the attraction that an atom has for the electrons orbiting about its nucleus.

Engineers use the term **electronegativity** to describe the measure of attraction that an atom has for its electrons. Conductors have low electronegativity. In other words, the atoms in a conductor have very little or a very weak hold on their electrons. The periodic table of the elements in Figure 14-2 has the conductors colored in green. Notice that conductors are located in the lower left.

Most conductors are metals. As we discussed in Chapter 8, a metal is a material that is a good conductor of heat and electricity. The copper atom has an atomic number of 29, which means that it contains 29 electrons (see Figure 14-5). The copper atom therefore has only one electron in its outer orbit. This one valence electron makes copper an excellent conductor of electricity.

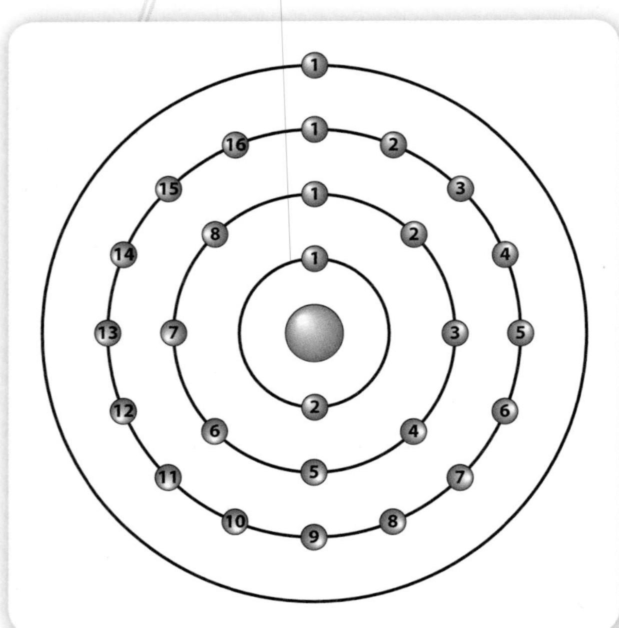

FIGURE 14-5 **The atomic structure of copper.**

Insulators

Materials that do not have free electrons are called **insulators**. Insulators resist the flow of electrons. Insulators are located on the upper right in the periodic table of elements. In

Figure 14-2, the insulators are colored red. Common insulators include wood, plastic, and rubber. Most materials that are nonmetals are also insulators. Engineers use insulators to stop the flow of electrons.

Figure 14-6 shows an argon atom. Argon, a gas in its natural state, has 18 total electrons and 8 electrons in its outer orbit, and thus no free electrons. This makes it very difficult for electrons to flow from one argon atom to another. Because argon gas is an excellent insulator, it is used to encase the filament of incandescent lightbulbs.

Semiconductors

Located on the periodic table of the elements in Figure 14-2 is an area between the conductors and insulators, or metals and nonmetals, and are the elements called **semiconductors**. This area is colored blue. Semiconductors typically contain four electrons in their outer orbit. Materials that are semiconductors include silicon and germanium. As shown in Figure 14-7, the silicon atom has four electrons in its outer orbit. By adding or subtracting electrons, engineers can modify semiconductors to perform a desired operation.

semiconductor

A semiconductor is a material that is neither a good conductor nor a good insulator; its electrical conductivity can be precisely altered by a manufacturing process.

Sources of Electricity

The flow of electrons can be found in nature in the form of **static electricity**. The discharge of static electricity can result when you walk across a carpeted

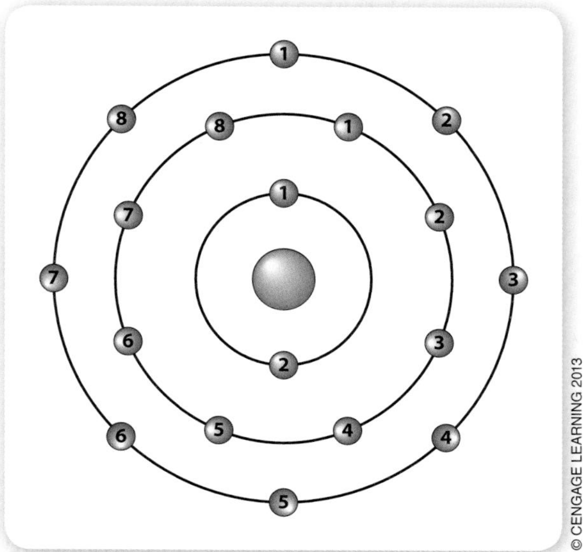

FIGURE 14-6 The atomic structure of argon.

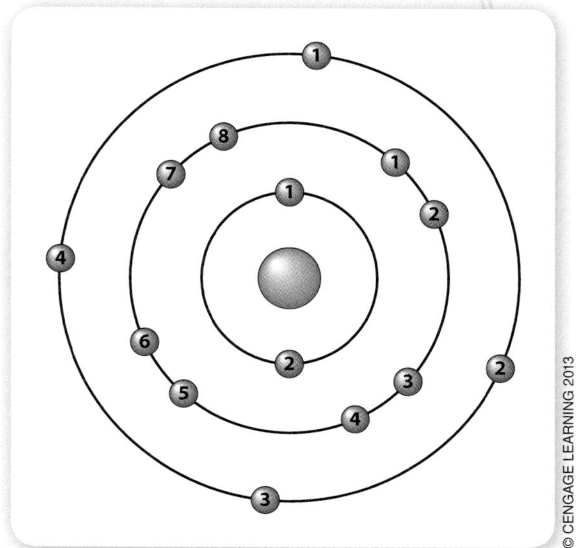

FIGURE 14-7 The atomic structure of silicon.

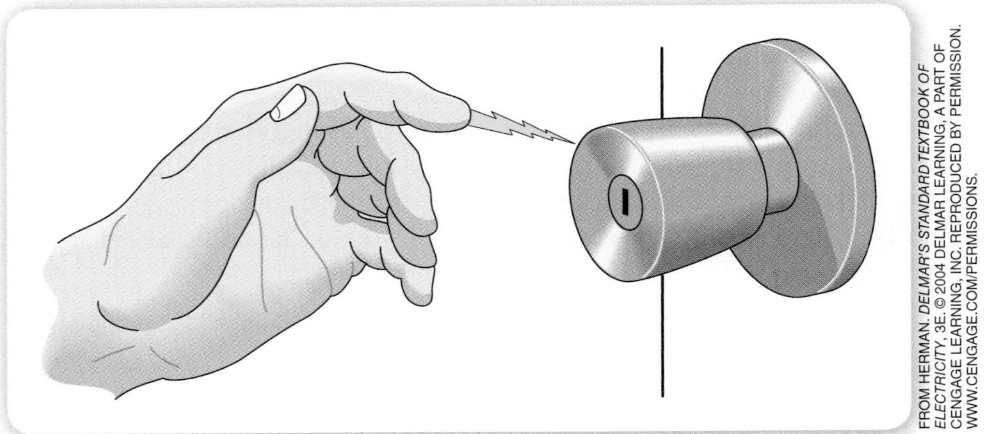

FIGURE 14-8 Static electricity in your home.

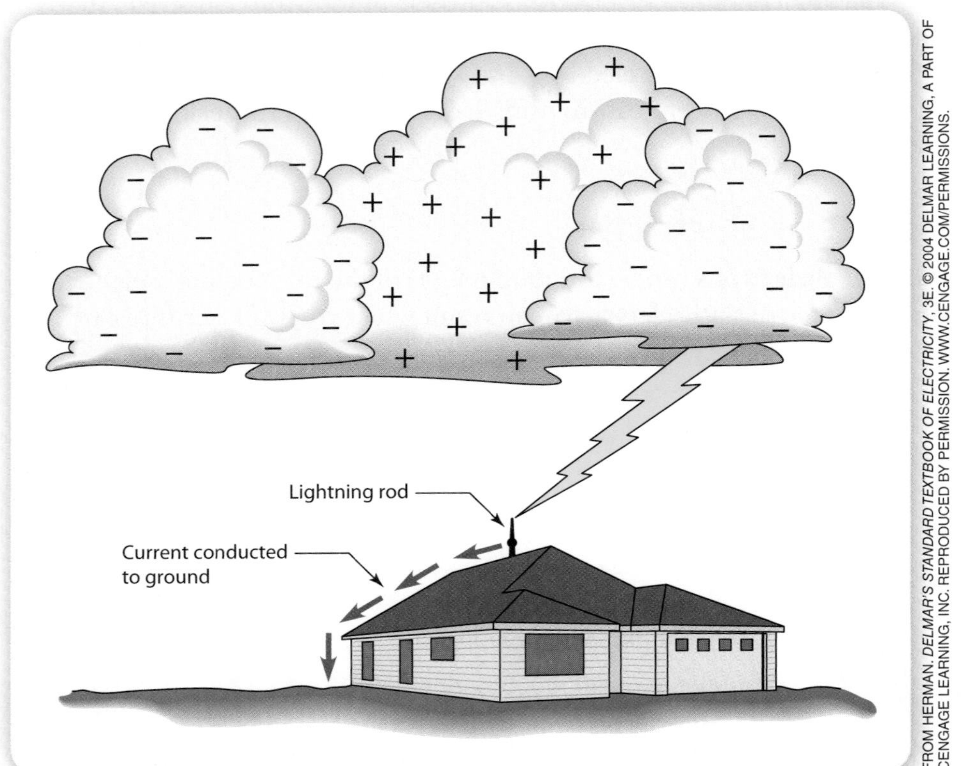

FIGURE 14-9 Lightning is a form of static electricity.

floor and reach out for a metal door knob (Figure 14-8). The valence electrons jump from your hand to the door knob, causing a "spark." Lightning is another form of static electricity that occurs in nature. As shown in Figure 14-9, the electrons can flow from the negatively charged cloud to the lightning rod of the house.

Chemical reactions can also be used to create the flow of electrons. The batteries used in everyday appliances (Figure 14-10) produce electricity through chemical reactions. The basic process of converting chemical energy to electricity takes place in a voltaic cell, which is a simple structure consisting of an **anode**, a **cathode**, and an **electrolyte**. These parts are shown in Figure 14-11.

FIGURE 14-10 Batteries make your life easier.

An electrolyte is usually a liquid, whereas the anode and cathode are typically metals. The electrons flow from the negatively charged cathode through the electrolyte to the positively charged anode. This flow of electrons will occur only when the pathway for the electrons outside the voltaic cell has been completed. You can make a simple voltaic cell using two unlike metals—zinc and copper, for example—and a fruit or vegetable to serve as the electrolyte (Figure 14-12). Two or more voltaic cells connected together create a **battery**. So the single household batteries like those in Figure 14-10 are not really batteries but single voltaic cells.

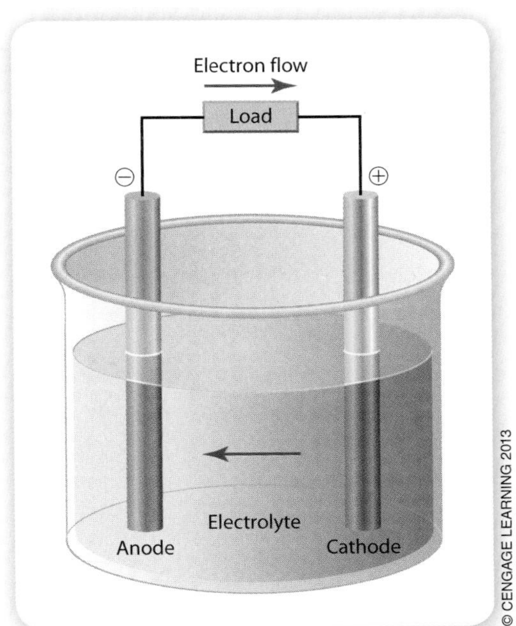

FIGURE 14-11 Components of a voltaic cell.

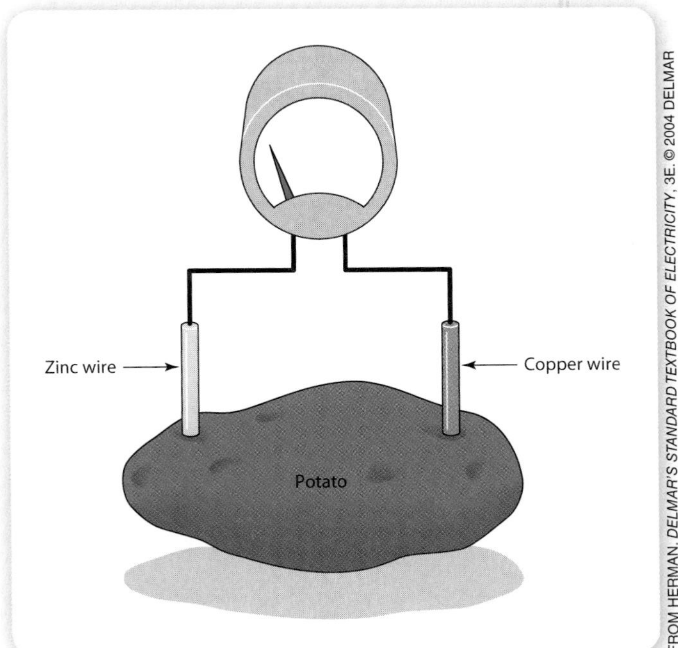

FIGURE 14-12 A simple voltaic cell can be made from a potato.

Science in Engineering

Hybrid vehicles, flashlights, and your laptop computer would not have been possible if not for an Italian physicist named Alessandro Volta. A professor at the Royal School in Como, Lombardy (Italy), Volta was interested in electricity and had studied the effects of dissimilar metals and electron flow. His studies led to the creation of the world's first battery, the voltaic pile (Figure 14-13). Volta placed alternating discs of zinc and copper in a wine goblet and then filled the goblet with saltwater. In 1801, Volta demonstrated his voltaic pile to Napoleon. Today we use the term *volt* to describe the pressure of electrons in a pathway.

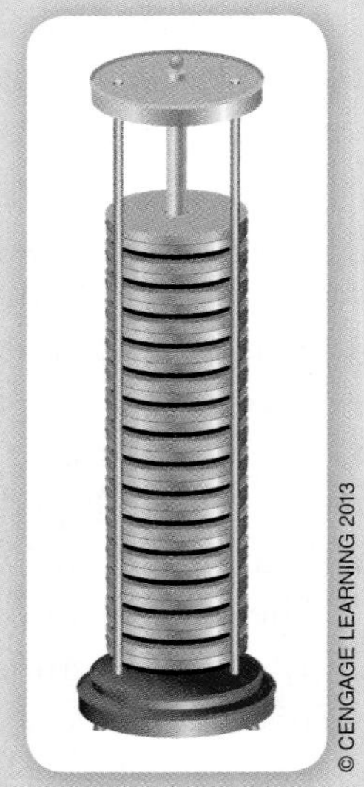

FIGURE 14-13 A voltaic pile.

SECTION 2: GENERATORS AND MOTORS

Laws of Magnetism

Understanding magnetism is essential in understanding electricity. A **magnet** is a device that attracts iron. Figure 14-14 depicts the force field that surrounds a magnet. This field is called a **magnetic field**. Magnetism occurs naturally on Earth. Navigators have used magnetism for centuries in their compasses to locate Earth's North Pole. Humans have created permanent magnets like those pictured in Figure 14-15. These magnets also have north and south poles just

Did You Know?

Toyota designed and produced the industry's first high-performance hybrid vehicle in 2004. The vehicle, which seats three people and has a carbon-fiber body, was named after the scientist who invented the first battery, the voltaic pile. The vehicle was called the Alessandro Volta.

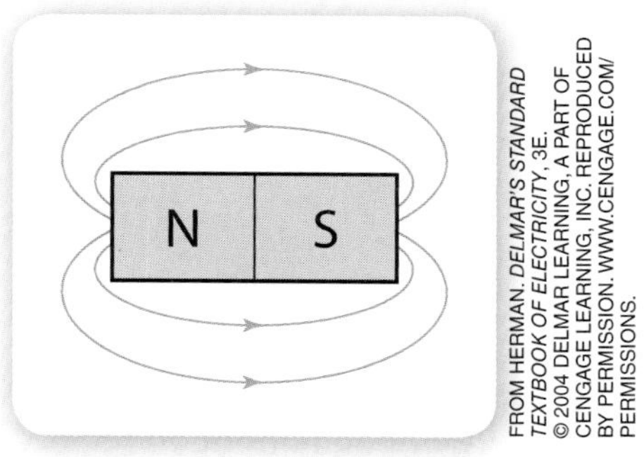

FIGURE 14-14 The magnetic lines of a magnet.

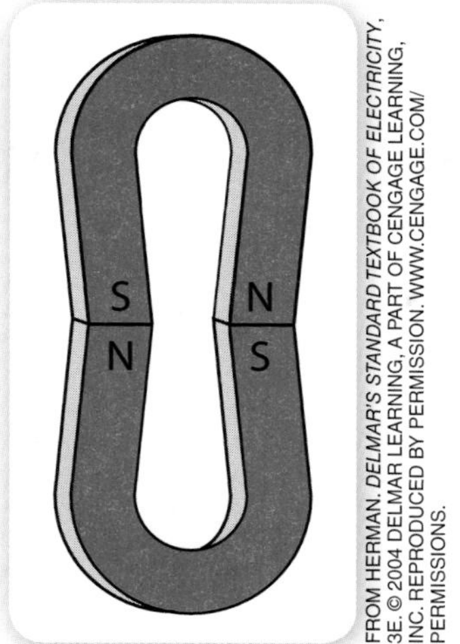

FIGURE 14-15 Common horseshoe magnets.

like Earth. The **laws of magnetism** note that unlike poles attract each other and like poles repel each other (Figures 14-16 and 14-17). Engineers can use the laws of magnetism to assist in the production of electricity.

An **electromagnet** is a magnet that is created by the use of electricity. When electrons flow through a conductor, a magnetic field is produced by the movement of the electrons. You can create a basic electromagnet by winding a wire

laws of magnetism

The laws of magnetism are the scientific principle that states that like magnetic poles repel and unlike magnetic poles attract.

electromagnet

An electromagnet is a metal core that is rendered magnetic by the passage of an electric current through a surrounding coil.

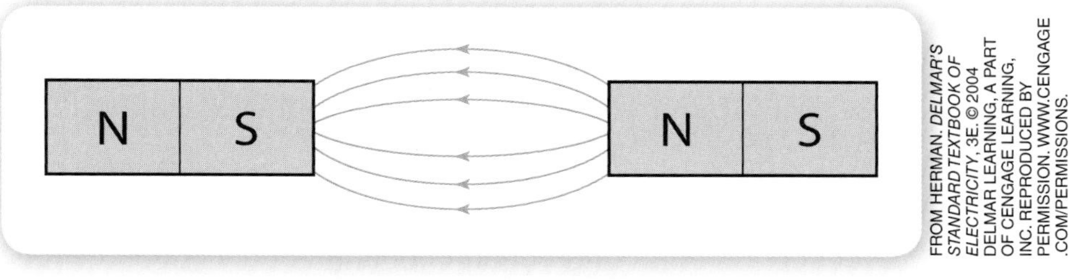

FIGURE 14-16 Unlike poles of a magnet attract each other.

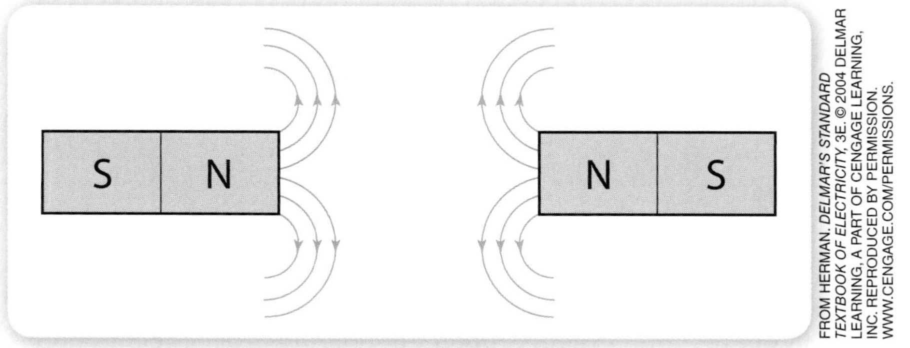

FIGURE 14-17 Like poles of a magnet repel each other.

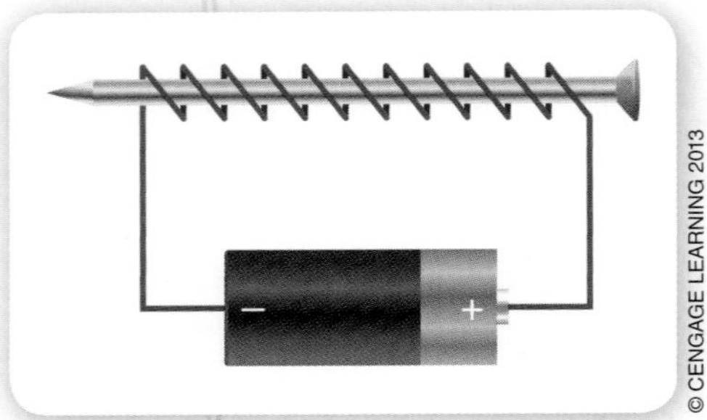

FIGURE 14-18 A simple electromagnet constructed with wire, a nail, and a battery.

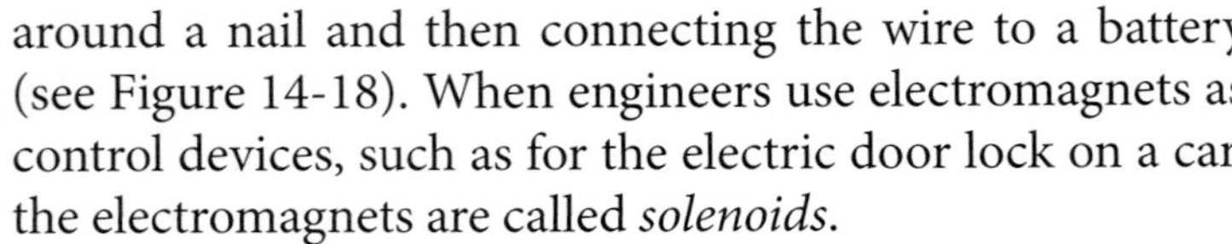

around a nail and then connecting the wire to a battery (see Figure 14-18). When engineers use electromagnets as control devices, such as for the electric door lock on a car, the electromagnets are called *solenoids*.

Generation of Electrical Energy

A **generator** is a device that converts rotational motion into electricity. You can mount a generator on your bicycle, as shown in Figure 14-19, to convert the bicycle's tire rotation into electricity for a light. Generators operate on the principle of **magnetic induction**, the generation of electron flow by a magnetic field. Magnetic induction is developed by moving a conductor through a magnetic field (Figure 14-20). The movement of the conductor past the magnetic field causes the valence electrons in the conductor to flow.

The parts of a generator include the **commutators**, the **brushes**, the magnets, and the rotating **armature**. These parts are shown in Figure 14-21. As the armature rotates past the magnets, magnetic induction causes the electrons to flow. The electrons flow to the commutators and then to the brushes. As the armature turns, the flow of electrons from the commutator switches to different brushes (Figure 14-22). This switching causes the direction of the electron flow to rotate every one-half revolution.

Electric Motor

The operation of an electric **motor** is the reverse of a generator. Electrical current is supplied to the motor and rotational motion is produced. As

FIGURE 14-19 A generator can convert the tire rotation of your bicycle into electricity to power a light.

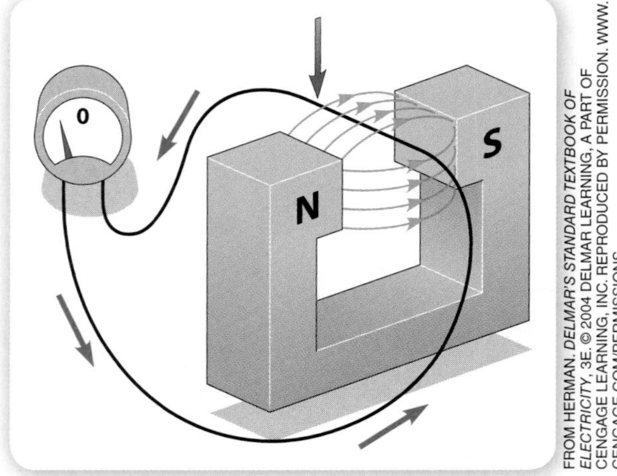

FIGURE 14-20 The principle of magnetic induction.

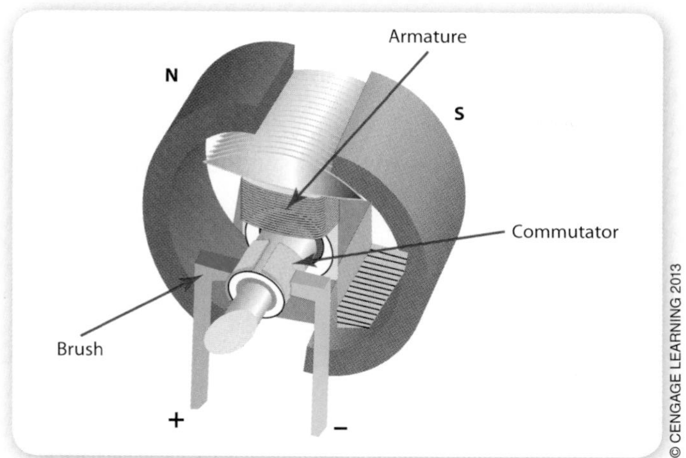

FIGURE 14-21 The parts of a generator.

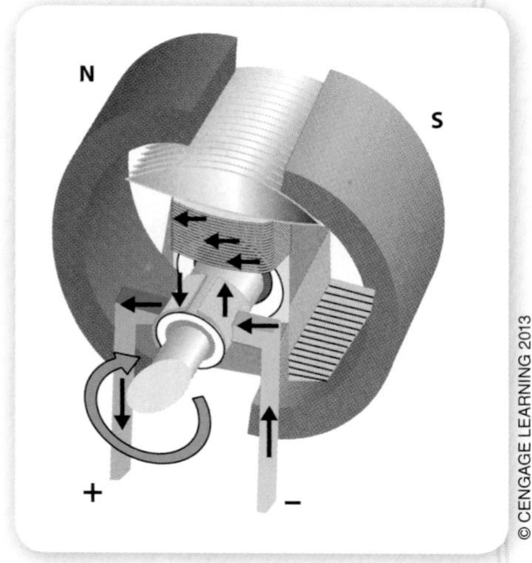

FIGURE 14-22 The flow of electrons in a generator.

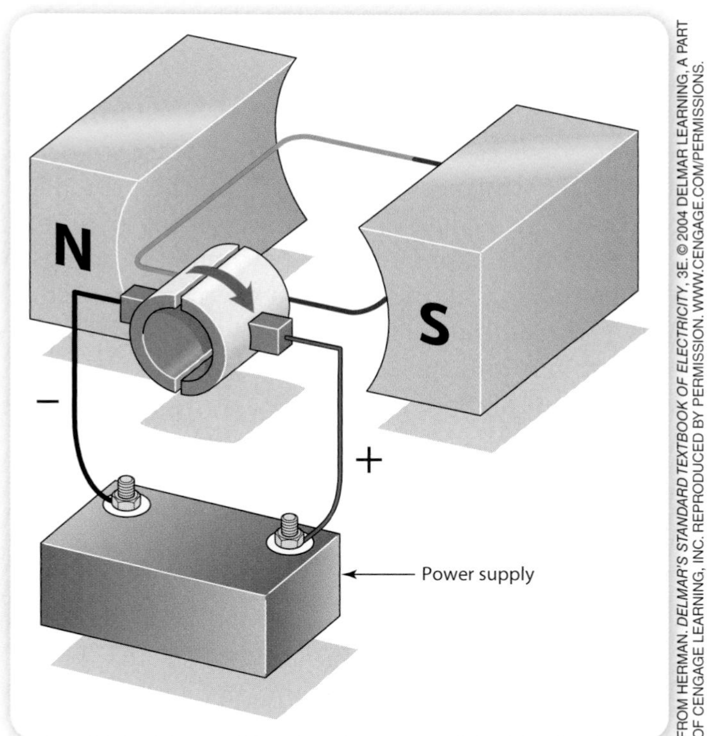

FIGURE 14-23 The magnetic induction of an electric motor.

shown in Figure 14-23, the electrons flow from the brushes to the commutators and then through the armature's windings. This flow of electrons develops a magnetic field in the armature. The armature's south magnetic field is attracted to the motor's north magnetic field, and the armature's north

Math in Engineering

Engineers use mathematical formulas to determine the number of windings required on the primary and secondary sides of a transformer. The ratio of these two sets of windings can be set up as a mathematical equation. In the equation, E stands for voltage and N represents the number of windings. The subscript p is for the primary winding and the subscript s means the secondary winding. The equation is noted below:

$$\frac{E_s}{E_p} = \frac{N_s}{N_p}$$

magnetic field is attached to the motor's south magnetic field. Because of the separation of the two commutators, the rotation of the armature causes the flow of electrons to reverse after one-half turn and hence reverse the armature's magnetic fields. This reversal of electron flow every half revolution causes the motor's armature to continue its rotation.

Reducing or Increasing Electrical Current Flow

Transformers are devices that are used to increase or decrease electrical current flow. Transformers also operate on the principle of magnetic
4 induction. A transformer has two separate sets of windings around an iron core. Electricity is supplied to the primary windings, which then develop a field of magnetic induction. This magnetic induction causes electron flow in the output, or secondary, windings. The ratio between the primary windings and the secondary windings determines the output voltage. For example, the transformer pictured in Figure 14-24 has

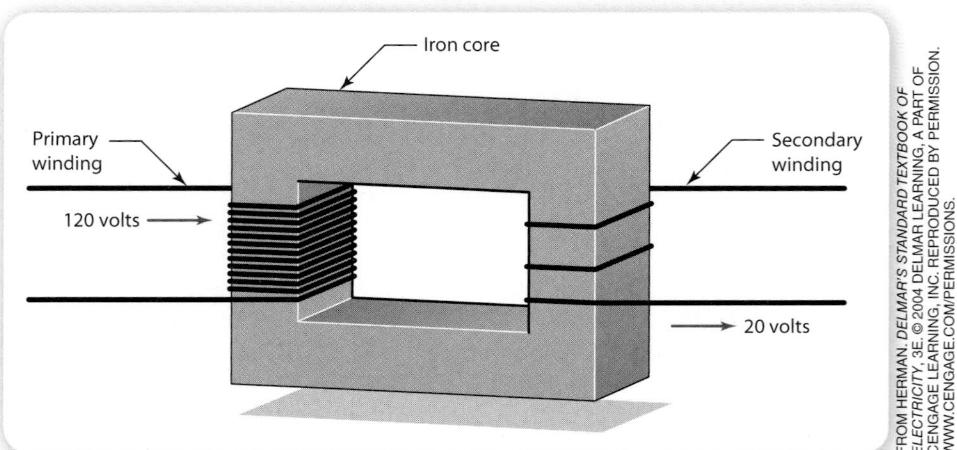

FIGURE 14-24 **The operation of a transformer.**

FIGURE 14-25 Computer power supply.

electromotive force
Electromotive force is the pressure on electrons to move them through a conductor that is supplied by the power source.

600 primary windings and 100 secondary windings, or a ratio of 6 to 1. If 120 volts is supplied to the transformer, the output voltage will be 20 volts. The computer power supply pictured in Figure 14-25 uses this ratio to convert your household current of 120 volts to the 20 volts required for your computer.

SECTION 3: MEASURING ELECTRICAL CURRENT

Multimeter

Engineers need to measure electron flow when they are designing and testing electrical devices. To perform this measurement operation, they use a tool called a *digital multimeter* (Figure 14-26). The most common measurements taken with the multimeter are voltage, resistance, and current flow.

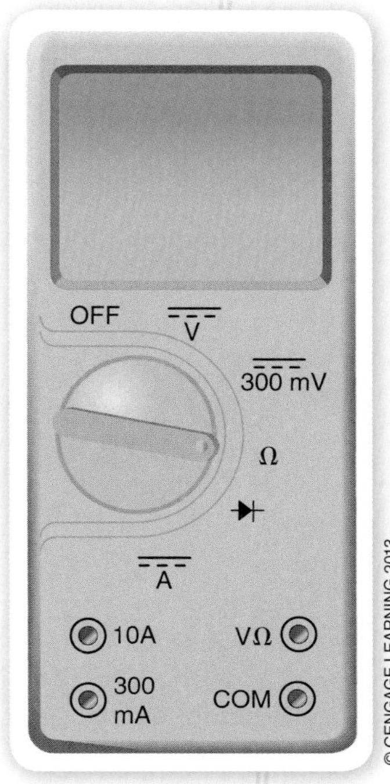

FIGURE 14-26 The digital multimeter.

Voltage

Voltage is the measurement of the force or pressure of the electrons in a pathway. Voltage is commonly referred to as electromotive force and is measured in **volts (E)**. The use of the multimeter to measure voltage is shown in Figure 14-27.

Resistance

The ability of a material to resist the flow of electrons is called **resistance**. Insulators have a high resistance, and good conductors have very little resistance. Resistance is measured in **ohms**. The Greek letter omega (Ω) is used by engineers to represent ohms. Figure 14-28 shows how to use the multimeter to measure resistance.

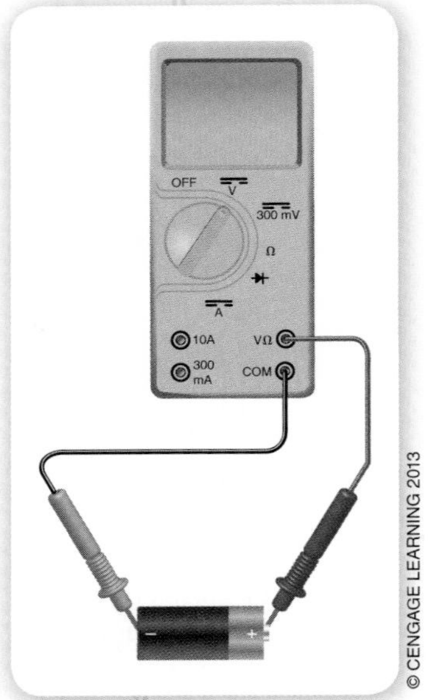

FIGURE 14-27 Using the multimeter to measure voltage.

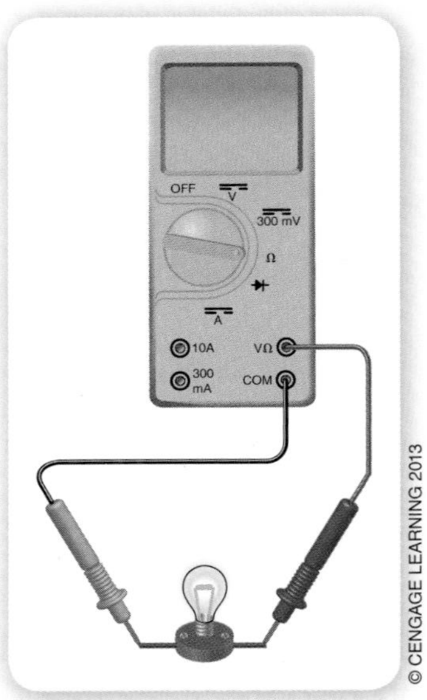

FIGURE 14-28 Using the multimeter to measure resistance.

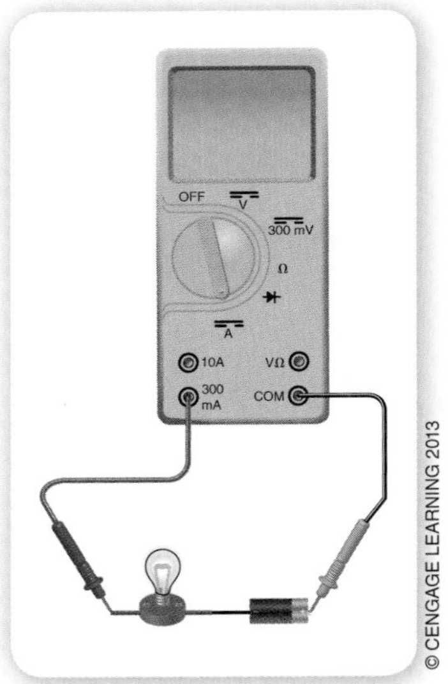

FIGURE 14-29 Using the multimeter to measure current flow.

Current

The rate of flow of the electrons along the pathway is termed **current**. Conductors allow the electrons to flow at a higher rate, whereas insulators slow down or stop the flow of electrons. Current is measured in **amperes** or **amp (A)**. To measure the flow of electrons along their pathway, the multimeter must be made a part of the pathway or circuit (Figure 14-29). The electrons must pass through the multimeter to be counted, just like the turnstile you walk through when entering a baseball stadium counts the number of fans.

Ohm's Law

In the nineteenth century, a German physicist named Georg Simon Ohm discovered that there is a directly proportional relationship among the voltage, resistance, and current flow in an electrical system. This mathematical relationship is called **Ohm's law**. Ohm's law states that voltage = current × resistance ($E = I \times R$). Engineers use the diagram in Figure 14-30 to visually demonstrate Ohm's law.

- E = the force of the electrons in volts
- I = rate of electron flow in amperes
- R = the resistance to electron flow in ohms

Ohm's law
Ohm's law is a mathematical formula that describes the relationship among voltage, current, and resistance: $E = I \times R$.

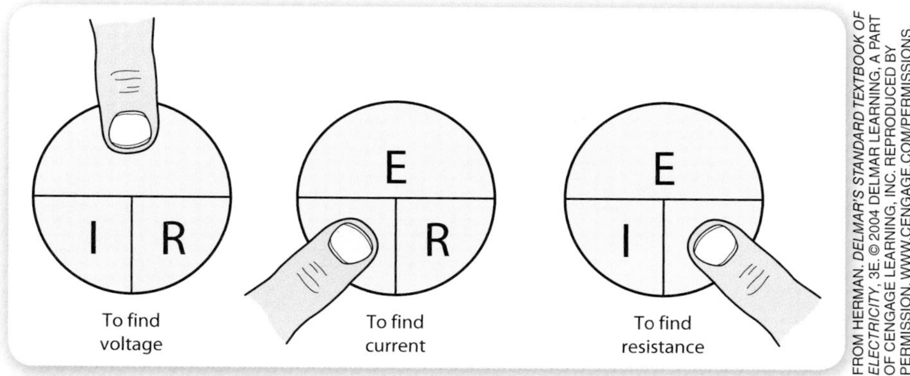

FIGURE 14-30 Using the Ohm's law diagram.

Engineering Challenge

ENGINEERING CHALLENGE 1

If an engineer knows two of the measurements of an electrical circuit, Ohm's law allows him or her to calculate the missing value.

- If an electrical circuit has a current flow of 2 amps and a resistance of 6 ohms, calculate the voltage.
- Next, if you know a circuit's voltage is 24 volts and the resistance is 6 ohms, you as the engineer can mathematically determine its current flow. What is the current flow?
- Finally, to calculate the circuit's resistance, you divide the voltage by the current flow. What is the resistance in an electrical circuit with 18 volts and a current flow of 3 amps?

Engineering Challenge

ENGINEERING CHALLENGE 2

- Using the digital multimeter, measure the resistance of various materials. Based on your measurement, determine whether these materials are conductors or insulators.
- Next, using the multimeter, measure the voltage of various batteries. Describe why the batteries do not all have the same voltage.

SUMMARY

In this chapter you learned:

- Electricity is the flow of electrons along a pathway.
- Some materials—conductors—allow for the electrons to flow easily. Other materials—insulators—resist the flow of electrons.
- The periodic table of the elements provides a graphic display of materials (conductors, insulators, and semiconductors).
- Electricity can be converted from the chemicals stored in a battery.
- Generators convert rotational motion into electricity, whereas motors convert electricity into rotational motion. Both devices use the principle of magnetic induction.
- The voltage, current flow, and resistance of an electrical pathway can be measured using a multimeter.
- If you know two of the three electrical pathway measurements (voltage, current flow, and resistance), you can calculate the third measurement using Ohm's law.

VOCABULARY

Write a definition for each terms in your own words. After you have finished, compare your answers with the definitions provided in this chapter.

Valence electron
Electricity
Electronegativity
Semiconductor
Laws of magnetism
Electromagnet
Electromotive force
Ohm's law

STRETCH YOUR KNOWLEDGE

Please provide thoughtful, written responses to the following questions.

- How has the development of electricity affected people's daily lives over time (both work and play)?
- How has this nation's dependence on foreign oil affected engineering and technology careers related to electrical theory?
- Describe how your life would be different if we did not have electricity.
- In the future, will engineering careers require a more focused area of specialization or will engineering careers call for a more interdisciplinary approach? Explain your thinking.

Onward to Next Destination

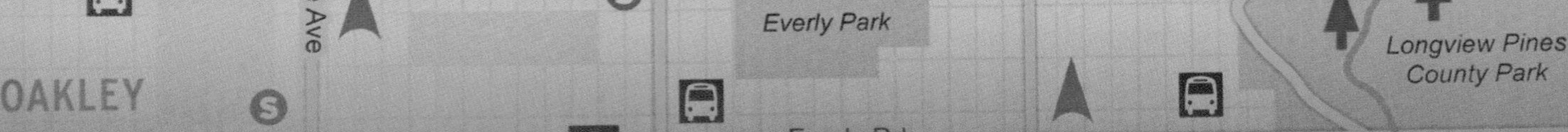

CHAPTER 15
Electrical Circuits

Menu

Before You Begin

Think about these questions as you study the concepts in this chapter:

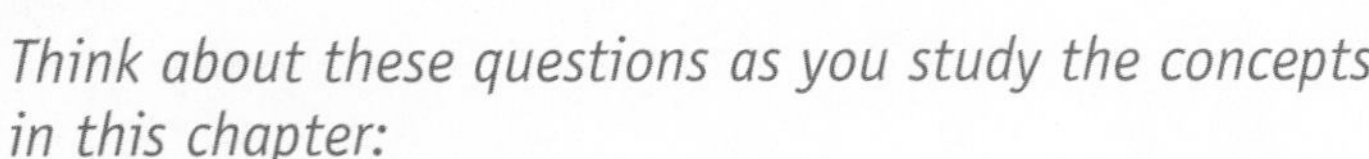

1. What are the components of an electrical circuit?
2. What are schematics, and how do engineers use them?

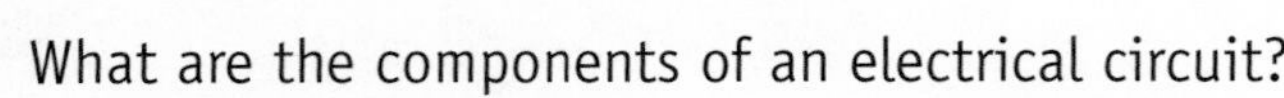

3. Why are fuses and circuit breakers wired in series?
4. What is the difference between a series circuit and a parallel circuit?
5. How do engineers use electrical circuits to help humankind?

© EMIN KULIYEV/SHUTTERSTOCK.COM

Engineering in Action

Gaze outside your home at night. What do you see? Street lights? House lights? Airplanes in the sky? None of these sights would have been possible without engineers controlling the flow of electrons. Through years of investigation and experimentation, scientists, technicians, and engineers have developed techniques to best control electricity and use this form of energy to help humankind.

Have you ever looked around when you were in a hospital or a medical doctor's office? Devices that use electricity are everywhere. These electrical devices make the work of doctors and nurses much easier. They also improve your health care. The basic understanding of how electrical circuits function and can be controlled is at the heart of these benefits to humankind.

SECTION 1: BASIC ELECTRICAL CIRCUITS

Introduction

Just as travelers use road maps to plan their trips, engineers use maps of the pathways that electrons will take on their journeys through electrical components. The route that electrons follow is termed a circuit. A typical electrical circuit contains the **pathway**, a **power source**, and a load device.

circuit
A circuit is an electrical pathway in which electrons travel.

load device
A load device is a component of an electrical circuit that consumes power.

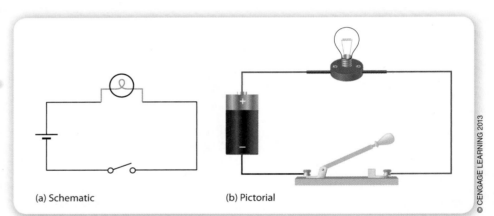

The resulting map, as shown in Figure 15-1, is called a **schematic** and uses symbols to represent electrical components along the pathway. The schematic in Figure 15-1a shows a battery, a light, and a switch. The corresponding electrical components are shown in Figure 15-1b. Figure 15-2 shows the symbols used for pathways, load devices, and power sources in schematic drawings, along with the pictures of their corresponding components.

(a) Schematic

(b) Pictorial

FIGURE 15-1 (a) The schematic diagram and (b) the corresponding pictorial diagram both show a battery, a light, and a switch connected in a certain sequence.

A set of standards governs the schematic symbols used for electrical circuits. The standards for electrical schematics were developed by the same agency that established the guidelines for dimensioning on engineering drawings, the American National Standards Institute (ANSI). The institute was originally formed in 1918 when five engineering societies and three government agencies merged.

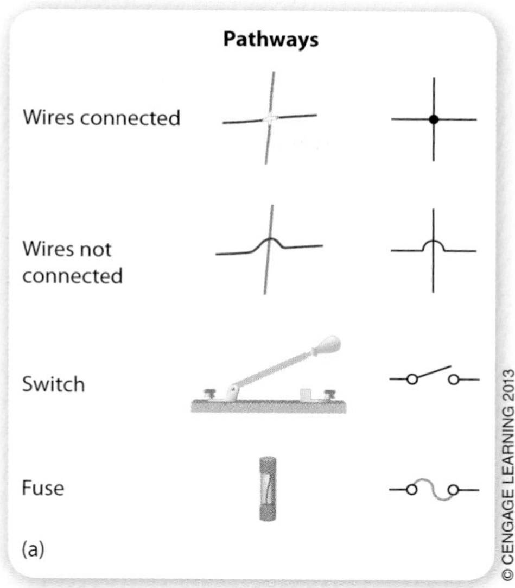

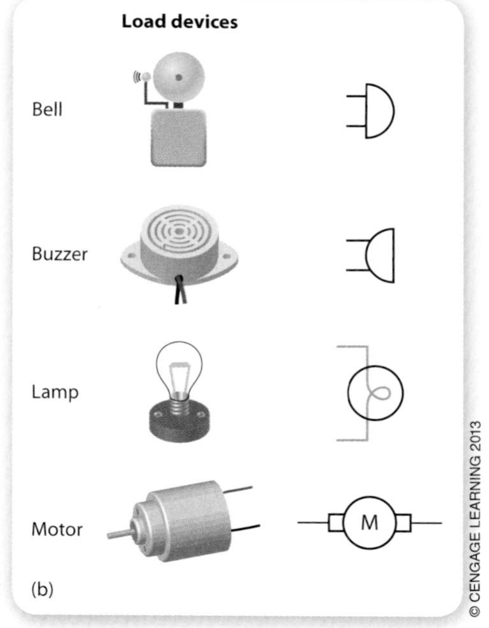

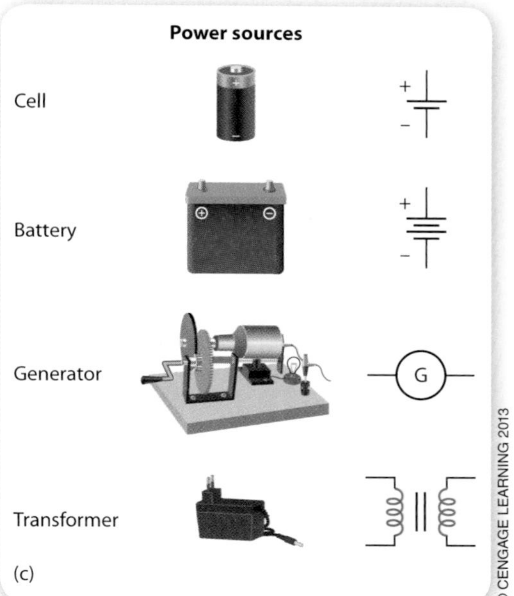

FIGURE 15-2 **Commonly used schematic symbols and their corresponding components.**

We typically use metal wire, a good conductor, as the pathway (see the section on conductors in Chapter 14). Usually, this metal wire is covered with an insulator such as plastic. Another type of pathway is a printed circuit (PC) board (see Figure 15-3). These PC boards have the pathway printed in copper on a plastic surface. PC boards make it easy for an electrician to assemble the electrical devices.

Most circuits also contain a **switch** that controls the flow of electrons along the pathway. Another component that may be included in the

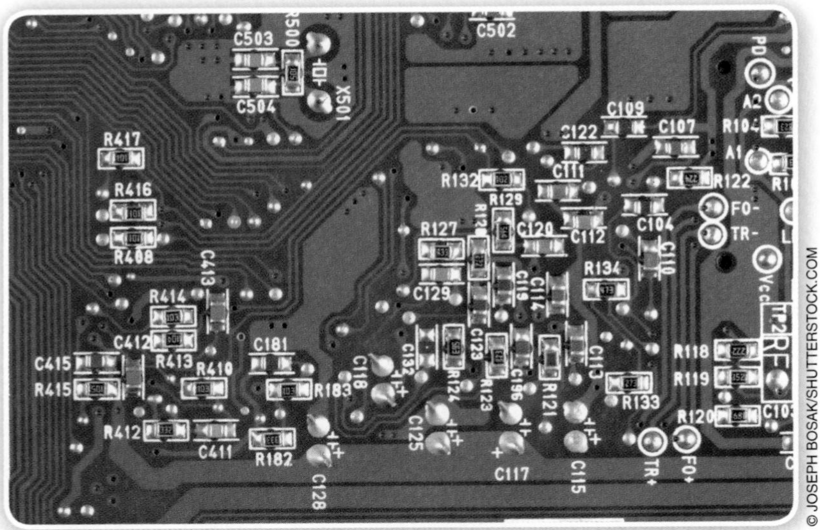

FIGURE 15-3 A printed circuit board.

circuit breaker
A circuit breaker is an electrical device designed to open when a current flow exceeds a limit.

pathway is a **fuse** or circuit breaker. Fuses and circuit breakers are safety devices. When the flow of electrons (amperage) exceeds the value of the fuse, then the extra current flow causes the fuse to overheat and melt, thus breaking the circuit. In a circuit breaker, the extra current flow initiates an electromagnet that opens a switch and interrupts the flow of electrons.

Power sources provide electron flow for the circuit. Voltaic cells and batteries are the simplest power sources. Generators and transformers are also power sources.

Load devices resist electron flow and perform work. These devices include lights, motors, buzzers, and anything else that transfers electrical energy into another form of energy (mechanical, heat, light, etc.).

If a circuit is open or not connected, we call it an *open circuit* (see Figure 15-4). Because of this opening in the circuit, the electrons cannot flow along the pathway and the load devices do not function. We call a circuit that does not contain a load device or has a "shortcut" past the load device—and thus no resistance in the circuit—a **short circuit**. Without resistance in the pathway, the electrons flow very fast, which causes the pathway to overheat (see Figure 15-5).

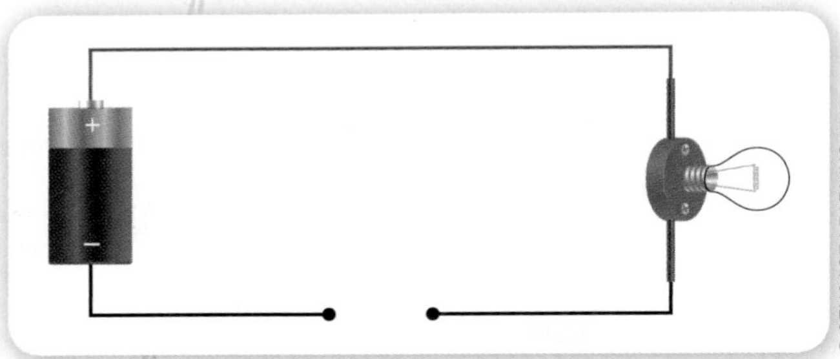

FIGURE 15-4 An open circuit.

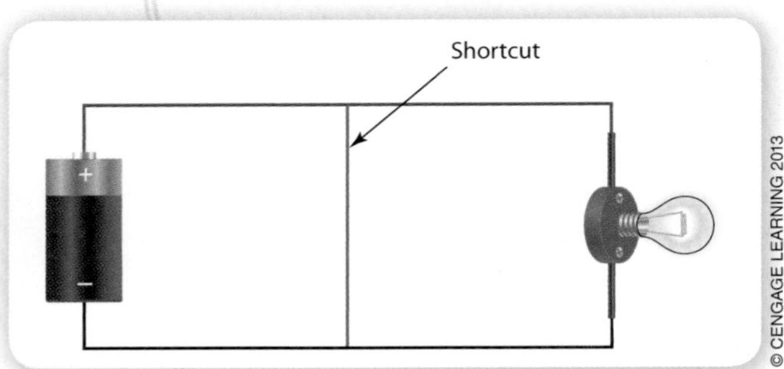

FIGURE 15-5 A short circuit.

Science in Engineering

As noted, electrical pathways are typically constructed with metal wire. Copper is commonly used as electrical wire because of its atomic number of 29. Copper has one valence electron, which makes it an excellent conductor. Silver, atomic number 47, is actually a better conductor than copper because its valence electron is farther from the nucleus. However, scientists and engineers have found that copper is more flexible and easier to work with than silver and is less expensive. That is why copper wire is used instead of silver wire in electrical circuits. In addition, the length and diameter of the metal wire can each affect the amount of electron flow.

SECTION 2: SERIES CIRCUIT

Construction

A **series circuit** is an electrical circuit in which the electrons have only one pathway to follow (see Figure 15-6). The electrons must flow through each load device in the circuit. For this reason, the current flow (amperage) is the same at any point in the circuit. Another characteristic of a series circuit is what happens when a load device fails (e.g., a lightbulb burns out) or is removed. If one load device stops functioning, all the devices in the series circuit stop. Imagine riding your bicycle down a one-lane path. If the riders ahead of you slow down, you will also slow your bicycle. If the group you are riding with speed up its pace, you will also peddle faster. If someone's bicycle breaks down and stops, everyone will be forced to stop. The same is true for electrons flowing along a series circuit.

series circuit
A series circuit is an electrical circuit that contains only one path for current to flow.

A fuse or circuit breaker is wired into a series. These devices are used to ensure circuit safety. If too many electrons try to travel the pathway, they will interrupt the route. A fuse is a thin piece of metal or wire. The smaller the size of the fuse material, the fewer electrons can pass through it. When too many electrons try to go through the fuse material, it overheats and then actually melts. With the fuse material melted, electrons can no longer flow, so the circuit is interrupted. A burned-out fuse must be discarded.

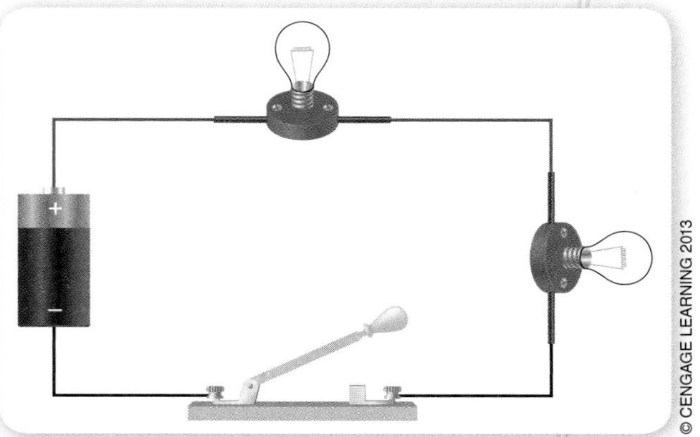

FIGURE 15-6 A series circuit with two lightbulbs.

The advantage of a circuit breaker is that it can be reused. The circuit breaker is a simple switch. On the circuit breaker, an electromagnet opens or closes the switch. We discussed electromagnets in Chapter 14. If too many electrons pass through the electromagnet, it develops additional pull and stops the flow of electrons. When either the fuse or the circuit breaker is triggered by excessive electrical current, electron flow is stopped throughout the entire circuit (see Figure 15-7).

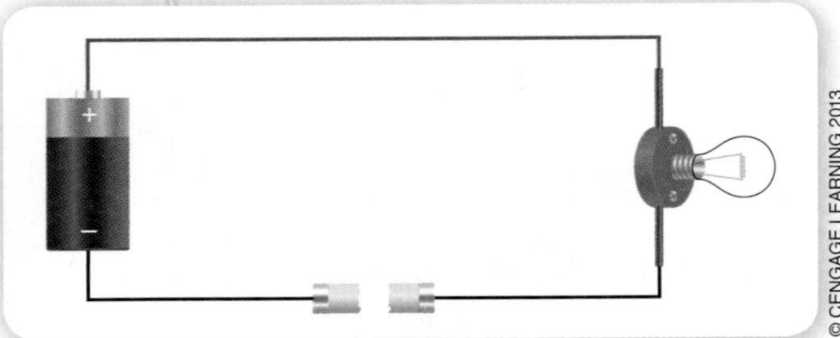

FIGURE 15-7 A series circuit with a burned-out fuse.

Characteristics

In a series circuit, the current flow (amperage) is equal across each load device in the circuit. In Figure 15-6, the current flow across each lightbulb is the same, but the voltage is divided between the bulbs. Each lightbulb receives only one-half of the circuit's total voltage. We call this decrease in voltage along the series circuit **voltage drop**. The voltage drop across each load device is **proportional** to the resistance of that load device. Because the lightbulbs are of equal resistance, the voltage across each is the same. Figure 15-8 shows a series circuit with three identical lightbulbs. In this series circuit, each bulb receives one-third of the total voltage.

Examine Figure 15-9. Even though the load devices are different with unequal resistances, the amperage across each load

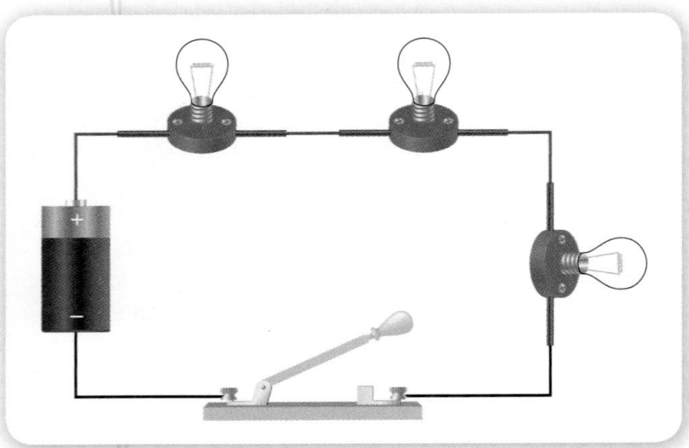

FIGURE 15-8 A series circuit with three lightbulbs.

proportional
Proportional is the quality that, given two values in a ratio, a change in one value will cause a corresponding change in the other value, keeping the same ratio. If the first value increases, then the second value also increases in the same ratio.

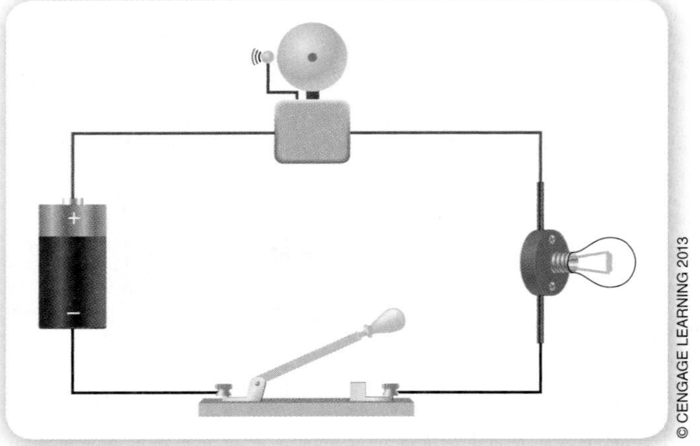

FIGURE 15-9 A series circuit with a lightbulb and a bell.

Math in Engineering

Engineers can use a mathematical formula to calculate the value of any one resistor in a series circuit, or they can calculate the total resistance of the series circuit.

The mathematical formula for series circuit resistance is:

$$R_T = R_1 + R_2 + R_3 + R_4$$

device is the same. However, the voltage for each load device is different and is based on its resistance. The resistance of the entire series circuit pictured in Figure 15-9 is the sum of the resistance of the lightbulb and the bell.

The characteristics of a series circuit are:

- The current flow (amperage) is equal throughout the circuit.
- The total circuit resistance is equal to the sum of the resistances of all the load devices.
- The voltage across each load device is proportional to that load device's resistance.

Engineering Challenge

ENGINEERING CHALLENGE 1

Ohm's law (discussed in Chapter 14) allows engineers to calculate values in a series circuit.

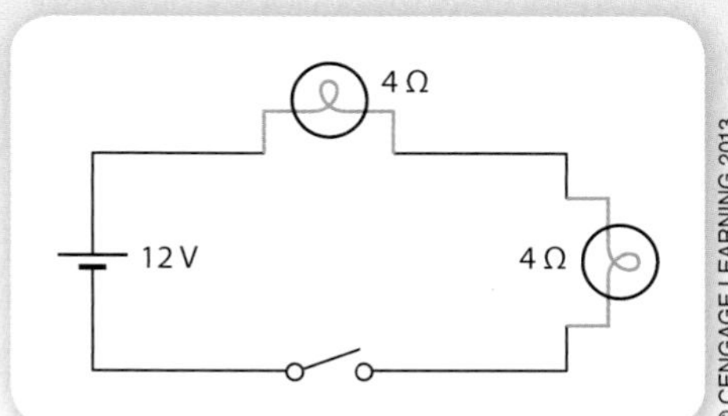

A schematic of Figure 15-6.

The schematic here is for the series circuit in Figure 15-6. The schematic indicates that the power source has 12 volts and that each lightbulb has a resistance of 4 ohms. Using Ohm's law, solve the following problems:

- What is the voltage at each lightbulb? Remember that a series circuit has voltage drop. (Hint: Because the lightbulbs have equal resistance, divide the total voltage by the number of lightbulbs.)
- What is this series circuit's total resistance? (Hint: Add the resistances of all lightbulbs.)
- What is the series circuit's total current flow (amperage)? (Hint: Use the total resistance you just calculated with the total voltage of 12 volts.)

Engineering Challenge

ENGINEERING CHALLENGE 2

The schematic here is for the series circuit in Figure 15-8. The schematic indicates that the power source has 12 volts and that each of the three lightbulbs has a resistance of 4 ohms.

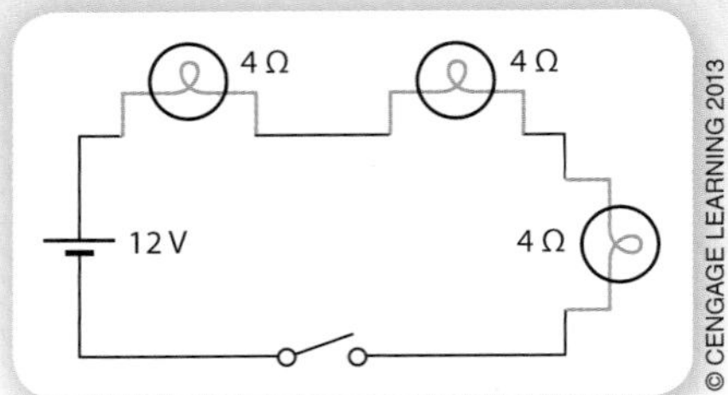

A schematic of Figure 15-8.

Using Ohm's law, solve the following problems:

- What is the voltage at each lightbulb? Remember that a series circuit has voltage drop. (Hint: Because the lightbulbs have equal resistance, divide the total voltage by the number of lightbulbs.)
- What is this series circuit's total resistance? (Hint: Add the resistances of all lightbulbs.)
- What is the series circuit's total current flow (amperage)? (Hint: Use the total resistance you just calculated with the total voltage of 12 volts.)

Engineering Challenge

ENGINEERING CHALLENGE 3

The schematic shown here is for the series circuit in Figure 15-9. The schematic indicates that the power source has 12 volts, the lightbulb has a resistance of 4 ohms, and the bell has a resistance of 8 ohms. Using Ohm's law, solve the following problems:

- What is this series circuit's total resistance? (Hint: Add the resistances of the lightbulb and the bell.)
- What is the series circuit's total current flow (amperage)? (Hint: Use the total resistance you just calculated with the total voltage of 12 volts.)

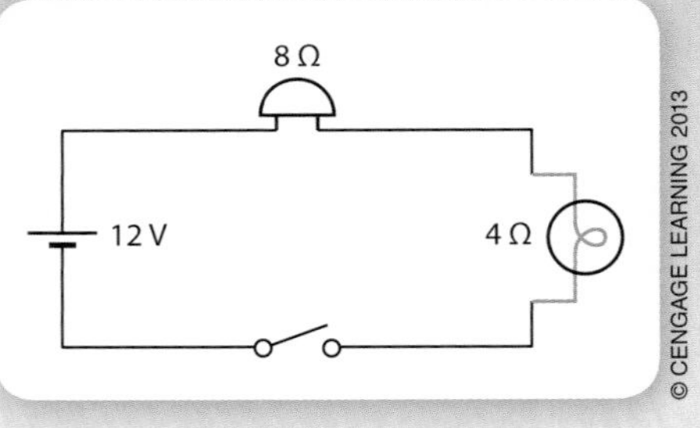

A schematic of Figure 15-9.

- What is the voltage at the lightbulb and at the bell? Remember that a series circuit has voltage drop. (Hint: The current flow in a series circuit is equal throughout the entire circuit, so you can use the total circuit's current flow and each load device's resistance to determine the voltage at that load device.)

SECTION 3: PARALLEL CIRCUIT

Construction

A **parallel circuit** is an electrical circuit in which the electrons have multiple pathways to follow. As shown in Figure 15-10 the electrons can travel through either lightbulb. If one of the lightbulbs burns out, the other bulb will still be on. Imagine that you are riding your bicycle in a city and you come to an intersection. You can ride straight ahead or turn right or left (multiple pathways). If the streets on your left and straight ahead are crowded, then you will probably turn right and take the path of least resistance. Electrons make the same decision. Electrons follow the path of least resistance.

parallel circuit
A parallel circuit is an electrical circuit that contains multiple pathways for current to flow.

Characteristics

The parallel circuit provides electrons with many options. Because the electrons follow the path of least resistance, the flow of electrons (amperage) varies across each load device. Amperage in a parallel circuit is proportional to the resistance of the load device. However, the electron pressure (voltage) is the same at each load device. Therefore, voltage in a parallel circuit is equal throughout the circuit.

Another interesting characteristic of the parallel circuit is that the resistance of the total circuit is less than the sum of the resistances of all the load devices together. This lower total resistance is a result of providing the electrons with more pathways to follow. Each additional load device in a parallel circuit provides another route for the electrons. The less the congestion of electrons, the less the resistance along the pathway. Electrical engineers can use this knowledge as they design electrical devices.

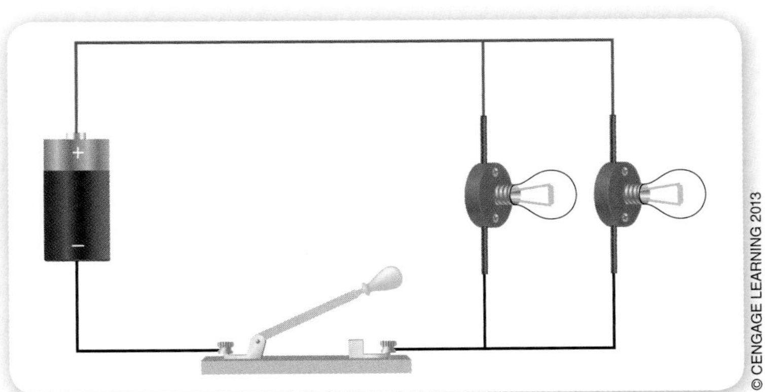

FIGURE 15-10 A parallel circuit with two lightbulbs.

Math in Engineering

STEM

Engineers can use a mathematical formula to calculate the value of any one resistor in a parallel circuit or to calculate the total resistance of the parallel circuit. The mathematical formula for parallel circuit resistance is:

$$\frac{1}{R_T} = \frac{1}{R_1} + \frac{1}{R_2} + \frac{1}{R_3}$$

The characteristics of a parallel circuit are:

- The voltage is equal throughout the circuit.
- The current flow (amperage) in each circuit branch is proportional to the resistance of the load device in that branch.
- The total resistance of the circuit is less than the sum of all the load devices' resistances.

Engineering Challenge

ENGINEERING CHALLENGE 4

Ohm's law (Chapter 14) allows engineers to calculate values in a parallel circuit.

This schematic shows the parallel circuit in Figure 15-10. The schematic indicates that the power source has 12 volts and that each lightbulb has a resistance of 4 ohms. Using Ohm's law, solve the following problems:

- What is the voltage at each lightbulb? (Hint: In a parallel circuit, voltage is equal throughout the circuit.)
- What is the current flow (amperage) at each lightbulb? (Hint: Use each bulb's resistance and the voltage.)
- What is the total parallel circuit's current flow (amperage)? (Hint: Add the current flow across each lightbulb.)
- What is this parallel circuit's total resistance? (Hint: Use the voltage and the total circuit's current flow.)
- How does the parallel circuit's total resistance compare to the resistance of each lightbulb?

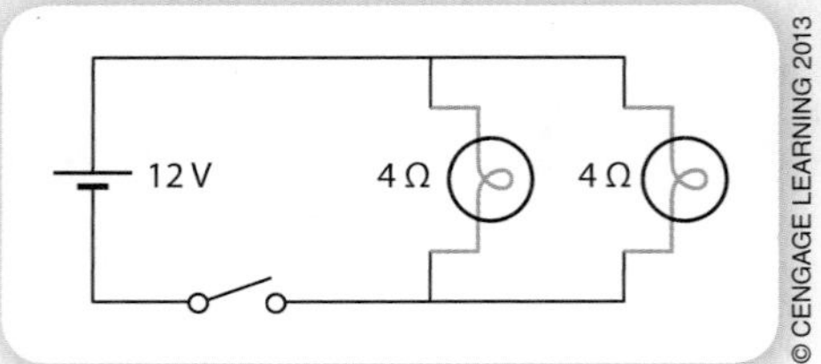

A schematic of Figure 15-10.

Career Spotlight

Name:
Jennifer Maffre

Title:
Senior Engineer, Advanced Research and Development, Biosense Webster Inc., a Johnson and Johnson Company

Job Description:
Maffre comes up with engineering ideas to help doctors treat their patients. "I get to work with physicians from all walks of life—from oncologists to obstetricians to heart specialists," she says. She also gets to travel a lot, meeting important doctors all over the world.

Maffre decided on an engineering career after her mother was diagnosed with cancer. It had not been caught early enough because the medical technologist had made mistakes. Maffre thought she would become a medical technologist herself until someone told her that machines would soon do their work for them. "I said, 'Who makes the machines?' And they said, 'The engineers do.' So that's what I decided to do."

Doctors tell Maffre what they need, and she dreams up solutions. For example, she designs devices to treat patients with abnormal heart rhythms. She works with a system that provides a 3-D image of the heart to identify trouble spots.

Maffre also helped develop a system for treating the foot sores of people with diabetes. She created a vial that contains a cream for the affected areas. "When you open the vial, it causes a chemical reaction, and the stem cells in the cream are activated," she says.

Education:
Maffre graduated from Rensselaer Polytechnic Institute with a bachelor's degree in electrical engineering and biomedical engineering. While studying at Rensselaer, she participated on the winning Ultra Rip Top challenge in her senior engineering-manufacturing design lab. Her team designed a top that could spin for 30 seconds. It was primarily assembled by a robotic arm that she programmed.

Advice to Students:
Maffre is an energetic advocate for the Hispanic community and an active member of the Society of Hispanic Professional Engineers.

"Don't be deterred if engineering is what you really want to do," she says. "With my background, you wouldn't think I would be an engineer. My dad drove a Yellow Cab and my mom was a hairstylist. It's easy to get caught up in jobs that your family has done. But if you're really inspired to be an engineer, it opens up a lot of doors to do worthwhile things."

SECTION 4: COMBINATION CIRCUIT

Construction

We call an electrical circuit that contains both a series circuit and a parallel circuit a **combination circuit**. As shown in Figure 15-11, the two lightbulbs are in parallel to each other. Both bulbs are connected in series with the bell. Electrical engineers use this combination of the characteristics of both a series circuit and a parallel circuit to design electrical systems.

combination circuit

A combination circuit is an electrical circuit that contains both series and parallel circuits.

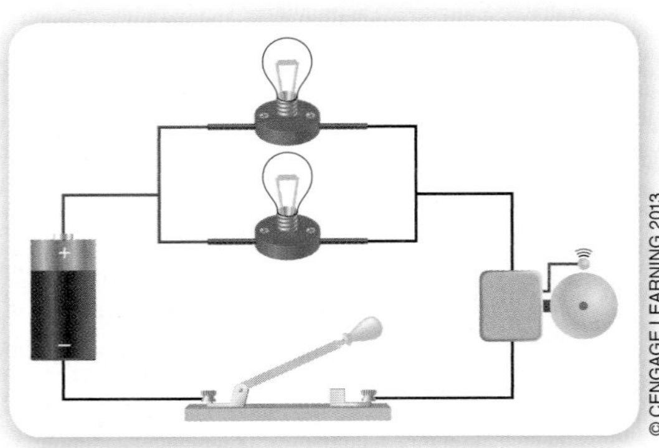

FIGURE 15-11 A combination circuit.

Characteristics

A combination circuit takes the characteristics of both the series and parallel circuits and combines them to accomplish the desired result. Most circuits constructed in today's electrical devices use the combination circuit.

SUMMARY

In this chapter you learned:

- Schematics are symbolic maps of electrical circuits.
- Electrical circuits contain a pathway, a load device, and a power source.
- Electrons follow the path of least resistance.
- The current flow in a series circuit is equal throughout the circuit.
- The voltage in a parallel circuit is equal throughout the circuit.
- Fuses and circuit breakers are safety devices used in electrical circuits.

VOCABULARY

Write a definition for each term in your own words. After you have finished, compare your answers with the definitions provided in this chapter.

Circuit
Load device
Circuit breaker
Series circuit
Proportional
Parallel circuit
Combination circuit

STRETCH YOUR KNOWLEDGE

Please provide thoughtful, written responses to the following questions.

1. Describe why electrical design engineers must understand the characteristics of different electrical circuits.
2. Compare and contrast the advantages of a series circuit and a parallel circuit.
3. Is it important to develop your mathematical skills in order to pursue a career in electrical engineering? Explain your thinking.

Onward to Next Destination

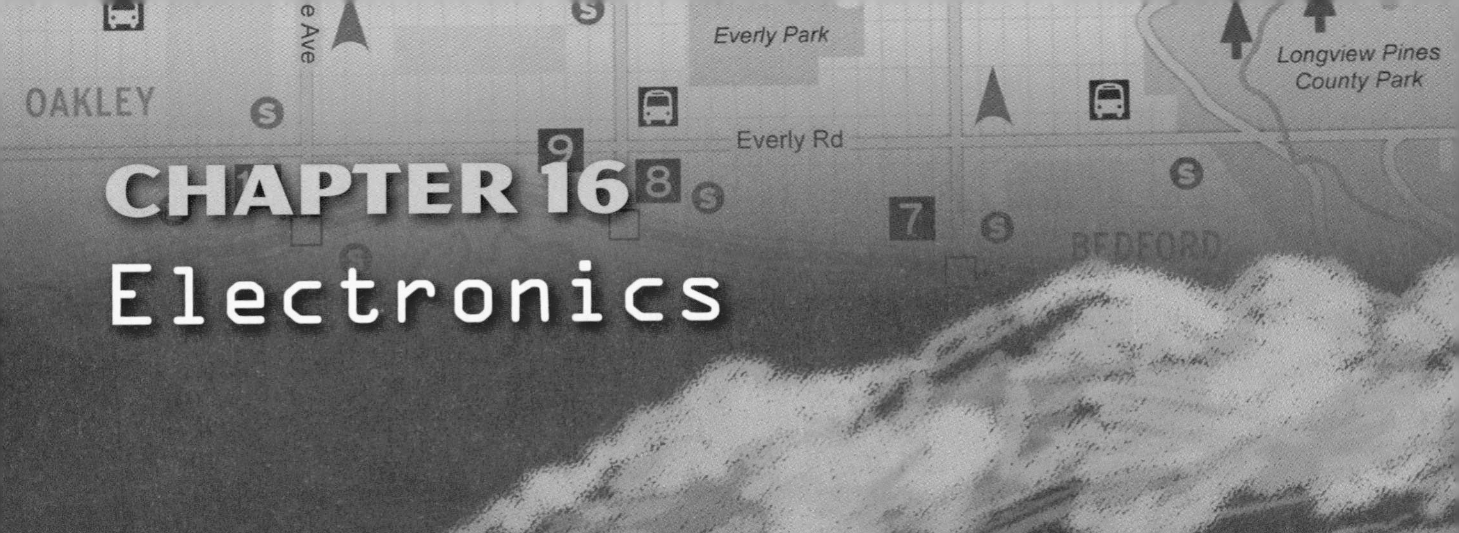

CHAPTER 16
Electronics

Menu

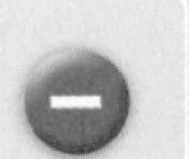

Before You Begin

Think about these questions as you study the concepts in this chapter:

1. What is electronics?
2. How do electronic devices, such as transistors, work?
3. What is meant by *digital electronics*?
4. What are logic gates, and how do they work?
5. What are integrated circuits?
6. How do we use electronics to improve our daily lives?

Engineering in Action

Did you play a portable electronic game last night? Did this electronic device produce sounds? Was there a video screen? You probably answered "Yes" to all these questions. But can you explain how this electronic game works? The use of electronic devices has become an integral part of our daily lives. Our homes, automobiles, appliances, and just about every other item we can name rely on electronics. Very few people, though, can explain how these devices actually operate. Designing and servicing electronic devices, such as your portable electronic game, require a knowledge of electronics.

SECTION 1: ELECTRONIC DEVICES

Electronics

transistor

A transistor is a device made by combining three semiconductors together. It uses a small amount of current to control a larger amount of current flow.

electronics

Electronics is the study and control of the flow of electrons, typically using low voltage.

doping

Doping is the addition of impurities to a semiconductor to modify its electronegativity.

When the transistor was invented at Bell Laboratories in 1947, it was one of the greatest inventions of its time (Figure 16-1). This small device helped change the world. Radios that were the size of our current televisions could now be handheld and portable.

A **transistor** is a device composed of materials called **semiconductors**. Semiconductors have the characteristics of both conductors and insulators. It was the transistor that started the digital revolution and the field of electronics. **Electronics** is the study and use of electrical devices whose operation is controlled by the flow of electrons. Typically, electronics use very small amounts of voltage to perform this controlling operation.

Silicon (Figure 16-2) is a common semiconductor. Silicon has an atomic number of 18. When silicon atoms combine, they form a lattice structure (Figure 16-3). Engineers use a process termed **doping** to add impurities to semiconductors such as silicon to develop the desired electrical properties. Some doping increases the number of electrons in the semiconductor and thus produces an **N-type semiconductor**. (The *N* stands for "negative.") Another doping process removes electrons and produces a **P-type semiconductor**. (The *P* stands for "positive.")

Scientists add small amounts of phosphorus to the silicon for N-type doping. Figure 16-4 shows the phosphorus atom. Phosphorus has an atomic number of 15 and has five valence electrons. Silicon has four electrons in its outer orbit. So when a phosphorus atom is placed in the silicon lattice,

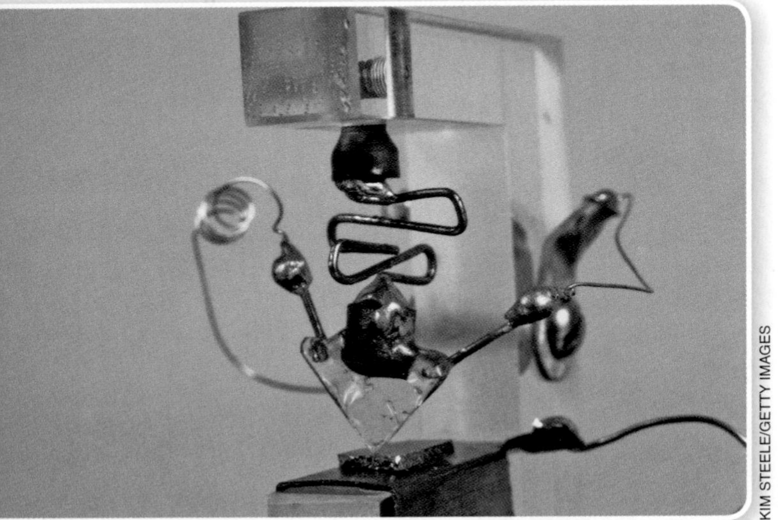

KIM STEELE/GETTY IMAGES

FIGURE 16-1 A replica of the first transistor, invented at the Bell Laboratories in 1947.

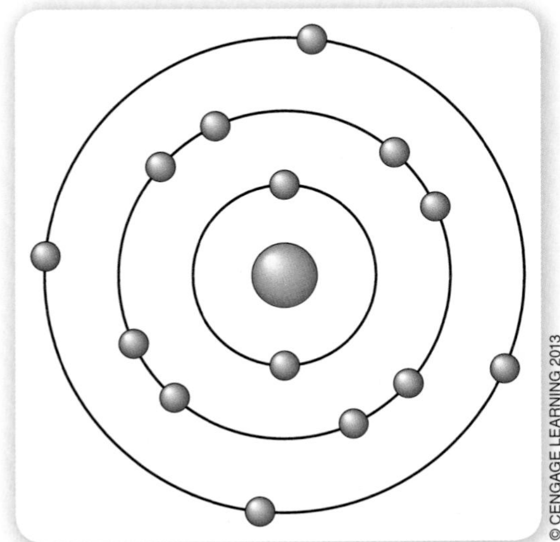

FIGURE 16-2 The atomic structure of silicon.

Science in Engineering

The investigation and experimentation of various types of materials, especially materials termed *semiconductors*, led to the development of electronics and the digital revolution. The semiconductor silicon has four electrons in its outer orbit. Silicon atoms placed together form a lattice structure. In this lattice structure (Figure 16-3), multiple silicon atoms team up to form full outer orbits with eight electrons. Doping adds impurities such as phosphorus or boron to the silicon.

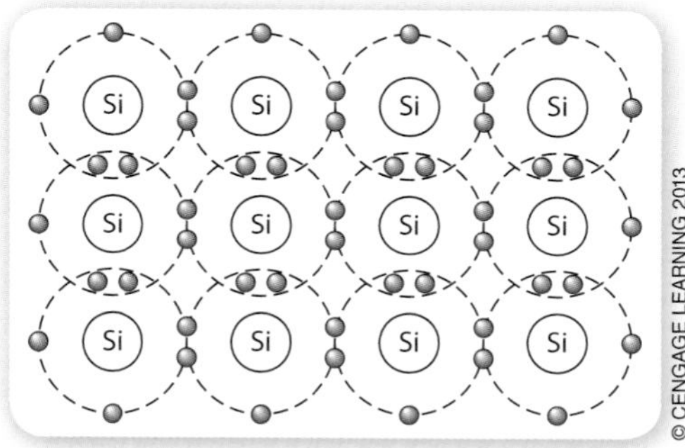

FIGURE 16-3 The lattice structure of silicon.

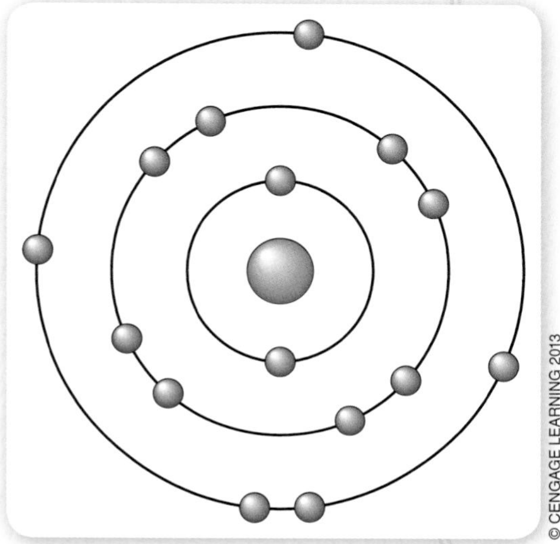

FIGURE 16-4 The atomic structure of phosphorus.

one of its five valence electrons is out of place. This fifth electron has nothing to bond to, so it is free to move. This free electron is what makes the phosphorus-doped silicon an N-type semiconductor.

Scientists add small quantities of boron to silicon in the P-type doping process. Boron has an atomic number of 5 (see Figure 16-5) and three valence electrons. Silicon has four electrons in its outer orbit, so the addition of a boron atom leaves a "hole" in the silicon lattice. This hole waits for an electron to fill it. This type of boron doping is called P-type because the material is short an electron.

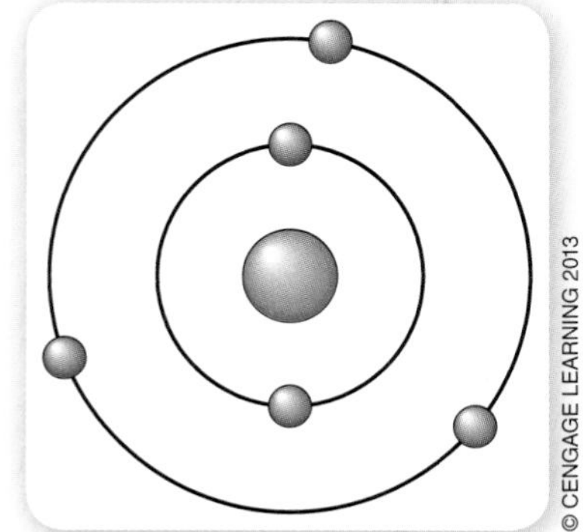

FIGURE 16-5 The atomic structure of boron.

Transistors

Transistors are the basic controlling or switching devices for today's electronics (Figure 16-6). They consist of three layers of semiconductor material. Transistors can be either an **NPN-type** or a **PNP-type transistor**. Figures 16-7 and 16-8 show these two types of transistors. The transistor has an electrical contact on each of the three layers. The center layer's contact is termed the **base**. The terminals on the two other layers are called the **collector** and the **emitter**.

2

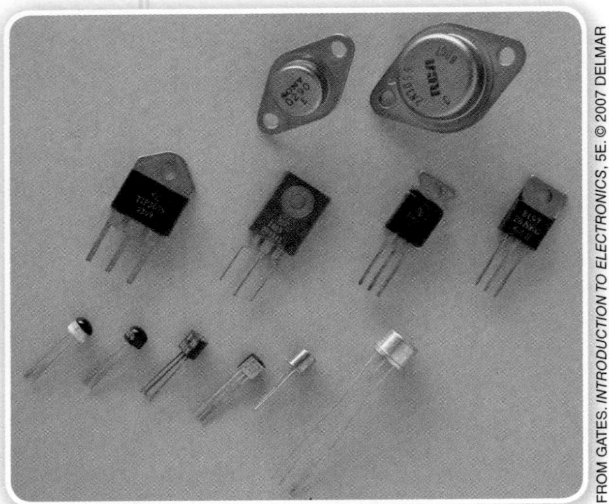

FIGURE 16-6 **Transistors of various shapes and sizes.**

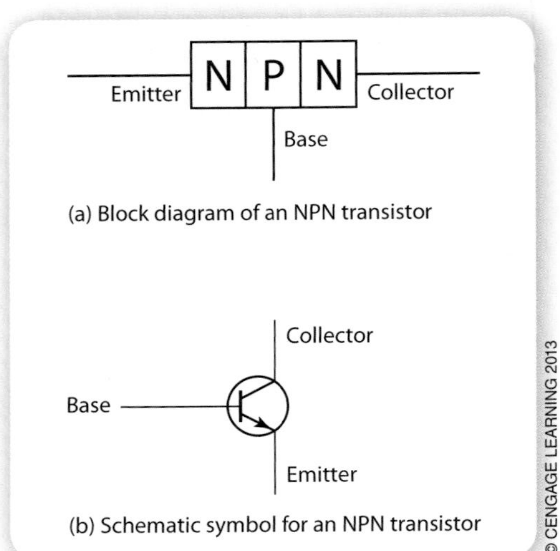

FIGURE 16-7 **A diagram of the NPN-type transistor and its schematic symbol.**

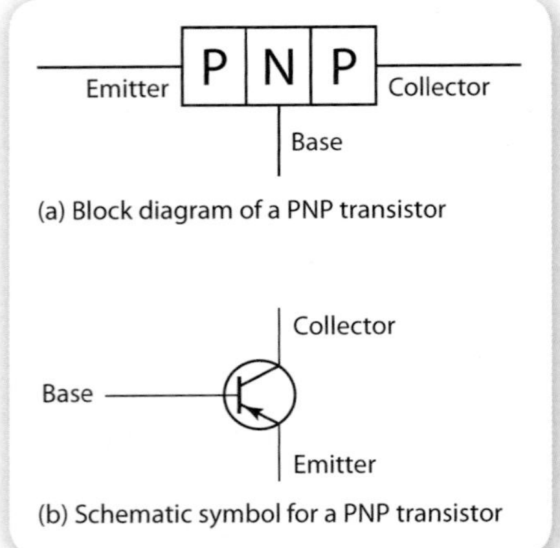

FIGURE 16-8 **A diagram of the PNP-type transistor and its schematic symbol.**

Electrons in the circuit "collect" at the collector terminal waiting for a signal. The signal is a small electric charge to the base terminal. This charge at the base of the transistor changes the conductivity of the base semiconductor, allowing the electrons to flow from the collector to the emitter and complete the electrical circuit. For an NPN-type transistor, a positive charge is required at the base terminal to open the circuit. For a PNP-type transistor, a negative charge is required at the base terminal to open the circuit.

Fixed Resistors

Resistors are some of the most common devices used in electronics. We use resistors to place opposition to the flow of electrons in the electrical circuit (we discussed resistance

Did You Know?

- The engineering team from Bell Laboratories was issued a U.S. patent (no. 02569347) in 1952 for its invention of the transistor.
- The Bell Laboratories engineering team was also awarded the Nobel Prize in physics in 1956 based on the impact of the transistor on society.

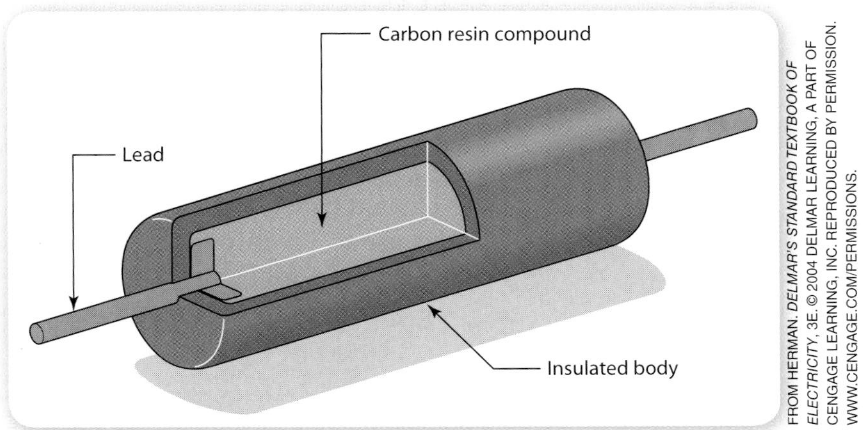

FIGURE 16-9 A fixed resistor.

in Chapter 14). Resistors convert electrical energy into heat energy, so heat is a by-product of using a resistor. Have you ever felt the back of your computer? You probably noticed that it is warm. This heat is a product of the resistors used in your computer. Resistors come in a variety of sizes, types, and resistance ratings to meet an engineer's needs in designing the electrical circuit.

The most commonly used type of resistor is the fixed resistor, in which the opposition to the flow of electrons provided by the resistor is fixed and cannot be adjusted. These fixed resistors are typicality made of carbon (Figure 16-9). The carbon resists the flow of electrons and converts the electrical energy

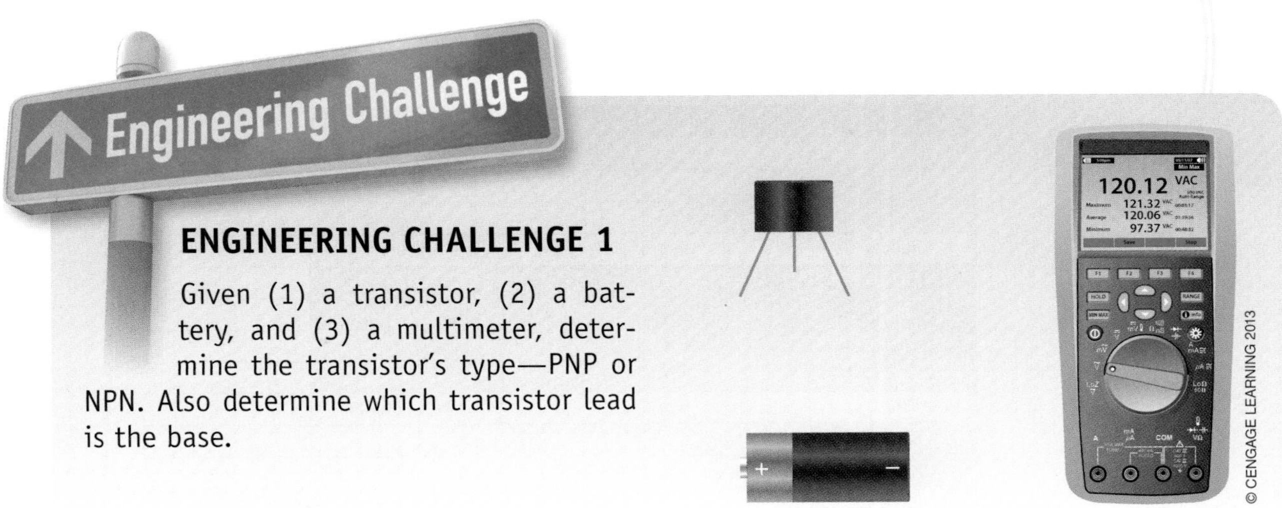

ENGINEERING CHALLENGE 1

Given (1) a transistor, (2) a battery, and (3) a multimeter, determine the transistor's type—PNP or NPN. Also determine which transistor lead is the base.

Decoding the transistor.

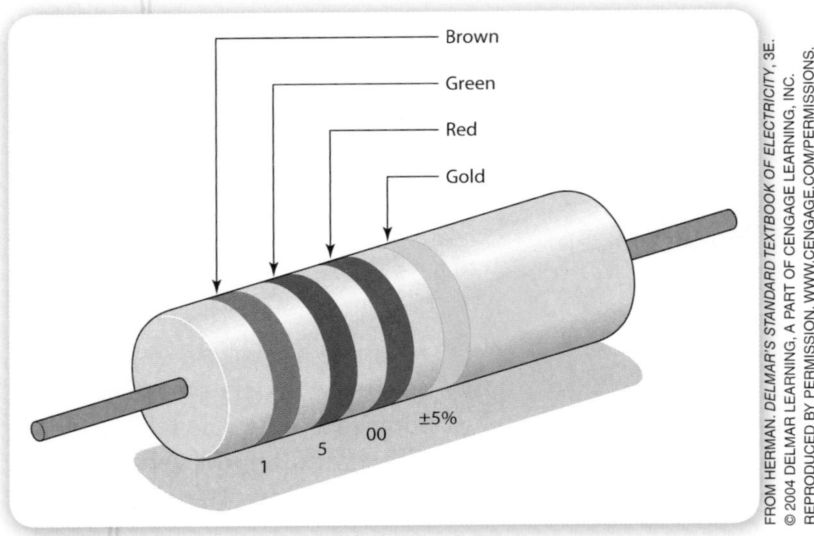

FIGURE 16-10 Decoding resistor color bands.

into heat energy. The value of a resistor—its resistance measured in ohms—is provided by the manufacturer using a color code.

Most resistors have four colored bands that an engineer can use to identify the resistor. The resistor in Figure 16-10 has a brown band, a green band, a red band, and a gold band. Bands 1, 2, and 3 provide the resistance value. Bands 1 and 2 are the first two digits in the resistance value. Band 3 is called the *multiplier*, and it indicates how many times the first two numbers are to be multiplied by 10. The fourth band indicates the **tolerance**, or range of the resistor. Tolerance is the amount of variation that is allowable from the established value. Figure 16-11 provides a table of the standard values for resistor color coding.

The resistor in Figure 16-11 thus has a value of 26 (red-blue) multiplied by 10,000 (yellow), or 260,000 ohms. The gold on the fourth band

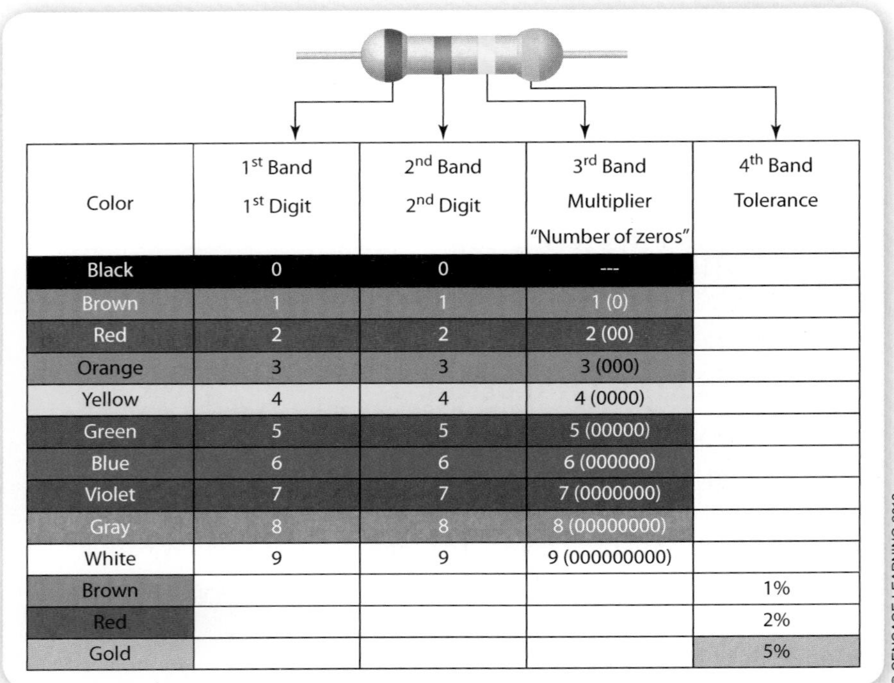

Color	1st Band 1st Digit	2nd Band 2nd Digit	3rd Band Multiplier "Number of zeros"	4th Band Tolerance
Black	0	0	---	
Brown	1	1	1 (0)	
Red	2	2	2 (00)	
Orange	3	3	3 (000)	
Yellow	4	4	4 (0000)	
Green	5	5	5 (00000)	
Blue	6	6	6 (000000)	
Violet	7	7	7 (0000000)	
Gray	8	8	8 (00000000)	
White	9	9	9 (000000000)	
Brown				1%
Red				2%
Gold				5%

FIGURE 16-11 Standard values indicated by resistor color bands.

tells the engineer the tolerance is five percent either above or below 260,000 ohms (273,000–247,000 ohms). Engineers use K to represent 1000 when measuring ohms. Therefore, we would note the resistor had a value of 260 KΩ.

Variable Resistors

Sometimes an engineer needs to be able to adjust the resistance in an electrical circuit. For this operation to be performed, a variable resistor is required. The potentiometer is a variable resistor that can be regulated by the operator or engineer. Figure 16-12 shows various types

FIGURE 16-12 Potentiometers of various shapes and sizes.

Engineering Challenge

ENGINEERING CHALLENGE 2

Engineers need to determine if electronic components such as resistors are serviceable for use in a circuit. That means they must ensure that the components meet their stated tolerances by calculating whether these components are manufactured within these tolerances. Given the measured readings taken on the fixed resistors in the table:

- Calculate each resistor's designated value.
- Calculate each resistor's tolerance range (highest acceptable value to lowest acceptable value).
- Determine whether each resistor is within its tolerance.

Measured value		1st Digit	2nd Digit	Multiplier	Calculated value	Tolerance %	Tolerance range
102 ohms							
1045 ohms							
53 ohms							
751 ohms							
337 ohms							

Calculating resistor values and tolerances.

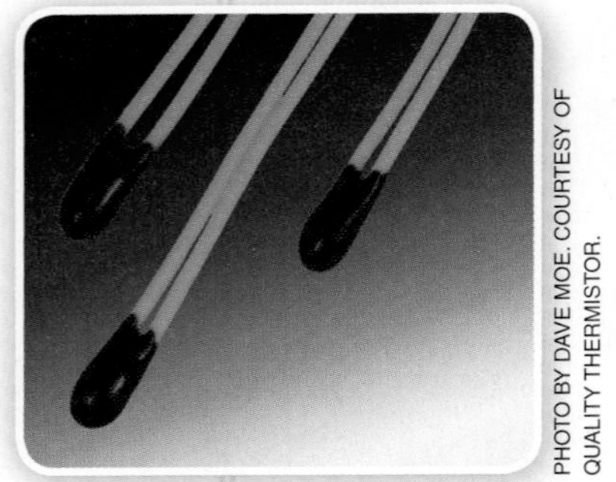

FIGURE 16-13 Thermistors of various shapes and sizes.

of potentiometers. By using a potentiometer, you can control the flow of electrons in a circuit. One example you may be familiar with is a dimmer switch used on ceiling lights. The dimmer switch is a potentiometer. It controls the flow of electrons to the lightbulb and therefore the brightness of the light. Engineers also use potentiometers to control the speed of electric motors.

Two other types of variable resistors use external conditions and not human input to control the electron flow of a circuit. The **thermistor** is a variable resistor that is regulated by temperature (Figure 16-13). Sometimes we call the thermistor a *temperature resistor*. When the temperature rises, electrons are able to flow with less opposition, so the resistance of the thermistor goes down. When the temperature gets colder, the resistance of the thermistor increases. Thermistors are used in electric thermometers and also in new electronic home thermostats. Thermistors are typically manufactured from semiconductors and metals such as manganese, copper, or nickel.

The **photoresistor** is another type of variable resistor. Its resistance depends on light. A well-lit area allows the electrons greater movement, so resistance decreases. When the environment is dark, the resistance of a photoresistor increases (Figure 16-14). Other names for the photoresistors are *photoconductor*, *photocell*, and *light-dependent resistor* (LDR). Photoresistors are also made from semiconductors. Photons from the light rays are absorbed by the semiconductors, freeing electrons and allowing current to flow.

Diodes

Like a transistor, a diode is an electronic device that uses P-type and N-type semiconductors. However, the diode has only two layers: one P-type and the other N-type. We call the area between the layers the *junction*. Because of this junction area, we sometimes call the diode a *PN junction diode*. This

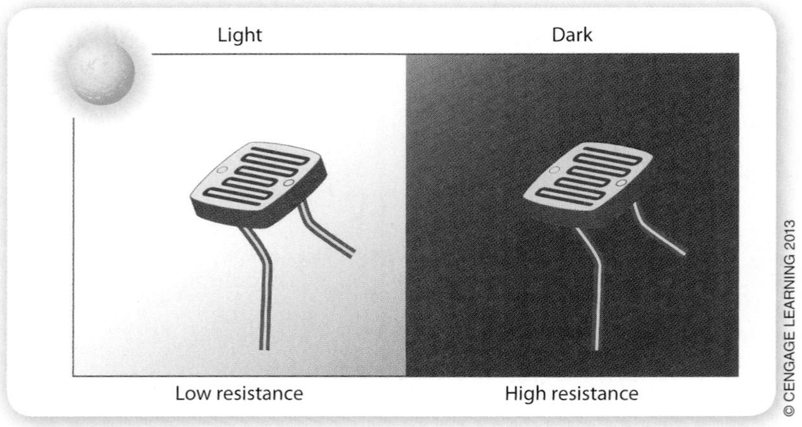

FIGURE 16-14 Photoresistors change values from light to dark.

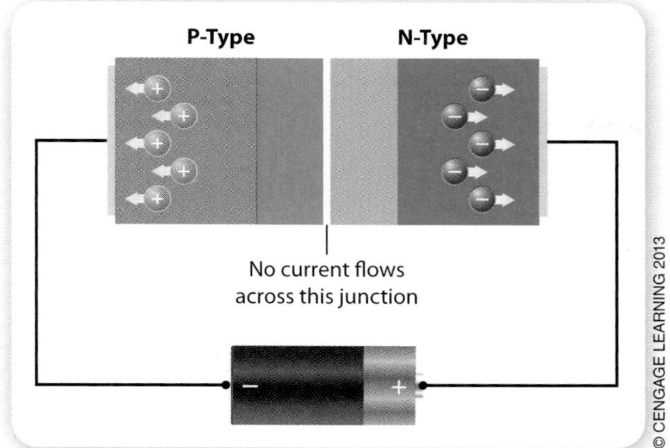

FIGURE 16-15 **Electrons are not flowing across this diode.**

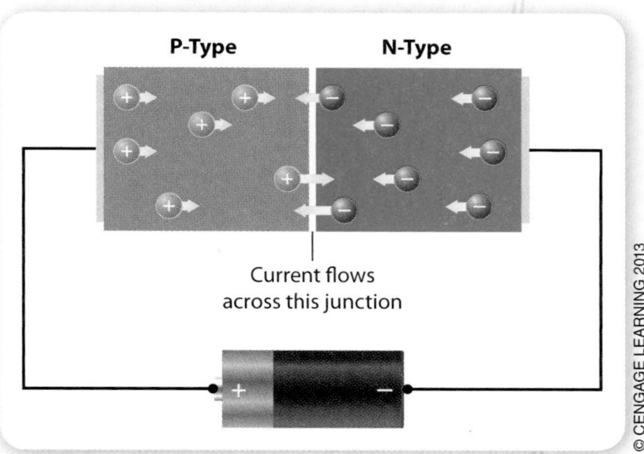

FIGURE 16-16 **Electrons are flowing across this diode.**

configuration allows the electrons to flow in only one direction through the diode. When the negative (N-type) terminal of the diode is connected to a positive current source, the electrons are attracted to the positive sources and the junction area is open. Likewise, when the positive (P-type) terminal of the diode is connected to a negative current source, the electrons are attracted to the negative sources and the junction area is open. Therefore, there is no current flow through the circuit (Figure 16-15). When the battery is reversed as in Figure 16-16, the electrons and protons are repelled into the junction area and current flows through the circuit.

There are many types of PN junction diodes. The most commonly used is the **light-emitting diode (LED)**. These come in various sizes and colors (Figure 16-17). The LED produces light when subjected to current flow; the electrons jump the junction area and convert electrical energy into light energy. Figure 16-18 shows the operation of an LED.

light-emitting diode (LED)
A light-emitting diode is an electronic device much like a small lightbulb that allows current to flow in only one direction.

Because the LED is a diode, connecting it to the correct positive or negative terminal is very important. Figure 16-19 notes two methods for determining how to connect the LED. The positive terminal is called the *anode* and has a longer lead. The negative side of the LED, termed the *cathode*, has a flat surface in the plastic cover. Because the LED does not have a filament that will burn out because of heat like the traditional incandescent lightbulb, the LED has an unlimited life span.

FIGURE 16-17 **Light-emitting diodes of various shapes and sizes.**

Capacitors

Sometimes engineers need to store electrical energy and then release it very quickly. We call an

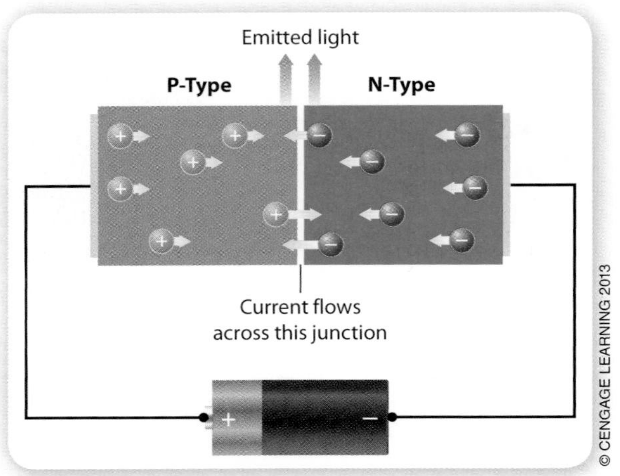

FIGURE 16-18 Electrons are flowing across this LED and causing it to light up.

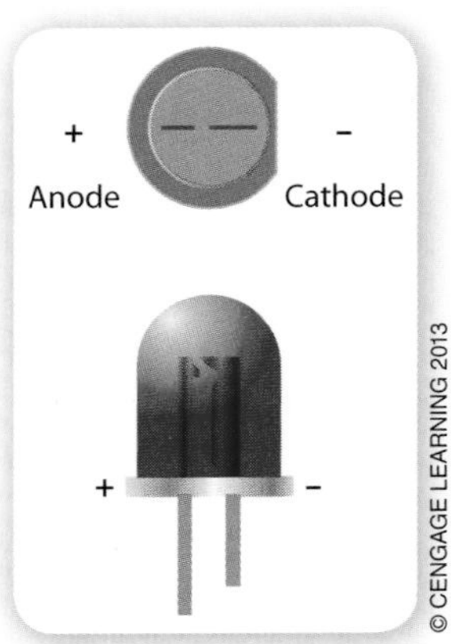

FIGURE 16-19 The anode and cathode terminals of an LED.

electronic device that allows such storage a **capacitor**. A capacitor is constructed by sandwiching two metal plates around an insulator plate. Capacitors come in various types and sizes, as shown in Figure 16-20. Most cameras use a capacitor to store electrons until they are needed for the flash.

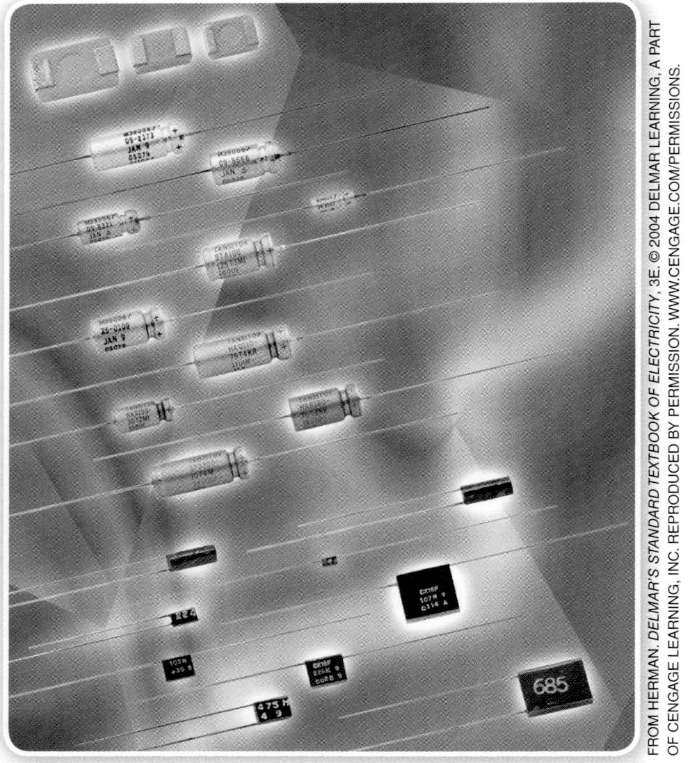

FIGURE 16-20 Capacitors of various shapes and sizes.

Science in Engineering

LEDs offer the lighting market a new and revolutionary lighting source that saves energy. Because LEDs operate on a lower voltage than regular incandescent lightbulbs, they use less electrical current, thereby costing less to operate and saving energy. In addition, LEDs last longer than incandescent lightbulbs because they do not burn out. Consumers will also be able to select LEDs that generate lighting with different colors. LEDs are one example of how science and technology help humans live better.

SECTION 2: DIGITAL ELECTRONICS

3

Digital Electronics

Before the invention of the transistor, all electrical devices operated under analog systems. These systems used inputs and outputs that could vary and were not always accurate. An example of an analog device is a rotary clock (Figure 16-21). Your school may have an old one. Most clocks today are digital, as are most electrical devices. **Digital electronics** is the use of inputs of either "on" or "off" to produce outputs of "on" or "off." Let us take a look at how digital electronics works.

digital electronics
Digital electronics is the use of digital logic to operate an electronic system.

Binary Number System

A number system is simply a code used to transfer information or numerical values. The number system that we use is based on the number 10 and is called the *decimal* system. We accomplish counting by using the numbers 1, 2, 3, 4, 5, 6, 7, 8, 9, and 10. We call the simplest number system the **binary system**. The binary system uses only two numbers, zero (0) and 1. The binary system is used in digital electronics because it is the simplest. Engineers can design decision-making circuits by selecting the binary numbers 0 or 1. Each 0 or 1 is called a **bit**. The term *bit* is a contraction of *b*inary dig*it*; each bit notes the memory location of the switch. A combination of eight bits is called a **byte**. This use of bits and bytes is codified and regulated by the American Standard Code for Information Interchange (**ASCII code**).

FIGURE 16-21 *A view of an analog clock.*

Math in Engineering

The decimal number system that we use every day contains 10 digits, 0 through 9. We sometimes call this number system the *base-10 number system*. The binary number system contains only two digits, 0 and 1. We call the binary system the *base-2 number system*.

Logic Gates

To use the binary numbers 0 and 1 in a decision-making circuit, an engineer uses a digital electronic device called a *logic gate*. **Logic gates** use inputs of 0 or 1 to signal outputs of 0 or 1. The binary numbers of 0 and 1 each indicate either "on" or "off." The number 1 represents an input or output of "on." The 0 notes "off."

$$1 = \text{On}$$
$$0 = \text{Off}$$

truth table

A truth table is a chart that represents the inputs and outputs for a desired outcome or electronic component.

The inputs and outputs of logic gates are shown on charts called **truth tables**. Each logic gate has its own truth table. There are four common types of logic gates: AND, OR, XOR, and NOT.

The schematic symbol for the **AND gate** is shown in Figure 16-22. The AND gate has two inputs; these are represented by the letters A and B. The A and B inputs can be either a 1 or a 0. The output (Q) is "on" or 1 only when both of the inputs are 1 or both are on. Engineers use truth tables to summarize the logic gate decision-making process. The truth table in Figure 16-23 graphically shows the AND gate's process.

The **OR gate** also has two inputs, A and B, and one output, Q (Figure 16-24). An OR gate produces an output (Q) of 1 when either of its inputs (A or B) is a 1. The truth table for the OR gate is provided in Figure 16-25.

A modified OR gate is the **XOR gate** or **exclusive gate** (Figure 16-26). The XOR gate uses two inputs (A and B) to generate its output (Q). To generate an output of 1 or "on," only one of the inputs can be a 1 or "on" (Figure 16-27).

The fourth common logic gate is termed the **NOT gate** (Figure 16-28). The NOT gate

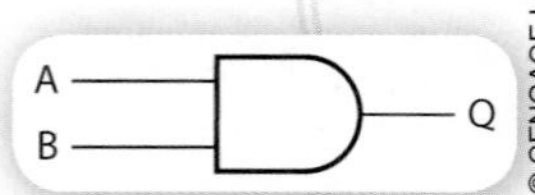

FIGURE 16-22 The schematic of the AND gate.

AND Gate		
Input	Input	Output
A	B	Q
0	0	0
0	1	0
1	0	0
1	1	1

FIGURE 16-23 The truth table of the AND gate.

only has one input (A). The NOT gate reverses the input as its output. An input of 1 generates an output of 0. An input of 0 produces an output of 1. Because of this action, the NOT gate is commonly called the *inverter gate*. Figure 16-29 shows the NOT gate truth table.

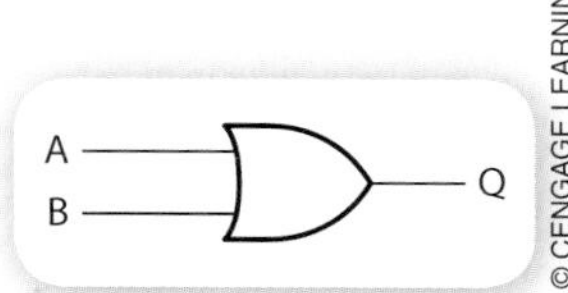

FIGURE 16-24 The schematic of the OR gate.

OR Gate		
Input	Input	Output
A	B	Q
0	0	0
0	1	1
1	0	1
1	1	1

FIGURE 16-25 The truth table of the OR gate.

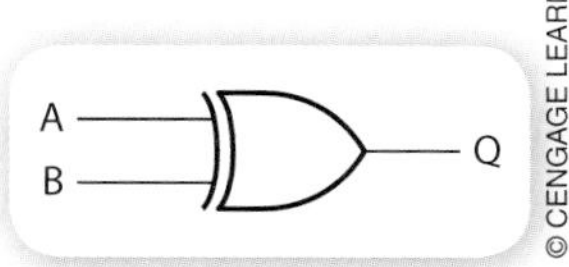

FIGURE 16-26 The schematic of the XOR gate.

XOR Gate		
Input	Input	Output
A	B	Q
0	0	0
0	1	1
1	0	1
1	1	0

FIGURE 16-27 The truth table of the XOR gate.

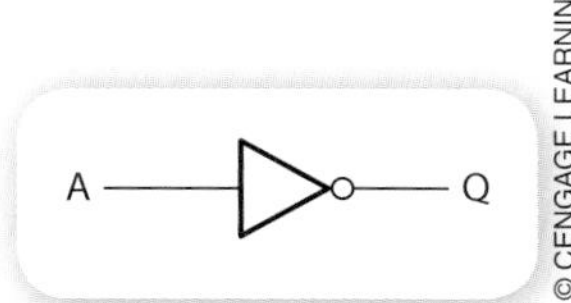

FIGURE 16-28 The schematic of the NOT gate.

NOT Gate	
Input	Output
A	Q
0	1
1	0

FIGURE 16-29 The truth table of the NOT gate.

Career Spotlight

Name:
Robert Noyce

Title:
Founder of Intel

Job Description:
Robert Noyce (1927–1990) is one of the inventors of the integrated circuit or microchip, a set of miniature electronic components etched onto a thin piece of silicon. Noyce's invention revolutionized the field of electronics and is used in almost all of today's electronic equipment.

Noyce invented the microchip while working at Fairchild Semiconductor, which he founded in the 1950s. In the 1960s, he founded Intel, where he helped develop another revolutionary device: the microprocessor. A microprocessor is a complete computation engine on a single chip. It is at the heart of today's computers.

Noyce established both Fairchild Semiconductor and Intel in the southern part of the San Francisco Bay Area. The location came to be called "Silicon Valley" because of all the high-tech businesses that concentrated there. Noyce was known as the "mayor" of Silicon Valley.

TIME & LIFE PICTURES/GETTY IMAGES

Noyce introduced a very casual working environment into his firms. This was not common at the time. So, in addition to his engineering genius with the integrated circuit, Noyce established our "casual Fridays."

Education:
Robert Noyce was born in a small Iowa town, where he enjoyed tinkering and figuring out how things work. When Noyce was attending Grinnell College, one of his professors received two of the first transistors that were developed at the Bell Laboratories. Noyce had to examine them and determine how they worked. His tinkering helped him gain valuable experience with these revolutionary electronic devices. He graduated with a bachelor's degree in physics and then attended the Massachusetts Institute of Technology for his Ph.D.

Impact on Society:
Robert Noyce had the same impact on the digital revolution as Henry Ford had on mass production and Thomas Edison had on the field of electricity.

SECTION 3: CONTROLLING YOUR WORLD

Integrated Circuits

The **integrated circuit (IC)** pictured in Figure 16-30 is an electronic device that incorporates an electrical circuit in a very tiny amount of semiconductor material. The circuit may include transistors, resistors, diodes, or capacitors. These components and their connections are etched into the semiconductor during the manufacturing process. Engineers design integrated circuits for specific purposes. We sometimes call this very small device a *chip*.

integrated circuit (IC)
An integrated circuit uses a single crystal chip of silicon that has all the components needed to make a functioning electrical circuit.

After the transistor, the integrated circuit was one of the most important inventions of the twentieth century. It would not have been possible without the invention of the transistor. Integrated circuits, chips, are produced in four common types: dual in-line packages (DIPs), pin-in grid arrays (PGAs), single in-line memory module (SIMMs), and single in-line packages (SIPs) (Figure 16-31).

FIGURE 16-30 Integrated circuits (IC) are used in computers.

We find integrated circuits in many different electronic devices including computers, cellular telephones, televisions, audio equipment, video recorders, and portable electronic games. The small size of the integrated circuit offers its greatest advantage. In addition, having complete circuits in one small unit reduces servicing costs for both consumers and industry. Integrated circuits can be produced as either a preprogrammed chip called a **microprocessor** or a memory chip that contains a blank memory. This option allows the engineer to select the chip for the desired process.

Typically, microprocessors contain more than one integrated circuit. You can find microprocessors in the central processing unit (CPU) of your computer.

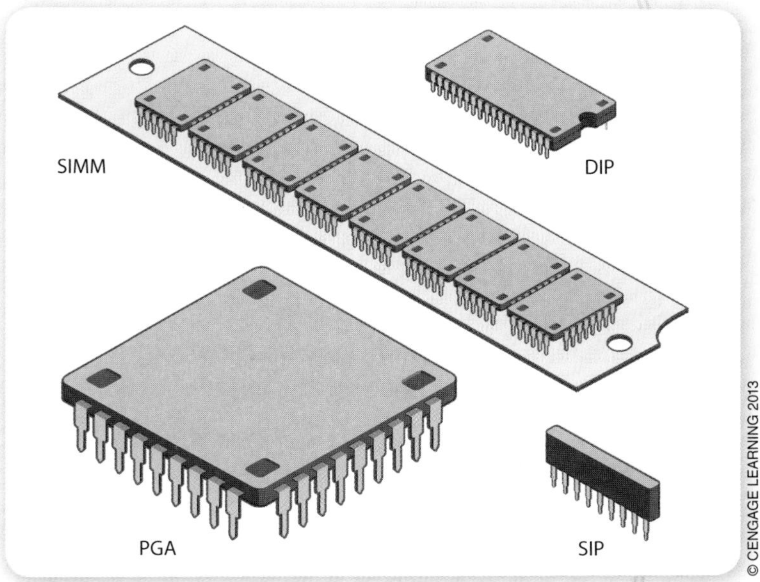

FIGURE 16-31 DIP, PGA, SIMM, and SIP integrated circuits.

Did You Know?

- Jack Kilby of Texas Instruments filed four U.S. patents—nos. 3138743, 3138747, 3261081, and 3434015—for a "solid state" circuit in 1959.
- Robert Noyce of Fairchild Semiconductor was awarded a U.S. patent in 1961 for a more complex integrated circuit termed the *unitary circuit*.

SUMMARY

In this chapter you learned:

- Electronics is the study and use of electrical devices that are controlled by the flow of electrons and typically involve low voltages.
- Transistors are small switching devices made of semiconductors.
- Semiconductors are typically manufactured by doping silicon with either phosphorus or boron.
- Both temperature and light can be used to control electrical current.
- LEDs are a type of diode that produces light.
- Digital electronics uses binary inputs and outputs to regulate an electrical circuit.
- Logic gates are used to control a digital electronics circuit.
- The integrated circuit or chip allows humans to control many systems.

VOCABULARY

Write a definition for each term in your own words. After you have finished, compare your answers with the definitions provided in this chapter.

Transistor
Electronics
Doping
Light-emitting diode (LED)
Digital electronics
Truth table
Integrated circuit (IC)

STRETCH YOUR KNOWLEDGE

Please provide thoughtful, written responses to the following questions.

1. Describe how your life would be different if the engineers at Bell Laboratories had not invented the transistor.
2. How has the development and expanded use of the integrated circuits affected the service and repair industry?
3. Some people have said that digital electronics is the language of the future. Do you agree or disagree? Explain your response.

Onward to Next Destination

CHAPTER 17
Manufacturing

Menu

Before You Begin

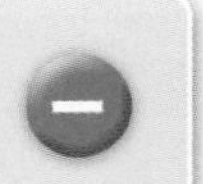

Think about these questions as you study the concepts in this chapter:

1. Why is manufacturing important?
2. What is manufacturing?
3. How has manufacturing changed over time?
4. What processes are used to shape materials into useful products?
5. What is quality assurance?
6. How are parts or products moved between stations in a manufacturing plant?

© TIHIS/SHUTTERSTOCK.COM

FIGURE 17-1 Skateboarding has become an internationally recognized sport partly because of improved materials and manufacturing techniques.

Engineering in Action

In the 1950s and 1960s, teens often had to make their own skateboards. This was done by taking apart a pair of metal roller skates and fastening each half to the end of a 2-inch by 4-inch piece of lumber. These homemade skateboards were crude versions of the manufactured skateboards available today (Figure 17-1).

Skateboard enthusiasts continued to make boards in their own garages or shops. As they did, they experimented with new materials and techniques. One early development was gluing thin strips of wood together using a technique known as *lamination*. Laminating has several advantages over using solid wood: it is easier to form shapes, it is stronger, and it is less likely to split the way a solid board will.

Improved materials and new manufacturing techniques have revolutionized the sport of skateboarding. Today's boards are often still made of wood, but other materials are also used, including composites, aluminum, nylon, and fiberglass. New board designs use a honeycomb material for the deck that is stronger but lighter in weight than wood (see Figure 17-2). The skateboard trucks (the part that holds the wheels) are usually made of metal. Polyurethane (a synthetic rubber polymer) is typically used for the wheels. The skateboards are often finished with a screen printing process for color and design.

(a) Laminating
(b) Forming
(c) Drilling
(d) Sanding
(e) Screen printing

FIGURE 17-2 Many processes are needed just to manufacture the skateboard deck.

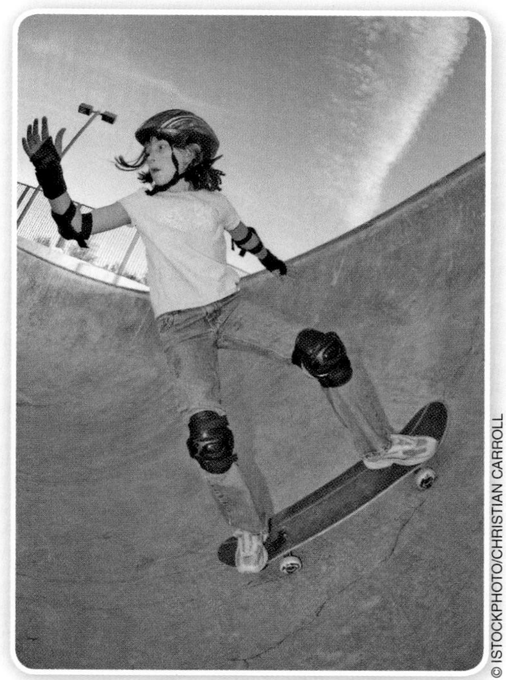

FIGURE 17-3 New products have been developed and manufactured to support skateboarding.

Skateboarding has created a demand for other new products such as helmets and wrist and knee pads. New manufacturing plants have opened to meet this demand. Manufacturing is an important part of our society. Society comes to depend on manufactured products when they are available (such as cell phones). But society also creates the demand for new products to be manufactured, as in the case of protective gear for skateboarding (Figure 17-3).

SECTION 1: THE IMPORTANCE OF MANUFACTURING 1

Have you ever gone to the store for a replacement part? Perhaps you needed to replace something as simple as a battery or a lightbulb. If a wheel on your skateboard wore out, would you be able to find a replacement? Your answer is probably, "Of course!" But did you know that it has been possible to find replacement parts in stores for only the last 100 to 150 years?

Until the early 1800s, most products were still made by individual craftsmen. Because each product was made individually by hand, each part was a little bit different. Often individual craftsmen had their own way of making things, so you could not always go to another craftsman for a replacement part.

Did You Know?

Although Eli Whitney is best known as the inventor of the cotton gin, he also invented many other things. History also credits Whitney with proving the value of interchangeable parts in America and launching the modern era of manufacturing.

In 1798, Eli Whitney signed a contract with the federal government to produce 10,000 rifles in two years. At the time he signed the contract, he did not have a rifle manufacturing plant. In fact, Whitney had never made a rifle in his life. In 1801, Eli Whitney took baskets of parts to President-elect Thomas Jefferson, who was able to assemble working muskets from the baskets of parts. This is the first documented case in the United States of mass-produced, interchangeable parts.

Eli Whitney did not have all 10,000 rifles made on time. However, the factory he set up to mass produce the rifles began the modern era of manufacturing. Eli Whitney used unskilled labor to operate machines to make thousands of duplicate parts. Before this, most products were made one at a time.

Modern-day manufacturing began in the early nineteenth century. The word *manufacturing* is actually based on two Latin words (*manu factus*) that together mean "made by hand." **Manufacturing** is the use of tools and machines to make things. Generally, manufacturing involves processing raw materials into finished goods.

Manufacturing is based on the idea of standardized parts. Standardized parts are also called **interchangeable parts**. The basic idea behind interchangeable parts is that hundreds or thousands of identical parts are made first and then products are assembled from the parts. Remember, before modern manufacturing, individual craftsmen made each product by hand, one at a time. **Mass production** is the manufacture of large numbers of identical pieces (interchangeable parts). These two developments—mass production and interchangeable parts—were the major factors that enable you to find replacement parts for products at different stores.

manufacturing
Manufacturing is the use of tools and machines to make things. Manufacturing converts materials into usable products.

The addition of computers and automation in manufacturing in the last decades of the twentieth century significantly boosted production and improved product quality. These improvements made things such as your video game possible, not to mention clothing, costume jewelry, contact lenses, and life-saving medical equipment.

Think about it for a minute. Look at what you are wearing from your shoes to your hair products. Your eyeglasses or contact lenses and your wristwatch are also examples of manufactured products. From the alarm clock

FIGURE 17-4 **Modern manufacturing processes have contributed to our standard of living. Can you tell what popular product is being manufactured in this photo?**

that woke you this morning to much of the food you ate at breakfast, you are surrounded by the products of modern manufacturing (Figure 17-4).

Without modern-day manufacturing, our lives would be quite different—and, it is probably safe to say, much less convenient. Modern manufacturing has contributed significantly to our standard of living and quality of life.

SECTION 2: MANUFACTURING PROCESSES 4

Manufacturing is the process of converting materials into usable products. This process is a technological system, as we learned in Chapter 2. Materials are the inputs, manufacturing techniques are the process, and finished goods are the outputs of the system. Customer satisfaction and sales are the feedback part of the system.

Manufacturers perform their operations in specialized facilities called *factories* or *manufacturing plants*. The materials to be processed, as described in Chapter 8, may be natural or synthetic materials. Typically, we think of manufacturing as the processing of materials into usable products for the consumer (Figure 17-5). As you might guess, we refer to these products as **consumer products**.

Sometimes, the manufactured product will actually be an input for another manufacturing operation. We refer to these products as **industrial products**. For example, an automated manufacturing facility that produces cell phones will assemble parts that have been manufactured at other plants (Figure 17-6). This manufacturing process is really more of an assembly process.

© YURI ARCURS/SHUTTERSTOCK.COM

FIGURE 17-5 Most of the things we use and enjoy are manufactured products.

In this chapter, we focus on the production processes that change the shape of materials to create usable products. To fabricate a prototype (see Chapter 8) or a product, engineers and technologists must use different processes to change the shape, size, or properties of the material. We can classify manufacturing processes into five common types: (1) forming, (2) casting or molding, (3) separating, (4) combining, and (5) conditioning.

Forming

Forming is a process that changes the shape and size of material without cutting it. We can form some materials when the materials are in either a hot or a cold state. We can form other materials only when they are hot. For example, most plastics must be heated to be formed. A few materials must be cold to be formed. Engineers determine whether to *hot form* or *cold form* based on the material's properties and the requirements of the design.

We may either hot form or cold form a metal, depending on the process. Forming the material will change its properties in the area where the forming occurs. For example, some metals may

forming
Forming is a family of processes that change the shape and size of a material without cutting it.

extrusion
Extrusion is a forming process that forces material through an opening in a die with a desired shape.

forging
Forging is a forming process that uses heat and force to shape metal.

© CENGAGE LEARNING 2013

FIGURE 17-6 Industrial products are components that are used to make other products.

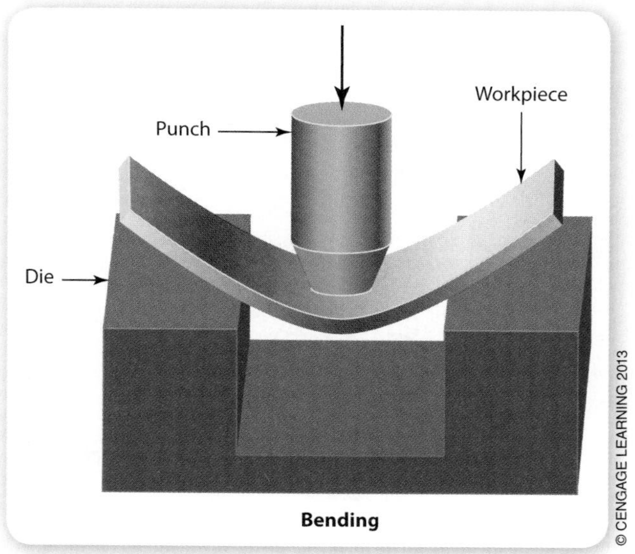

FIGURE 17-7 **Sheet metal and bar stock can be cold formed.**

FIGURE 17-8 **Extruding is a process that shapes material by forcing it through an opening of the desired shape. The machine in this photo is extruding trout bait.**

become more brittle after bending (see Chapter 8 for a description of material properties).

Cold forming metal could simply be bending the metal or forcing it into a shape using a pattern (Figure 17-7). Products such as shelf brackets and sheet metal tool boxes are cold formed. Sheet metals are formed using a machine called a *brake*.

Extrusion is a type of forming process. It forces material through an opening in a die that holds a desired shape (Figure 17-8). We often use extrusion to produce long objects, usually as a metal hot forming process. The metal is heated to make it softer and easier to force through the form, but it is not melted (Figure 17-9). Plastics may be heated, or even melted, before extrusion. Some food products are extruded, such as pasta shapes and some dry cereals.

Forging is a forming process that uses heat and force to shape metal. Forging is the process used by blacksmiths, a hot forming process done by hand (Figure 17-10). The metal is heated and then hammered into shape. Because of the heat and force required, forging is now completed by large industrial presses (Figure 17-11).

Forging actually compresses metal grains closer together. As a result of compressing the metal, it is actually stronger. The primary advantage of forging is the increased strength of the finished part. We also use forging to create decorative wrought iron products.

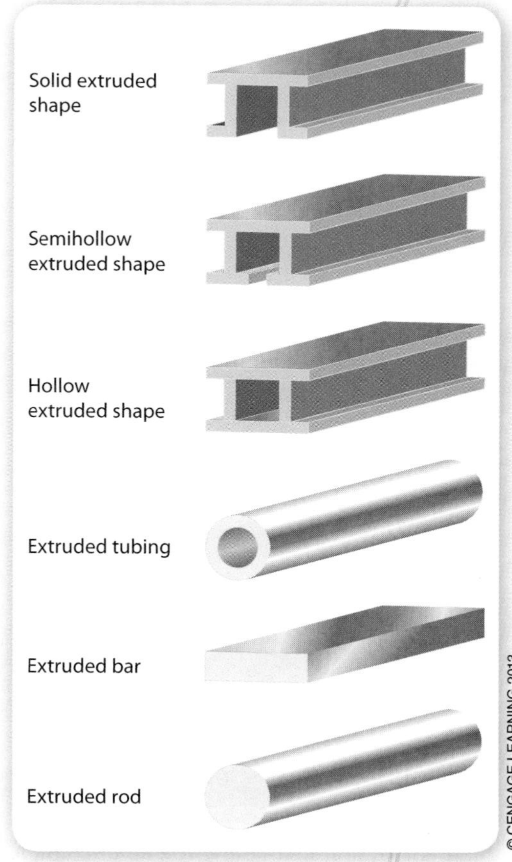

FIGURE 17-9 **Common extruded metal shapes.**

© PATRICIA MARROQUIN/SHUTTERSTOCK.COM

FIGURE 17-10 Blacksmiths use a forging process.

COURTESY OF ALCOA

FIGURE 17-11 Industrial forges can be enormous machines.

casting

Casting is a forming process in which the material is first made into a liquid and then poured into a mold.

molding

Molding is a process that forces plastic into a mold or cavity of the desired shape.

Casting and Molding

Casting is a forming process in which the material is first made into a liquid and then poured into a mold (Figure 17-12). The material solidifies in the mold and is then removed. Some molds are permanent, but others are used once and must be destroyed so the part can be removed.

Making ice cubes is similar to a casting process. In manufacturing, the materials are typically heated to a liquid state and then poured into the mold and allowed to cool at room temperature (Figure 17-13). In the ice cube example, the water is a liquid at room temperature, so it must be placed in a freezer to solidify.

© ISTOCKPHOTO/KEVIN BROWN

FIGURE 17-12 Casting involves pouring a molten metal into a mold.

Molding is very similar to casting and is used to force plastic into a new shape. The plastic is first heated to a liquid or softened state and then forced into a mold or cavity with the desired shape. Then the plastic is allowed to cool in its new shape. Plastics can generally be formed much more quickly than metal.

There are many different types of molding processes for plastic. **Injection molding** is one of the most common processes (Figure 17-14). The plastic starts out in the form of small pellets. After being melted to a liquid state, the plastic is forced into a mold, known as a *die*, under pressure. Injection molding is used to form game pieces, dice, and similar products.

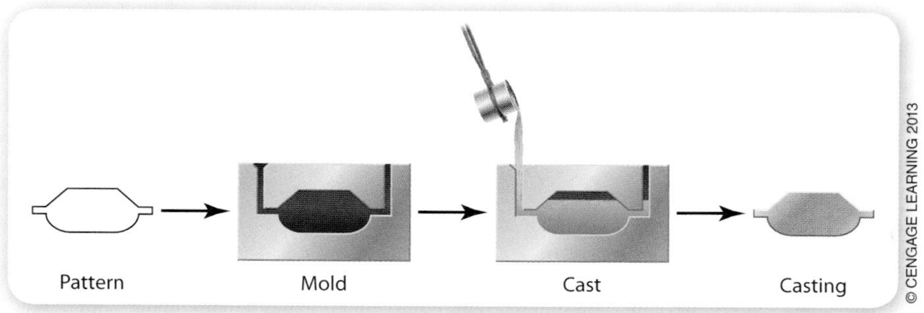

FIGURE 17-13 Casting often uses a hollow form called a mold.

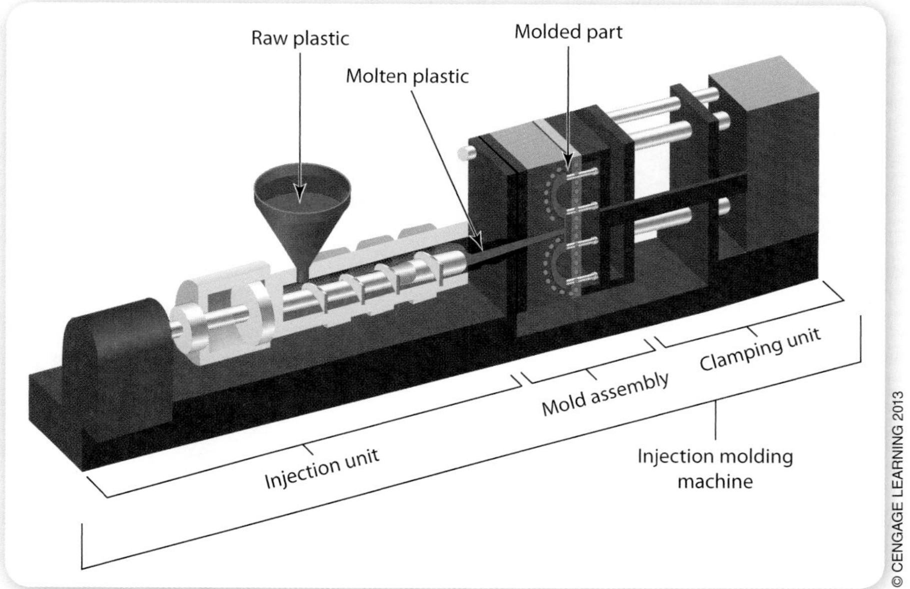

FIGURE 17-14 A plastic injection-molding machine.

Thin plastic sheets may be formed into trays by **vacuum forming** (Figure 17-15). This process is different than injection molding. The plastic sheet is laid across a form and then heated to a softened state, but not melted. Vacuum suction pulls the softened plastic sheet down into the mold.

Blow molding is another molding technique for plastic. It is used to make things such as milk jug containers, soft drink bottles, shampoo bottles, and even hard-sided plastic cases for tools (Figure 17-16).

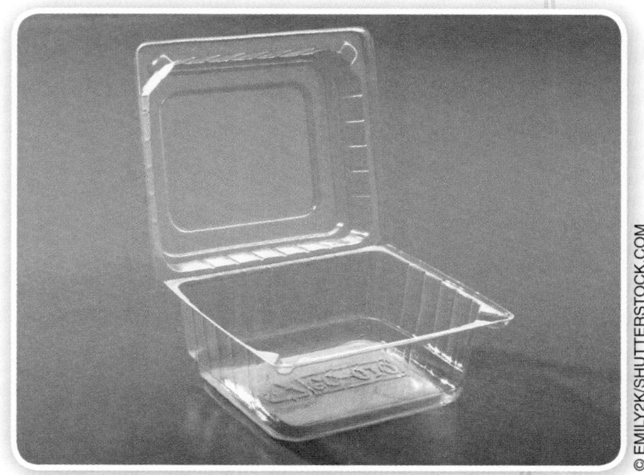

FIGURE 17-15 Food containers are familiar examples of vacuum forming.

separating
Separating is a family of processes that remove excess material to change the size, shape, or finish of a product.

Separating

Separating is a family of processes that remove excess material to change the size, shape, or finish of a product. These separating processes include sawing, drilling, turning, shearing, and sanding. One of the most common separating processes is **sawing**, which uses a set of teeth to remove material (Figure 17-17). We call the groove left by the sawing process the **kerf**. Engineers determine the type of saw to use based on the properties of the material to be cut. For example, for soft materials larger teeth are used to remove more material on each cut. For hard or brittle material, like metal, small teeth must be used (Figure 17-18).

Drilling is a separating process that produces a round hole. The hole may or may not pass all the way through the material (Figure 17-19).

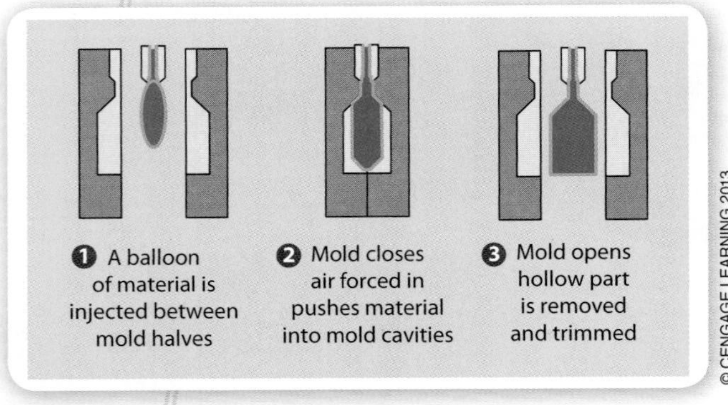
❶ A balloon of material is injected between mold halves
❷ Mold closes air forced in pushes material into mold cavities
❸ Mold opens hollow part is removed and trimmed

FIGURE 17-16 Blow molding is a fast and efficient method for making hollow shaped containers, such as milk jugs or shampoo bottles.

DELTA

FIGURE 17-17 Most middle schools use a scroll saw to cut materials.

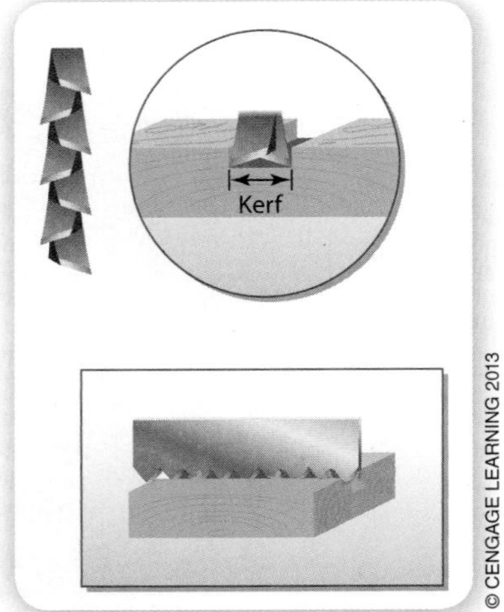
Kerf

FIGURE 17-18 The shape and size of a saw's teeth depend on the material to cut.

FIGURE 17-19 Drill bits are used to produce round holes in many types of materials.

A **drill bit** is used as the cutting tool. Drill bits may be used in portable electric hand drills or in a machine called a **drill press** (Figure 17-20).

To produce a cylindrical object such as a softball bat, a table leg, or a bowl, technicians use a process called **turning**. We use a machine called a **lathe** for turning operations (Figure 17-21). During the turning process, the material turns against the cutting tool (the material turns, and the tool remains stationary). This is the opposite of both sawing and drilling, in which the cutting tool turns against the material (the tool turns and the material does not). Small lathes can be hand operated, whereas industrial machines are usually automated (Figure 17-22).

To cut a piece of paper, you would use a pair of scissors. The scissors shear paper. **Shearing** is a process that separates material without producing waste. There are hand shears, like tin snips, for cutting sheet metal. Hydraulic shears can cut flat sheet metal or metal strips.

Using sandpaper to smooth an object is also a separating process. **Sanding** uses small abrasive grain particles to remove material from a softer surface. The sawdust produced from sanding is the material that has been separated.

FIGURE 17-20 Drill presses make drilling easier and more accurate than hand drilling.

Combining

Combining is a process of joining two or more materials together. Using **mechanical fasteners** such as nails, staples, screws, bolts, and rivets is a common method of combining materials. One advantage of mechanical fasteners is that they can join together different types of materials. Mechanical fasteners also allow for easy assembly and disassembly of a prototype.

FIGURE 17-21 Lathes are used to make cylindrical (round) products such as softball bats or bowls.

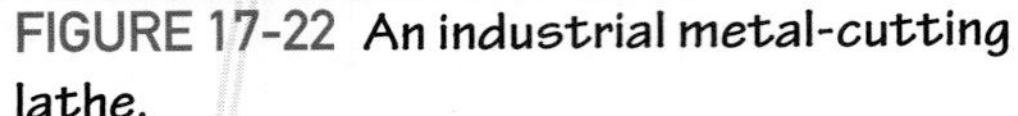

FIGURE 17-22 An industrial metal-cutting lathe.

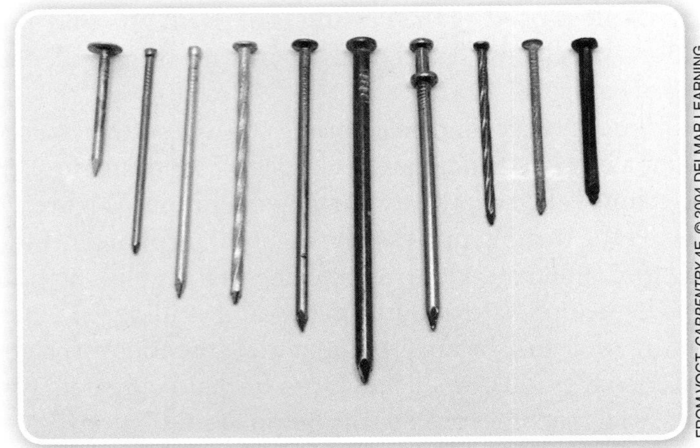

FIGURE 17-23 Common nail types.

A **nail** is a wedge made of metal used to join lumber together. Nails are produced in various sizes and types (Figures 17-23 and 17-24). Each type of nail has a specific use. The technician selects the appropriate type and size of nail for the application. Nails are frequently used in construction.

A screw is used to assemble products and prototypes. It has greater holding power than a nail and can be easily installed and removed. **Wood screws** are tapered and used in lumber but not metal. A **machine screw** is a small bolt typically used with a nut. Screws are available in many different head shapes, but the most common are called *round*, *oval*, and *flat* (Figure 17-25).

Penny (D) sizes
2D 3D 4D 6D 8D 10D 12D 16D 20D 30D 40D 50D 60D
1/2" 1" 1 1/2" 2" 2 1/2" 3" 3 1/2" 4" 4 1/2" 5" 5 1/2" 6"
Length

FIGURE 17-24 Common nail sizes.

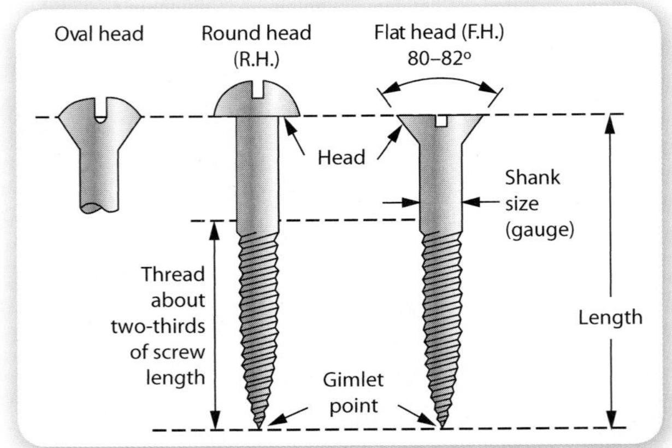

FIGURE 17-25 Common screw types.

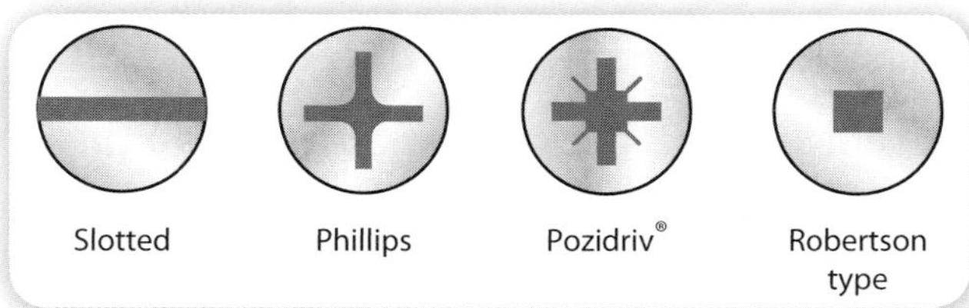

FIGURE 17-26 *Common types of screw drive slots.*

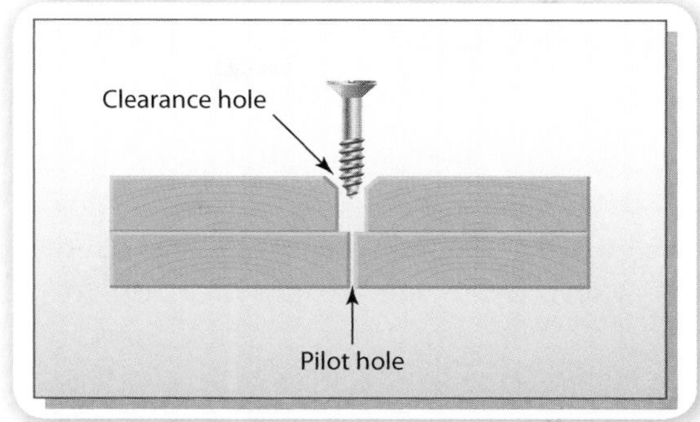

FIGURE 17-27 The material must be countersunk to receive a flat-head screw.

Metal and some plastic materials can be threaded to receive a machine screw so a nut is not be needed. There are also different types of head slots for driving screws. Two of the most common are slotted (straight slot) and Phillips (looks like a plus sign) (Figure 17-26). When installing a flat-head screw, the hole must be countersunk (Figure 17-27).

Another method to combine materials is to join them with an **adhesive** such as glue (Figure 17-28). Each type of adhesive is designed to be used with specific materials. Wood glue is used when building a prototype with lumber. It bonds quickly and has good strength.

Metals can also be combined using **welding** or **soldering**. In welding, the materials are heated and "melted" together (Figure 17-29). The two pieces literally become one at the point where they are joined. Soldering is a process in which two pieces are joined by a molten metal that adheres to the base metal. In soldering, the base metal is not melted, as it is in welding. Soldering is used for electrical connections (Figure 17-30).

FIGURE 17-28 Glue is an adhesive used to join materials.

FIGURE 17-29 Welding is a combining process that melts two pieces of metal together.

Science in Engineering

Understanding properties of materials is critically important to engineering. Each processing technique will have some effect on the material, sometimes in dramatic ways. When metals are heated, for example, their properties change. Sometimes we deliberately heat the metal to cause this change to happen, such as when hardening. But other times the change is a by-product of the process—for example, when welding the heat from welding can cause the metal to warp or become brittle.

FIGURE 17-30 Soldering is a combining process used for electrical connections.

Conditioning

Conditioning changes the internal characteristics of the material. Engineers typically use three types of conditioning for metal: hardening, tempering, and annealing. All three of these processes involve heat-treating the metal in a furnace (Figure 17-31). **Hardening** is a process that makes the material harder. **Tempering** makes the material more flexible. **Annealing** makes the material softer and easier to work. Engineers decide which process is needed based on how the product will be used.

FIGURE 17-31 An industrial heat-treating process.

SECTION 3: ORGANIZATION OF A MANUFACTURING SYSTEM

As described earlier in this chapter, manufacturing is a technological system. Systems are organized to accomplish a task as efficiently as possible. In Chapter 2, you learned about seven technological resources. Do you remember what they are? In this section, you will read about a few of these resources as they relate to the organization of a manufacturing system.

Corporate Ownership

Most manufacturing firms today are corporations. A corporation is a legal entity with rights under the law. A company may sell stocks to many people (called *shareholders*). This business is then referred to as a publicly owned corporation. Selling stock is the primary method of raising capital (one of the key resources) for the operation of the business.

Corporate Personnel

The shareholders (the people who bought the stocks) are the owners of the corporation. The shareholders elect a board of directors to represent their interests in the company. The board is responsible for making sure the company is run as efficiently as possible and makes a profit. The board of directors hires the top manager of the company, often called the *chief executive officer* (CEO).

Manufacturing Management

Managing a manufacturing business is a complex process. The success of a business is largely determined by how well it is managed. Companies must be able to accurately produce consistent products on time and within budget. Consumers want to know that when they go to the store to buy something it will be available. They also want to know that it is safe and of high quality. We can divide the management of a manufacturing company into two parts: production (Figure 17-32) and sales.

Workforce

A skilled workforce is essential for successful manufacturing. Many unskilled jobs have been replaced by robots and automation. More jobs are high-tech, requiring some training beyond high school. Technicians are skilled workers who have usually trained at community colleges. Engineering technologists may have been trained at either a community college or a university. We consider both positions skilled labor (Figure 17-33).

FIGURE 17-32 Production management involves understanding the entire manufacturing process.

FIGURE 17-33 Skilled technicians must be able to service complex machines.

Quality Assurance

5 **Quality assurance** is the process of making sure the *entire* manufacturing process is the best it can be. Quality assurance includes everything from obtaining the raw materials to shipping the finished product. It is both a planning function of management and a technical aspect of production. Management and skilled and unskilled labor work in teams called **quality**

Math in Engineering

When a manufacturing company is making millions of parts, it is not possible to measure each part. For this reason, engineers and engineering technologists use a type of mathematics called *statistics* to guarantee that parts are made to the right tolerances. This involves *sampling*. For example, the quality control technician might measure only 1 out of every 100 pieces. Using statistical process control programs allows the company to predict with certainty the overall quality.

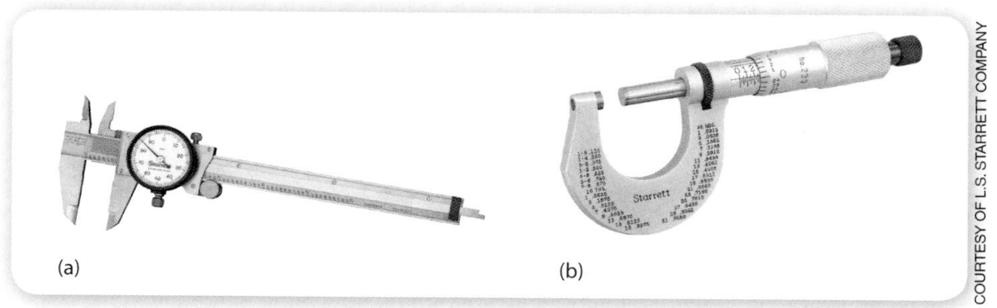

FIGURE 17-34 (a) A dial caliper and (b) a micrometer are precision measuring instruments.

circles. Workers at all levels are charged with identifying problems and developing creative solutions. Everyone has a stake in making sure the product is the best it can be. Teamwork is one of the most important traits that managers look for when hiring new employees.

Quality Control **Quality control** is the process of making sure each individual part and manufacturing process is of the highest possible quality. Manufacturing is a very complex process. Many different quality control systems must be in place to make sure that each part meets the customers' expectations.

Manufacturers often check the parts with a variety of precision measuring tools. Part inspection may be done by hand with micrometers or calipers when accuracies of 0.001 to 0.0001 of an inch are adequate. (Figure 17-34). When accuracy greater than 0.0001 of an inch is required, very sophisticated machines take measurements. These measuring machines may use lasers for very precise measuring.

Product Transportation

A **transportation system** is a system designed to move people or goods from one place to another as efficiently as possible. Product transportation is very

Engineering Challenge

Engineering technologists use precision measuring instruments. They call the most common instruments *micrometers* and *calipers*. They regularly measure objects with an accuracy of one-thousandth of an inch (0.001) or even one ten-thousandth of an inch (0.0001). For comparison, the average human hair is about two-thousandths of an inch in diameter. This is written as 0.002.

The dial caliper can accurately measure objects to thousandths of an inch. It takes practice to read this instrument. This is a versatile measuring device because it can measure either outside or inside dimensions. The long jaws measure the outside dimension of an object. On the opposite side are two shorter jaws that are designed to measure inside a hole or a piece of tubing.

Engineers must be able to use both English units of measure and the metric system. In nanotechnology, engineers measure in nanometers. One nanometer (nm) is one billionth of a meter (m). As a decimal number, this is written as 0.000000001. Although that looks like a very large number, it actually represents a very, very small number.

How do you know how many places to the right of the decimal to use when writing such small numbers? Each decimal number is a placeholder for a value, just as when counting in whole numbers to the left of the decimal place. When writing numbers smaller than 1, the numbers to the right of the decimal place are as follow:

The first decimal place is the *tenths,*
the second decimal place is the *hundredths,*
the third is the *thousandths,*
the fourth is the *ten-thousandths,*
the fifth is the *hundred-thousandths,*
the sixth is the *millionth,*
the seventh is the *ten-millionths,*
the eighth is the *hundred-millionths,* and
finally, the ninth is the *billionths.*

1. Write the following numbers as a decimal:
   a. one-tenth of a meter
   b. one-hundredth of a meter
   c. one-thousandth of a meter
   d. one-millionth of a meter
2. Measure each of the parts in the basket provided by your teacher. Record your measurements in the table provided by your teacher. The following is just an example.

Object	Diameter	Length or thickness
Marble		
Skateboard ball bearing		
Copper tubing	Inside: Outside:	
Miniplug from earphones		
CD		

3. From your teacher, obtain 10 nails of the same size and 10 wood screws.
   a. Record the diameter of each in the following table.

b. Remember that wood screws are tapered, so measure each one in the same location, right below the head.
c. Calculate the average diameter for each.
d. Which product has the greater variation in size: nails or screws?
e. Why do you think this variation exists?

Nail	Diameter	Screw	Diameter
1		1	
2		2	
3		3	
Etc.		Etc.	
Average		Average	

important in manufacturing for three reasons. First, materials need to be delivered to the plant on time as needed. Second, parts need to be moved around inside the plant to be at the right station at the right time. Third, the finished product must be delivered to customers on time. To be competitive in today's global economy, each of these three transportation processes must be accomplished efficiently and on time.

We call the process of making sure materials and products are delivered on time **supply chain management**. Another term commonly used is *logistics*. Supply chain management is the process of planning, implementing, and controlling the movement of materials as efficiently as possible during the entire manufacturing cycle. Supply chain management covers everything from the acquisition of raw materials to part movement in the plant and delivery of finished goods to consumers.

supply chain management
Supply chain management is the process of making sure materials and products are delivered on time from raw material to finished product.

Moving Parts Inside the Manufacturing Plant Inside the manufacturing plant, parts are commonly moved by conveyor belts. Large belts or moving chains move the part to the next station. We use the generic term **conveyor belt** to describe a continuous looped system to move parts (Figure 17-35). Sometimes parts are hung on hooks from an overhead chain, as in the case of Harley Davidson motorcycles. The bike frames move through the plant while hanging from a moving chain. The bike frames even go through a series of chemical baths to get them ready for painting as they hang.

Engineers and engineering technologists must carefully plan the movement of parts and how much time a part will need to be at each station. In manufacturing, the adage "time is money" really is true. Companies have huge financial investments in the raw materials they purchase. They cannot afford to order too much of a product and have it sit in a warehouse waiting for production. Sometimes companies are paying millions of dollars in interest payments, so each day that materials sit idle may cost the company thousands of dollars.

FIGURE 17-35 Parts are often moved on conveyor belts in a manufacturing plant.

SUMMARY

In this chapter you learned:

- Manufacturing is the process of converting materials into usable products.
- Manufacturing processes (sometimes called materials processing) change the shape, size, and properties of materials.
- Interchangeable parts (also called standardized parts) are hundreds or thousands of identical parts made to the same size and specifications.
- We classify manufacturing processes into five common types: forming, casting or molding, separating, combining, and conditioning.
- Most manufacturing firms today are corporations. A corporation is a legal entity with rights under the law.
- Quality assurance is the entire process of making sure the manufactured product is the best it can be.
- Supply chain management is the process of planning, implementing, and controlling the movement of materials as efficiently as possible during the entire manufacturing cycle.

VOCABULARY

Write a definition for each term in your own words. After you have finished, compare your answers with the definitions provided in the chapter.

Manufacturing
Forming
Extrusion
Forging
Casting
Molding
Separating
Supply chain management

STRETCH YOUR KNOWLEDGE

Provide thoughtful, written responses to the following questions and assignments.

1. Why is it important to understand manufacturing?
2. What are some of the major changes that have affected manufacturing in the past 200 years?
3. Research how a particular product is made, beginning with obtaining the raw materials. Construct a time line showing each significant process in the manufacturing of the product. Your time line should be suitable for posting on the class bulletin board. Present your time line to the class.
4. Describe the quality assurance process.
5. Research and present to the class common products made from any one of the manufacturing processes described in this chapter.

Onward to Next Destination

CHAPTER 18
Robotics

Menu

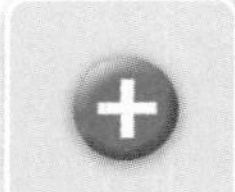

Before You Begin

Think about these questions as you study the concepts in this chapter:

1. What does the word *robotics* mean?
2. What kinds of jobs do robots perform?
3. How do robots make our life easier?
4. How are robots used in manufacturing?
5. How are robots powered?
6. What type of engineer programs robots?
7. When were robots invented?

COURTESY OF iROBOT

FIGURE 18-1 Roomba by iRobot was the first widely sold functional personal robot.

Engineering in Action

S T E M

ENGINEERING IN ACTION: ROOMBA AND YOU

Roomba by iRobot was the world's first widely sold robot for personal use (Figure 18-1). Roomba is designed to automatically vacuum an entire room. Its designers programmed it to "learn" the layout of a room. Built-in sensors cause it to back up and turn whenever it encounters an obstacle such as furniture or a wall; they also keep it from falling down stairs. Roomba keeps track of where it has been in the room and keeps searching for a path to cover the entire room. Having a robot vacuum for you is a great labor-saving device, but a robot that would clean the bathroom or wash dishes would be even better.

A company called iRobot first developed Roomba in 2004. More than 1 million Roomba vacuums were sold the first year. Roomba is significant because there were more Roomba robots sold in the world in two years than all of the industrial robots sold since the 1960s, when they were first introduced into manufacturing. Colin Angle, Helen Greiner, and Rodney Brooks, roboticists from the Massachusetts Institute of Technology, started the iRobot company in 1990. The company develops robots to perform dirty, dull, or dangerous tasks to make the world better for people. Its robots have disarmed explosives in war zones, explored Mars, and even explored the far corners of an Egyptian pyramid.

SECTION 1: ROBOTS IN YOUR WORLD

robot
A robot is a mechanical device that acts in a seemingly human way and that is most often guided by automatic (usually programmable) controls.

robotics
Robotics is the science of designing and using robots.

We can define a **robot** as a mechanical device that acts in a seemingly human way and that is most often guided by automatic (usually programmable) controls. The important points are that it is a mechanical device (not living) and usually programmable. A robot's primary purpose is to relieve humans of work that is tedious, repetitive, hazardous, or dirty or that requires precise accuracy. Can you think of a job that you would like to have a robot do for you?

The term **robotics** refers to the science of designing and using robots. There are many types of robots. Some are mobile; most of these are on wheels, but a few actually have the ability to walk. Other robots are stationary and do repetitive work on assembly lines. Most industrial robots are fixed in one place and dedicated to one specific job. Industrial robots are often nothing more than an arm with a gripper or some type of tool on the end of an arm.

We can classify robots as personal (household), assistive, medical, industrial, military or security, or space exploring. Each type of robot has a different use, but all robots have similar systems. Although many people think industrial robots are the most common type of robot, the fact is that robots have become commonplace in many aspects of your life. The introduction of Roomba was the beginning of widespread personal robots for home use. But what does the future hold for personal robots? Personal robots are becoming much more sophisticated. Perhaps one day soon you will have a personal assistant like ASIMO, featured later in this chapter.

GETTY IMAGES

FIGURE 18-2 C3PO and R2D2 from *Star Wars*.

Personal Robots

The *Star Wars* movie series that featured R2D2 and C3PO made the idea of a personal robot assistant popular (Figure 18-2). Most people had a general idea about robots but really did not know what kind of robots existed or what they could do. Their understanding of robots was influenced by popular movies, not reality. Movies such as *I, Robot* added to their misperceptions.

Until recently, personal robots were primarily for entertainment. Today, many types of robots are available to students in technology education classes, including kits that you can build. These robots are not only fun but also educational tools. Many stores sell a variety of robotic kits for home use, too.

Math in Engineering

Learning to program robots, even those that appear to be toys, is an excellent way to develop the logic required in advanced mathematics. Yes, you will become better in math if you practice programming robots.

The Lego Mindstorms NXT robotic kit combines LEGO TECHNIC elements with ultrasonic, sound, light, and touch sensors (Figure 18-3). This combination allows you to make an intuitive robot. The light sensors can detect both color and intensity of light. The sound sensor allows the robot to respond to sound patterns and tones. This robot uses ultrasonic "eyes" to measure distance and movement. Students can program the robot to perform a variety of tasks.

Many middle school technology education classes include robots, like the Vex robot shown in Figure 18-4. What will your robot design look like? What purpose will it serve?

The Roomba robot featured in the opening of this chapter is capable of vacuuming several rooms of your house. Wouldn't it be great if someone would design a robot to pick up your room and make your bed? Although we do not have robots yet with that capability, we undoubtedly will in the very near future.

There are, however, personal robots such as the Robomower that can mow your lawn (Figures 18-5). This robot responds

FIGURE 18-3 Lego Mindstorm robotic kits are in many classrooms (and are available for home use, too).

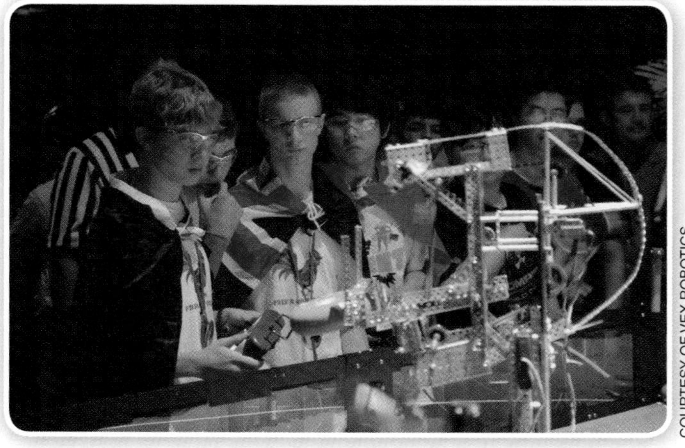

FIGURE 18-4 Middle school students practice with their robot at a national competition.

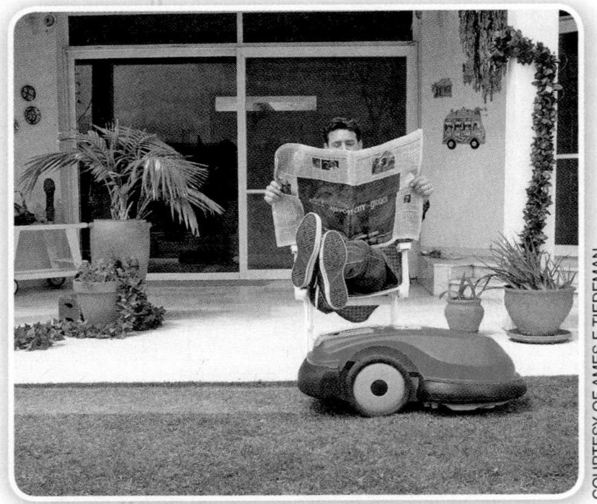

COURTESY OF AMES F. TIEDEMAN

FIGURE 18-5 Robots can help you relax in the shade while they mow your lawn.

COURTESY OF iROBOT

FIGURE 18-6 A robot to clean the leaves out of your gutters.

to a wire perimeter around a yard. It will work its way back and forth across the yard, making several passes. When it is finished mowing, the robot will return to its home base to recharge its battery. There are even robots designed to clean the leaves out of your rain gutters (Figure 18-6). Clearly, robots are designed to do tasks that most people do not want to do.

Personal robots have advanced considerably in the past decade. Most robots that are available to homeowners are not really personal assistants. But what does the future hold? Will you be the one to design a truly affordable personal assistant robot? Remember, before Henry Ford revolutionized mass production, most people could not afford an automobile. Today we cannot imagine living without one. Will the same be true for personal robots?

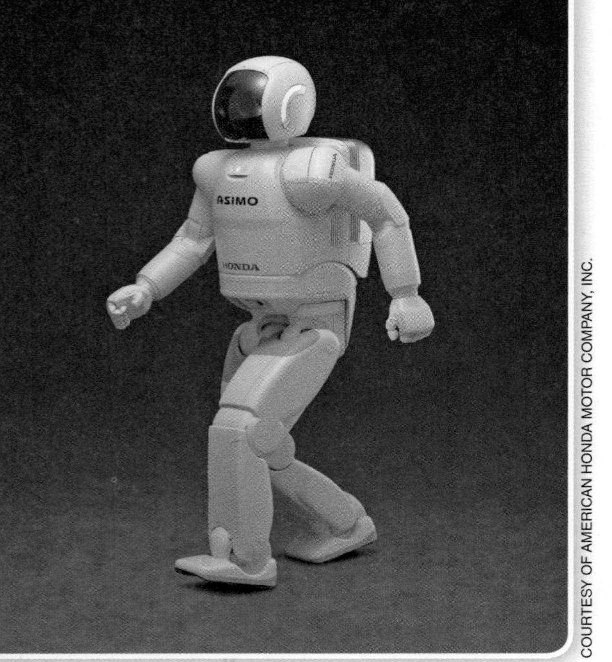

COURTESY OF AMERICAN HONDA MOTOR COMPANY, INC.

FIGURE 18-7 ASIMO, Honda's humanoid robot.

Honda's Humanoid ASIMO

ASIMO is an acronym that stands for Advanced Step in Innovative Mobility. In 1986, Honda engineers set out to create a walking robot. Honda's engineers have spent more than two decades revising their designs, and ASIMO has undergone 10 distinct revisions. The latest version of ASIMO is capable of running approximately 4 miles per hour, something earlier versions could not do. ASIMO can also walk up stairs (Figure 18-7).

Science in Engineering

Scientists at the Massachusetts Institute of Technology are researching robots that can help stroke victims and children with cerebral palsy develop motor skills by using video games. A patient's arm and hand are strapped to a robotic arm. If the patient cannot move his or her hand quickly enough for the video game, the robot moves and teaches the patient how to coordinate his or her hand–eye movements. This research has been very successful so far.

A humanoid robot is a very complex machine with sophisticated computer programming. ASIMO is made of lightweight materials and has 34 servomotors throughout its body. A **servomotor** is a special type of motor that provides very precise position feedback. The motor moves in small increments to certain positions. This enables ASIMO to imitate many human actions. It is designed to be able to reach for things and pick them up, walk on uneven surfaces, and even climb stairs. With sophisticated computer software and camera eyes, ASIMO is able to recognize people's faces and voices. ASIMO can even comprehend and respond to simple voice commands. "He" has the ability to map his environment, avoid stationary objects, and even avoid moving objects as he moves through his environment. ASIMO is designed to function in the real world and truly assist humans.

servomotor
A servomotor is a special type of motor that provides very precise position feedback used in robotics and automation. The motor moves in small increments to certain positions.

Honda engineers take ASIMO around the world to encourage and inspire students to study science and engineering (see Figure 18-8). In the very near future, ASIMO may be able to serve as the eyes, ears, hands, and legs for people with disabilities. As a personal assistant, ASIMO might assist the elderly or someone confined to a bed or a wheelchair. And ASIMO might also be able to complete tasks that are too dangerous for people, such as cleaning up toxic spills or fighting fires. What a wonderful assistant that would be. Can you think of other dangerous jobs robots could perform?

COURTESY OF AMERICAN HONDA MOTOR COMPANY, INC.

FIGURE 18-8 ASIMO helps a group of students cross the street safely.

Assistive Robots

Robots can already assist people with a wide variety of disabilities. Many types of injuries or conditions can prevent a person from walking. Many older people suffer from a stroke and need rehabilitation to be able to walk again. Spinal cord injuries can also leave a person unable to walk. The AutoAmbulator robot was developed to help injured people learn to walk again.

People who suffer severe disabilities have had access to powered wheelchairs for many years. Robotic technology was only recently combined with a powered wheelchair to provide more freedom to people with severe disabilities. The Raptor was the first commercially available assistive robotic wheelchair that was approved by the Food and Drug Administration.

Medical Robots

There are different types of medical robots. Early robots in hospitals were used to transport medical supplies and deliver drugs to doctors and nurses from the pharmacy. Engineers have significantly refined these robots. One of the newest versions, SpeciMinder, has a simple push-button control panel that allows lab technicians to select several destinations for the specimens. The SpeciMinder will calculate the most efficient route and stop at all assigned locations (see Figure 18-9). When its route is finished, the robot will automatically return to its charging dock.

CCS ROBOTICS

FIGURE 18-9 A fully autonomous specimen transport system.

Today, robots are also being used to perform delicate surgery. This surgery may be performed even deep in the ocean or in outer space, or wherever humans are stationed and have telecommunication capabilities. The doctor performing surgery never has to leave her office. In conventional operating rooms, robots have much more precise movements than a human hand. One model of a surgical robot is called the da Vinci Surgical System (Figure 18-10).

Combining doctors' skills with robotic technology allows surgeons to perform delicate surgeries better than ever before. There are several benefits of robotic surgery for the patient as well, including:

- reduced trauma to the body because the robotic procedure is very precise;
- less blood loss, which means fewer transfusions are necessary;
- less pain after the surgery because a smaller part of the body was impacted by the surgery; and
- less risk of infection.

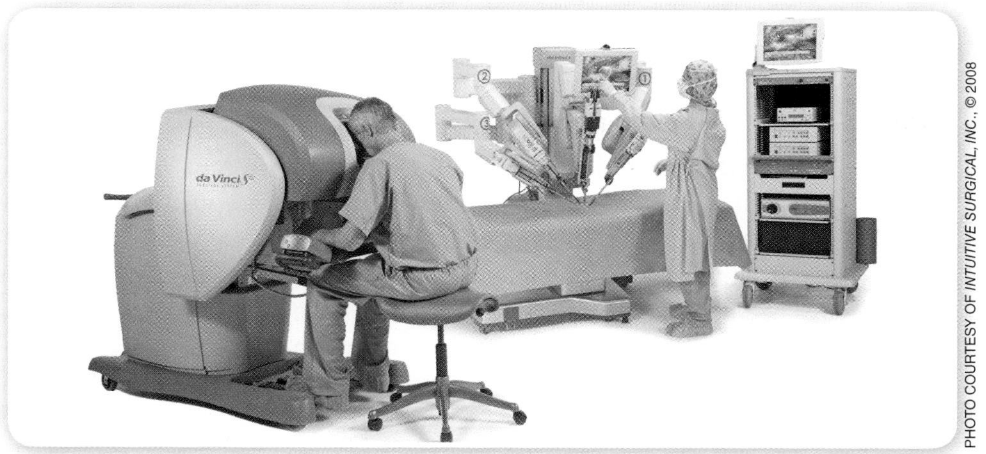

FIGURE 18-10 The use of surgical robots allows doctors to perform delicate procedures accurately.

These benefits add up to a shorter hospital stay and a quicker recovery, things that everyone wants.

Other robots in hospitals enable patients and care providers in remote areas to have access to specialists anywhere in the world. Two-way telecommunications enable doctors in one location to see and converse with a patient and doctor in another location. Other robots have been programmed with sophisticated medical information about specific injuries. Rural emergency room personnel can consult this robot when they do not have a specialist on staff (a neurosurgeon, for example). Robots clearly have an important role to play in the medical field.

4

Industrial Robots

When robots were first introduced into manufacturing in the 1970s, many people were afraid that robots would replace human workers. They were afraid these workers would lose their jobs and be unemployed. Because of this resistance, robots were introduced slowly into manufacturing. Now robots are very common in manufacturing (Figure 18-11).

FIGURE 18-11 Industrial robots are used extensively in manufacturing today.

Robots do jobs that are dirty, repetitive, or dangerous for people. It is true that robots replaced some humans in these less desirable jobs. However, many new jobs were also created. Robots need to be manufactured, programmed, and serviced. Entire new businesses were developed to support automated manufacturing. There is a growing need for engineering

© ISTOCKPHOTO/GIANLUCA BARTOLI

FIGURE 18-12 An articulated robot.

technologists and skilled technicians to program and service industrial robots.

An **industrial robot** is a computer-controlled, reprogrammable, and multipurpose unit that can move in three or more directions. This is similar to length, width, and height (think 3-D). Imagine the movements your arm would go through if you wanted to touch each corner of your desk. Most industrial robots are basic "arm" robots. The technical name for this type of robot is an **articulated robot** (Figure 18-12). An articulated robot has two or more joints that swivel. An articulated robot mimics human motions. Many industrial robots are fixed in one place, swivel at the waist like a human, and reach out with an arm to either move a part or complete a task.

Typical industrial applications for robots include welding, painting, and assembly. Industrial robots can also be used for measuring and inspection in quality control procedures. Industrial robots may be programmed to work in teams to accomplish complex tasks (Figure 18-13). Other uses for robots in manufacturing include transporting parts around a factory and inventory storage and retrieval.

articulated robot
An articulated robot is a robot that has two or more joints that swivel.

Industrial robots are commonly used for "pick and place" operations. That means they can be used to place or remove parts on a conveyor belt or pick them up and place them in a carton for shipping (Figure 18-14). Robots can handle everything from fresh meat and dairy products to packaging games and puzzles. The gripper (hand) at the end of the robot's arms must be adjustable and have pressure sensors to avoid crushing soft products and to handle irregular shapes.

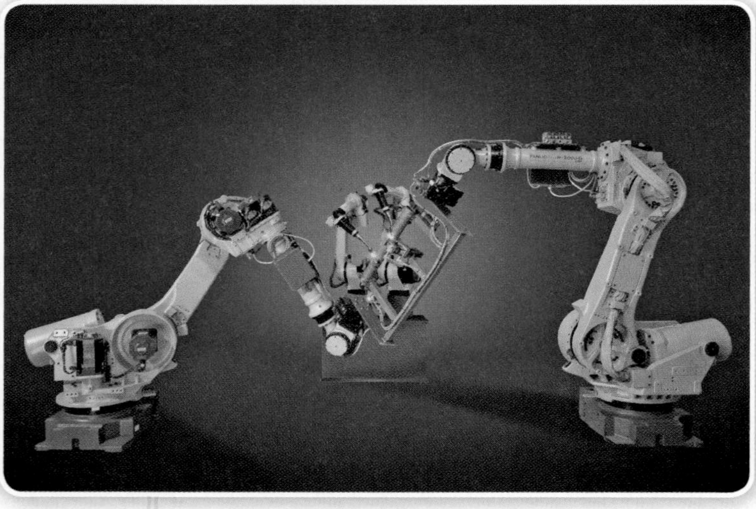

COURTESY OF FANUC ROBOTICS AMERICA, INC.

FIGURE 18-13 An articulated material-handling robot holds a part while another robot welds it. These two robots have to be programmed to work together.

Many factories have wires buried in the concrete floor along each aisle. These wires can send an electronic signal. Robots can be programmed to pull carts with raw materials or parts from one location in the factory to another following these wires. These robots are referred to as **automated guided vehicles (AGV)** (Figure 18-15). The newest AGVs can "learn" the layout of the plant. When programmed to make multiple stops, they can calculate the most efficient route and do not need to follow the wire in the floor.

Other robots are used to store or retrieve materials and parts from designated storage areas (Figure 18-16). Racks or shelves can store hundreds or thousands of bins that can be stacked several stories high. Each bin is numbered, and its location is programmed into a robot's memory. When it receives orders to place or retrieve parts, the robot can quickly find the right bin.

COURTESY OF FANUC ROBOTICS AMERICA, INC.

FIGURE 18-14 This robot is packing freshly baked goods for shipment.

Advantages of Robots in Manufacturing

There are many advantages to having robots work in manufacturing. Robots are able to work long hours and never get tired. They do not need lunch or coffee breaks like humans. Industrial robots are also able to move more quickly and more accurately than humans. These robots can repeat their programmed operation with a great deal of accuracy. The addition of robots into manufacturing has improved both the quality and the quantity of products manufactured.

COURTESY OF JBT CORPORATION

FIGURE 18-15 An automated guided vehicle (AGV).

COURTESY OF THE UNITED STATES DEPARTMENT OF DEFENSE

FIGURE 18-16 An automated storage-and-retrieval robot.

Robots in Military and Homeland Security

The military has made extensive use of robots. Robotic aircraft are called **drones**. Drone aircraft are able to fly over enemy airspace undetected because they are smaller and quieter than conventional piloted aircraft (Figure 18-17). As a result, they are able to photograph and videotape enemy movements on the ground.

© ISTOCKPHOTO/ALLAN TOOLEY

FIGURE 18-17 Drones are a type of robot that can provide aerial reconnaissance.

Security robots with video surveillance were first used to patrol a large-scale sporting event at the World Cup in Berlin, Germany, in 2006 (Figure 18-18). The onboard video and thermal imaging sensors can send a live feed to a security office. The robot, OFRO, also uses global positioning system satellites to steer and keep accurate records of its exact location. Manufactured by Robowatch, OFRO was the world's first widely sold outdoor mobile surveillance robot. OFRO and its sensors can function independently of weather conditions.

Robots in Space

Perhaps two of the world's most widely watched robots were the Mars Exploration Rovers *Opportunity* and *Spirit* (Figure 18-19). Designing robots to travel on the Martian soil was an incredible engineering challenge, but getting them there was an even bigger challenge. NASA engineers had to figure out not only how to get a spacecraft millions of miles across space to Mars but also how to get the robot safely to the Martian surface. Their solution was very creative: they used a very large balloon-like structure to completely surround the robot (see Figure 18-20).

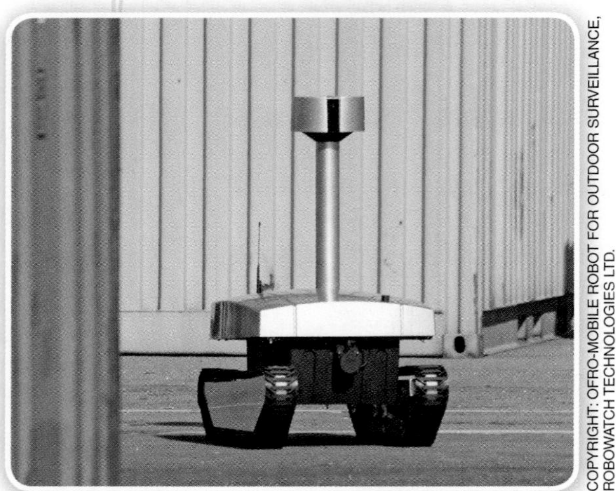

COPYRIGHT: OFRO-MOBILE ROBOT FOR OUTDOOR SURVEILLANCE, ROBOWATCH TECHNOLOGIES LTD.

FIGURE 18-18 OFRO, a surveillance robot, was used to patrol the World Cup in Berlin in 2006.

The mission raised lots of questions for NASA scientists. What would happen if the balloon-enclosed rover bounced when it hit the ground? What if it landed upside down? How would the robot get free of the fabric material? What kind of a robot could survive in the harsh Martian terrain, where nighttime temperatures drop to 250° below zero Fahrenheit? Also, Mars has the largest dust storms in our solar system. Would the dusty Martian soil clog the wheels

COURTESY OF NASA/JPL-CALTECH

FIGURE 18-19 An artist's rendering of the Mars Exploration Rover *Spirit*.

COURTESY OF NASA

FIGURE 18-20 NASA's solution to safely landing a robot on Mars.

and drive mechanism? What if the robot sank into the soil? Maybe the easiest problem the engineers solved was the power source: solar panels and rechargeable batteries. But could the batteries survive the long nights and the extreme cold temperatures?

NASA engineers and scientists learned a great deal from the Opportunity and Spirit robotic explorers. They applied that knowledge to design the next robotic explorer, Curiosity (Figure 18-21). Curiosity landed on Mars August 6, 2012, and immediately began transmitting images of the red planet. The engineers designed the rover to determine whether Mars has ever supported life, or if it could do so now. Due to Curiosity's size, NASA engineers could not use an airbag design like they did before. Instead, they designed a sky crane to lower Curiosity to Mars' surface (Figure 18-22).

COURTESY NASA/JPL-CALTECH

FIGURE 18-21 An artist's rendering of the Mars rover Curiosity.

COURTESY NASA/JPL-CALTECH

FIGURE 18-22 NASA engineers designed a tethering system to lower Curiosity to the Martian surface.

Being an engineer is exciting and rewarding work. In every facet of designing and building robots, engineers are faced with many challenges to overcome. Engineers are people who like solving problems and who like challenges. Sometimes their solutions are based on trial and error, but most often they apply their knowledge of mathematics and science to solve design problems. What problems or challenges would you like to solve? What will you design?

SECTION 2: ROBOTIC SYSTEMS

Remember from Chapter 2 that systems can be either open loop or closed loop. Robots are considered a closed-loop system. A robot can respond to changes and adjust its inputs or outputs. Sensors can provide the inputs to the program, or the feedback, that controls the robot's movement or actions.

Anatomy of a Robot

Robotic systems are similar to human systems. They are alike physically, but they do not have emotions or understand right from wrong. Physically, you have a bone structure to keep your body rigid. You have joints that hold your bones together and allow you to move. You have sensory perception through your eyes and ears and have a sense of touch and a sense of smell. You have muscles to power your movements. And you have a brain to tell your body how to move based on the input you receive from your sensory perception, usually your eyes.

Robots typically have the same general parts and functions your body has. They also have limbs, joints, sensory perception, a power source, and a computer for a brain. Most robots imitate human functions. Some robots are on wheels, some move up and down tracks in a fixed path, a few have legs for walking, and many are fixed in one place for a specific job in manufacturing.

Robots must be able to perform tasks to be useful. Typically, these are tasks that humans would have done. So a robot's anatomy often mimics human anatomy. An industrial robot must be able to swivel at the waist, raise and lower its arm, and operate a gripper or tool.

Robotic Arm The main body of most industrial robots is stationary. The robot has a joint that allows the body to rotate in place. The shoulder is the first joint connected to the main body, and it allows the arm to raise and lower. An industrial robotic arm with six joints can imitate the motion of a human arm. Fortunately, we need only a shoulder, elbow, and wrist to accomplish the same movements. This is because some of our joints can move in two or three directions, whereas each robotic joint can only move in one direction.

End Effector The purpose of your arm is to move your hand. Similarly, the purpose of a robotic arm is to move an **end effector**, an attachment that is designed for a specific task. It is placed at the end of a robot's arm. The end effector is attached to the wrist joint at the end of the arm. It may be a type of gripper, like your hand, or a tool (like a drill, welding rod, or paint sprayer). The end effector is what makes a robotic arm adaptable to many different tasks.

end effector
An end effector is the attachment at the end of a robot's arm that is used for a specific task such as holding a tool, painting, or welding.

One challenge for engineers has been to design grippers that have the sensitivity of a human hand. You can sense how tightly to grab something so that you will not drop it or squeeze it so hard that it breaks. NASA engineers have designed a robotic hand with fingers, but this is not yet available in factories. Many robotic grippers have two stationary fingers and one mobile finger (or thumb).

The end effector could also be a magnet. The robot would energize the magnet to pick up metal objects and then shut off the current to release the object. An end effector could also be a syringe that dispenses liquids such as perfumes into bottles. Other end effectors resemble small forklifts designed to lift and move pallets. There are literally dozens of types of end effectors, each designed to handle a certain type of product.

An end effector might also be a tool such as a welding gun. Welding is one of the most common uses of robots in industry. The automotive industry makes extensive use of robots in welding operations. Another very common use of robots is painting automobiles. In this case, the end effector is the paint-spraying apparatus.

Movement As we saw in Chapter 12, *pitch*, *yaw*, and *roll* describe movement in space. In this case, we are concerned with the movements of a robotic arm (Figure 18-23). Up and down motion is referred to as *pitch*. Movement to the left or right is called *yaw*. Rotating movement is called *roll*. Your shoulder can move in all three directions: it has pitch, yaw, and roll. Your elbow, however, has only pitch. Your wrist can move up and down, side to side, and can twist a small amount.

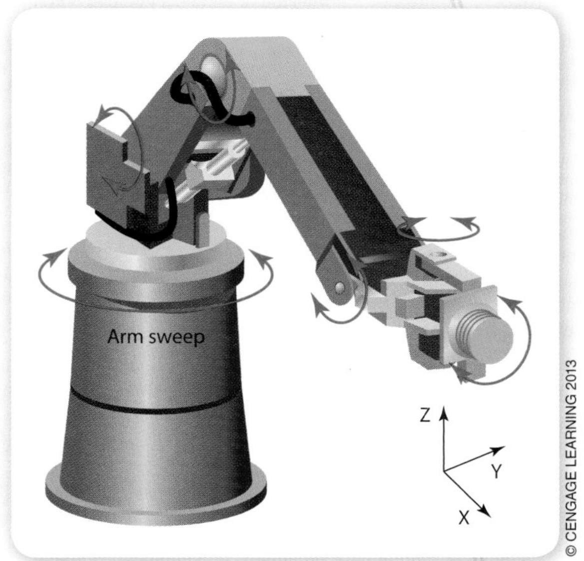

FIGURE 18-23 Robotic arm movements: pitch, yaw, and roll.

work envelope
A work envelpe is the area of normal reach for an industrial robot.

The term **work envelope** is used to describe the normal reach a robot has, typically referring to industrial robots that are fixed in one place. The work envelope is important to know when designing a manufacturing process. It is also important to know for safety reasons. Safety zones around robots need to be clearly marked to prevent injury to humans and damage to the robot.

Power for Robots

Robots are often a combination of electrical motors and fluid power (either pneumatic or hydraulic). Most industrial robots are connected to an electrical supply because they are stationary. Robots on wheels must have batteries. The Robomower and Roomba robots described earlier have batteries. When the batteries begin to run low, their program has them return to their base for charging. The NASA robots used to explore Mars had powerful batteries that could be recharged from solar panels.

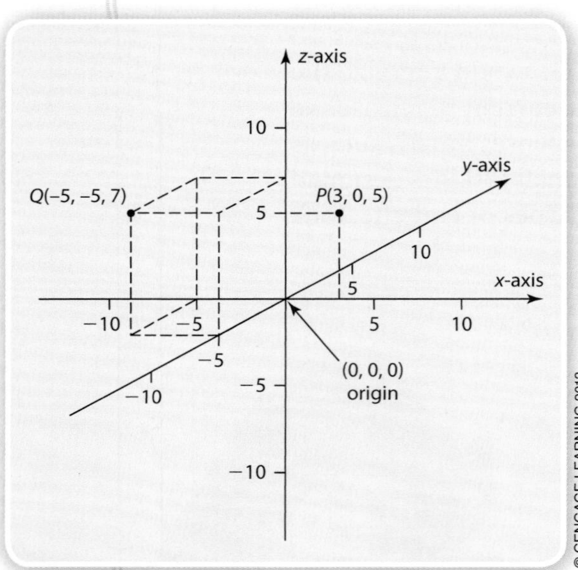

FIGURE 18-24 The positive and negative quadrants of the Cartesian coordinate system.

Coordinate System

Automated machine tools and robots move through three-dimensional space. We describe this space with the Cartesian coordinate system. Using your desk as an example, the desktop has width and depth and is some height off the floor. We commonly refer to the width as the *x*-axis, the depth as the *y*-axis, and the height off the floor as the *z*-axis.

Any point in space can be identified by these three axes. This requires having a common point of reference to measure from (Figure 18-24). This reference point is called the *origin*. Mathematically, it is the 0, 0, 0 point. It is customary to refer to the points to the right or above the origin as positive numbers. Points to the left or below the origin are referred to as negative numbers. These

Did You Know?

The Cartesian coordinate system is named after René Descartes, a French mathematician and philosopher, who first developed the idea of this mathematical system in 1637.

designations are crucial for controlling robots and machine tools. When an engineering technologist programs a robot to perform a task, each location the end effector moves to is identified as a point in space in the Cartesian coordinate system.

Controlling Robots

Robots are controlled by a series of commands in a computer program. Sometimes the program is a set procedure that is followed repeatedly. Other times the robot must react to information obtained through sensors. In both cases, a series of instructions is embedded in computer code.

Autonomous robots are robots that react to their environment to complete their tasks. For example, parts coming down a conveyor belt might be lying at different angles. An autonomous robot has some type of vision system that enables it to adjust its pickup position to accommodate each part.

autonomous robot
An autonomous robot is able to react to its environment to complete its task.

Robots are often programmed by engineering technologists who may have two- or four-year college educations. Many community colleges and universities offer programs in robotics. It is important to have a solid background in mathematics to understand how robots are programmed.

ENGINEERING CHALLENGE: YOUR ROBOTIC ARM

Purpose

To write a robotic program using the Cartesian coordinate system.

Goal

The goal of this activity is to move a penny from the origin and drop it into a cup.

Background

If you were to hold a penny over your desk, how would you describe its precise location? We use the Cartesian coordinate system to describe the exact location of objects in space. Robots move in three-dimensional space, just as we do. This system tells the robot each movement to make. It is customary to list the *x*-axis first, then the *y*-axis, and then the *z*-axis. The *z*-axis is usually the vertical distance.

Materials required

To complete this activity you will need a partner, three paper cups, masking tape, a penny or small coin, and a tape measure. Be sure to obtain your teacher's permission before you begin this activity.

Procedures

1. Use the masking tape to lay out an *x*-axis and a *y*-axis across the middle of your desk. Your tape should look like a large plus (+) sign.
2. Label the tape that crosses from left to right the *x*-axis; label the tape that runs top to bottom the *y*-axis. We will use a tape measure for the *z*-axis.
3. The origin is the point where the *x*- and *y*-axes cross. It is the (0, 0) point. Using the tape measure, mark off 1-inch increments along each piece of tape from the origin. Remember that numbers are positive above and to the right of the origin but negative to the left and below.
4. Have your partner place three cups at different locations on your desk. Place a penny on the origin point (0, 0).
5. Begin with the penny. Your goal is to pick the penny up and drop it into each cup, starting from the origin. Do not forget that your first movement should be to raise the penny off the desk, or on the *z*-axis, say, 5 inches. Use the tape measure to measure the *z*-axis.
6. Record each movement of your "robotic arm" in Cartesian coordinates to write a program.
7. Have your partner record each movement as you call it out. The following table provides an example.

Step number	Coordinates (*x*, *y*, *z*)	Explanation
0	0, 0, 0	Starting point. The penny is sitting on the origin.
1	0, 0, 5	Lifts the penny 5 inches off the desk on the *z*-axis.
2	6, 0, 5	Moves the penny 6 inches to the right on the *x*-axis.
3	6, 2, 5	Moves the penny 2 inches on the *y*-axis.
4		Open gripper and release the penny.
5	6, 2, 0	Penny is successfully dropped into the first cup.

8. Repeat the activity for the two remaining cups, starting the penny at the origin each time.
9. Your partner is responsible for checking the accuracy of your program.
10. Reverse roles with your partner and repeat the activity.

Career Spotlight

Name:
Stephanie Horne Swindle

Title:
Operations and Maintenance Manager, Southern Power

Job Description:
Swindle manages eight employees at a Georgia power plant. The plant is a "peaking plant," which means it runs only when the weather is really hot or really cold and the power grid needs a boost. "We always say we're on 'hot standby,'" she says. "It's almost like a last resort. We can run the units within 30 minutes so people will be able to flip their switch and have power."

Ms. Swindle is in charge of maintenance at the plant, and that involves some serious problem solving. She cannot fix things if the plant is running or is about to run. She must make hard decisions about when to act and when to wait.

Ms. Swindle combines a technical background with business skills. "When I started getting into a management role, I realized how important it was to understand accounting and economics," she says. "I have to maintain a very large budget at the plant. I'm always making business decisions: Do we hire this person or not? Do we replace this piece of equipment this year?"

Swindle has worked in many places besides management, including the instrument control shop, the mechanical shop, and operations. "Just because I'm in mechanical engineering doesn't mean I'm isolated in that one area," she says.

Education:
Swindle earned her bachelor's degree in mechanical engineering from the University of Alabama and her master's degree in business from the University of Alabama–Birmingham.

Advice to Students:
Swindle encourages students to get both a bachelor's and a master's degree in engineering. Anyone interested in management should get a master's degree in business as well. She recommends that students obtain the advanced degrees while they are young. "If people have been out of school too long, they have trouble getting back to it," she says. "They have spouses, kids, and other obligations. Before you have children, get all your education behind you as soon as possible."

Swindle says that being successful requires hard work. "I love my job," she says, "so it's not an issue for me to put in long hours. I believe in what I do and wouldn't want to do anything different."

SECTION 3: HISTORY OF ROBOTICS

No one knows for sure when the first robot was invented. Al-Jazari, an Arab Muslim inventor, made a number of automatic machines around A.D. 1200. One of his inventions was a boat with four automated musicians. The boat was simply used for entertainment at parties. He used a series of pegs in a drum that pressed little levers to move the musicians that played drums. Because the pegs could be placed into different holes, the rate of motion could be controlled. For this reason it is considered a programmable robot.

A Japanese inventor, Hisashige Tanaka, designed and made many complex mechanical toys. One of these mechanical toys was even capable of drawing arrows from a quiver and firing them. No wonder he has been called "Japan's Edison." He published some of these designs in 1796.

The word *robot* was first used in a play written by Czechoslovakian Karel Capek in 1920. The play was called *Rossum's Universal Robots*. The word *robot* is derived from the Czech word *robota*, which means "forced labor."

Isaac Asimov, a science fiction writer, used the word *robotics* in 1941 to describe the technology of robots. He was the first to predict that robotics would become a powerful new industry. In 1942, Asimov wrote "Runaround," which is a story about robots. In this story, he developed the three laws of robotics (Figure 18-25). These "laws" have been widely referenced by roboticists ever since they were first published.

The Second Industrial Revolution

George Devol invented the first industrial robot arm in 1954, but it was not until 1961 that the Unimate began working in a General Motors factory. It was a simple pick-and-place type of robot. By the 1980s, industrial robots had become so widespread that it could be considered the second Industrial Revolution.

Remember that the first Industrial Revolution, discussed in Chapter 2, had profound effects on labor, the economy, and social life. The introduction of robots into manufacturing had the same type of effects.

- A robot may not injure a human, or, through inaction, allow a human being to come to harm.
- A robot must obey the orders given it by human beings except where such orders would conflict with the First Law.
- A robot must protect its own existence as long as such protection does not conflict with the First or Second Law.

FIGURE 18-25 The three laws of robots.

The Future of Robots

Today, robots have affected almost every aspect of our lives. The egg-shaped Miuro is an example of how a robot can be both fun and functional. (Miuro stands for Music Innovation based on Utility Robot technology.) Miuro was designed to roll around your house playing tunes from your iPod player wirelessly wherever you are (Figure 18-26). This is just one example of how robots will soon become our personal companions.

Remember Honda's ASIMO? Honda engineers take ASIMO around the world to encourage and inspire students to study science and engineering. In the very near future, ASIMO may be able to serve as the eyes, ears, hands, and legs for people with disabilities. As a personal assistant, ASIMO might be able to assist the elderly or handicapped in and out of bed or even a wheelchair. And ASIMO might also be able to complete tasks that are too dangerous for people, such as cleaning up toxic spills or fighting fires. What a wonderful assistant that would be!

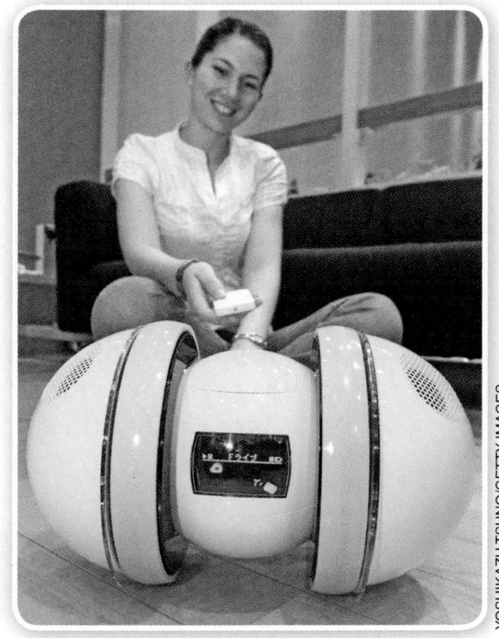
YOSHIKAZU TSUNO/GETTY IMAGES

FIGURE 18-26 Miuro will follow you around while playing your favorite tunes from your iPod.

SUMMARY

In this chapter you learned:

- There are more personal robots in homes than there are robots in industry.
- A robot is a mechanical device that acts in a seemingly human way but is most often guided by automatic (usually programmable) controls.
- Robots may be classified as personal (household), assistive, medical, industrial, military or security robots, and space exploration robots. Each type of robot has different uses, but all robots have similar systems.
- The primary purpose of a robot is to relieve humans of work that is tedious, repetitive, hazardous, dirty, or requires very precise accuracy.
- Robots can often perform some types of work more accurately than humans, including medical surgery.
- The addition of robots into manufacturing has improved both the quality and the quantity of products manufactured.
- Robotic systems are similar to human systems.
- Autonomous robots are robots that are able to react to their environments and adjust their actions.
- The introduction of robots has affected manufacturing and society as profoundly as the Industrial Revolution.

VOCABULARY

Write a definition for each term in your own words. After you have finished, compare your answers with the definitions provided in the chapter.

Robot
Robotics
Servomotor
Articulated robot
End effector
Work envelope
Autonomous robot

STRETCH YOUR KNOWLEDGE

Provide thoughtful, written responses to the following questions and assignments.

1. What advantages and disadvantages do you think we will see if we develop personal robot assistants? Will the advantages outweigh the potential problems?
2. Should limitations be placed on the use of robots?
3. What impact do you think robotics will have on your future?
4. Why have manufacturing companies replaced humans with robots? What are the advantages of using robots?
5. Conduct research to create a time line of significant robotic developments. Present your findings to your class through either a presentation or a time-line poster.

Onward to Next Destination

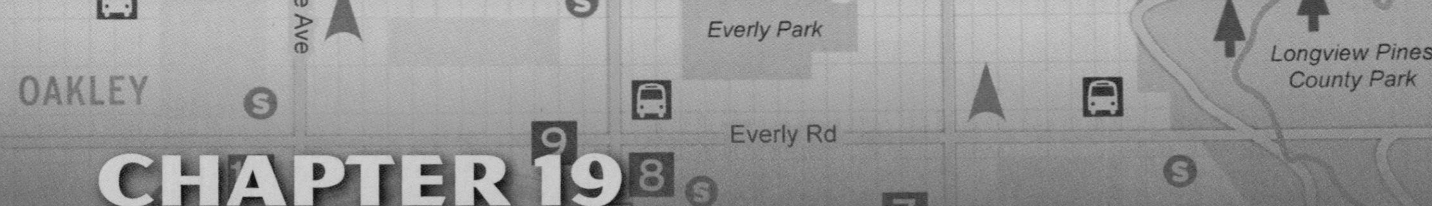

CHAPTER 19
Automation

Menu

Before You Begin

Think about these questions as you study the concepts in this chapter:

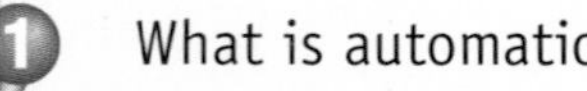

1. What is automation?

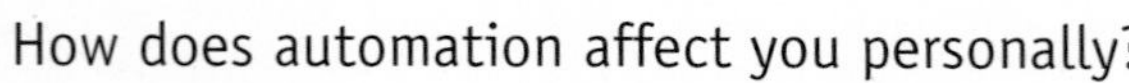

2. How does automation affect you personally?
3. What does the term *automated manufacturing* mean?
4. What is CAD, and what is CAM?
5. What does a workcell look like?
6. What is flexible manufacturing?

© ISTOCKPHOTO/RTIMAGES

FIGURE 19-1 Text messaging involves an automated process.

Engineering in Action

When you text message a friend, do you ever wonder how that message magically gets to your friend? We take many daily activities for granted without realizing that they are automated processes that have been designed to make our lives more enjoyable. Downloading music and new ringtones, using cell phones for a dozen things other than talking, playing the latest video game, and surfing the Web are all examples (Figure 19-1). Thank an engineer if you enjoy any of these activities.

Many of these daily activities are based on automation. Our lives would certainly be harder, perhaps even boring, without engineers designing new ways to automate daily tasks. Soon after being introduced to the market, the magical nature of the latest automation becomes commonplace, and we cannot imagine living without it. Have you ever seen old movies in which early automobiles had cranks on the front and someone had to physically crank the engine to get it started? How much easier our lives are with keys or push buttons to start our cars. In cold climates, many people even have remote starters so their cars will be warm when they get into them.

SECTION 1: AUTOMATION IN YOUR WORLD

Automation is a system in which some action happens as a result of another event and without human intervention. We use the term to refer to robots and the control of machines in manufacturing, but automation is much more than just manufacturing. Automation is present in many aspects of your personal life on a daily basis. Our lives are better as a direct result of automation.

Automation requires a closed-loop system. Often a sensor provides an input signal, usually to a computer program of some sort. The automated device performs a task or routine until another input signal tells it to stop when a certain condition has been met. These sensors can be such things as light sensors, thermometers, timers, motion detectors, and pressure sensors. Automation in many forms saves lives and makes life easier (Figure 19-2).

Your cell phone is an example of automation at work. When you are traveling in a car, you move out of the range of one cellular tower and into the range of another. Your phone will connect to the next tower, which sends a signal to the phone company to confirm your identity as a customer. All of this happens in an instant without you even being aware of it while you continue to talk (Figure 19-3).

FIGURE 19-2 Automatic garage door sensors prevent children and pets from being crushed under a closing door.

automation
Automation is a system in which some action happens as a result of another event without human intervention.

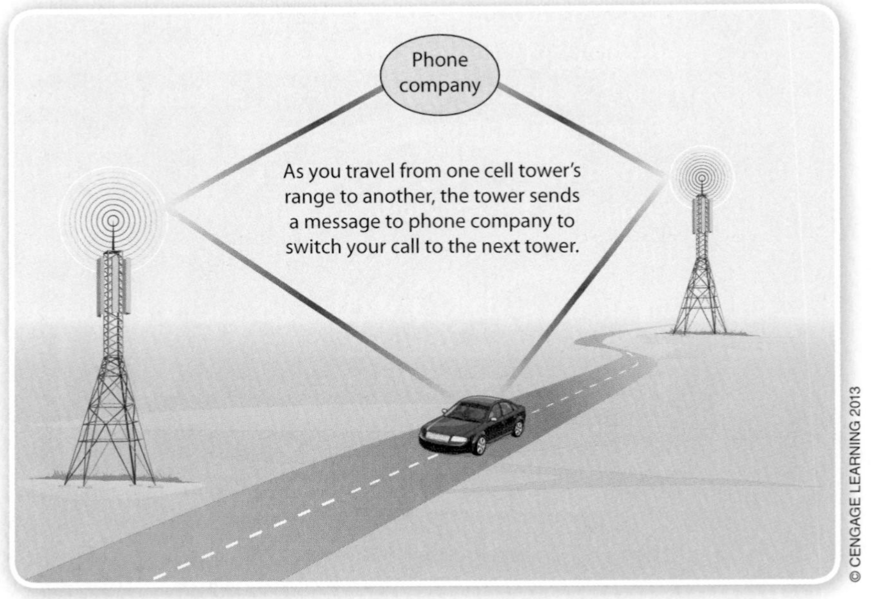

Phone company
As you travel from one cell tower's range to another, the tower sends a message to phone company to switch your call to the next tower.

FIGURE 19-3 Cell phone networks automatically switch your call from tower to tower as you travel.

Science in Engineering

An understanding of physical science principles is very important in designing automatic devices. For example, physical sciences explain how various sensors work. Many sensors are based on wave theory. Sound and light, for example, travel in waves. Sensors are devices that detect conditions in the environment (such as sound, light, or motion) to provide input or feedback to an automatic system.

Automation in Your Home

Do you ever wonder how your house stays at a constant temperature? You have a thermostat, usually located on an interior wall, that triggers the furnace to come on when the temperature drops 1 or 2 degrees. Then the thermostat will shut the furnace off when the room reaches the desired temperature setting (Figure 19-4). Why do you suppose the thermostat is usually located on an interior wall?

FIGURE 19-4 Thermostats are an example of an automated system.

Does your house have a basement or do you know someone whose house does? Have you ever known someone whose basement has flooded? Most basements have a sump pump, a type of pump that is designed to sit in a hole below the basement floor. As groundwater runs into the hole, the sump pump is activated and pumps the water up and out of the hole to the outside ground (Figure 19-5). What causes the pump to turn on and then shut off?

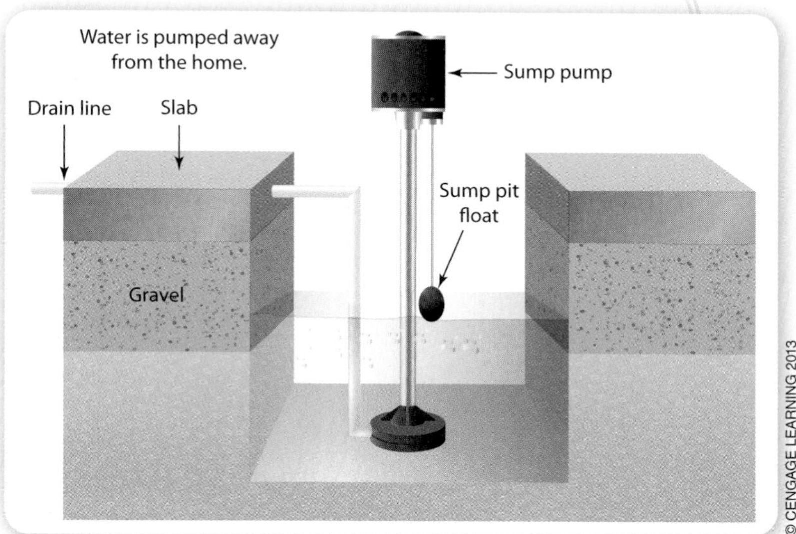

FIGURE 19-5 Sump pumps automatically pump groundwater out before it can seep into a basement.

Automation in Your Car

Many cars are equipped with automatic door locks. As soon as your car reaches a certain speed, the doors will lock automatically. Many cars today are also equipped with antilock brakes. These brakes are designed to keep the wheels from skidding when the breaks are applied. Sensors on each wheel release the brake for a fraction of a second when that wheel starts to skid. How do you think the sensor can tell the difference between when the wheel is skidding and the car is stopped?

Can you think of other everyday examples of automation? Perhaps when you are riding in the car you have noticed that some traffic signal lights provide a left-turn arrow only when a car is waiting to turn left. How does the signal know when a car is waiting to turn left? Or you may have noticed that streetlights come on automatically at dusk. How do the streetlights know when to turn on and off?

Automation and Your Health

Diabetics must monitor their blood sugar levels many times each day. Although some diabetics can control their disease with diet alone, those who have severe cases must take regular insulin shots to help their bodies process sugar. Engineers have now designed insulin pumps that diabetics can wear on their belts; the insulin is delivered as needed through catheters that the diabetics insert into their abdomens (Figure 19-6). An individual

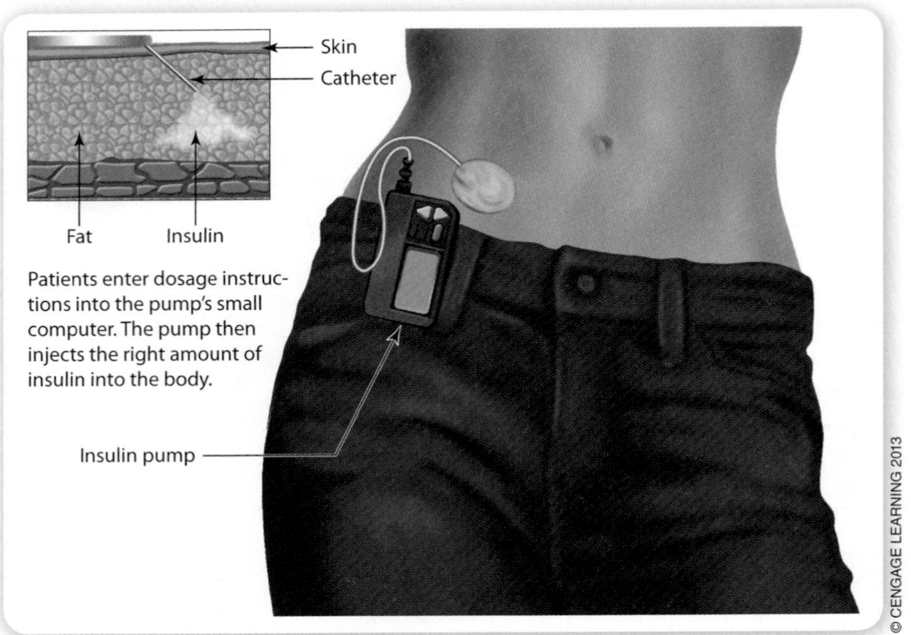
Skin
Catheter
Fat
Insulin
Patients enter dosage instructions into the pump's small computer. The pump then injects the right amount of insulin into the body.
Insulin pump
© CENGAGE LEARNING 2013

FIGURE 19-6 An automated insulin pump helps diabetics lead more worry-free lives.

Math in Engineering

Digital electronics is what makes all kinds of miniature devices possible, from watches to pacemakers. Mathematics is the language of digital electronics. The language is a very useful tool that allows engineers to design lifesaving devices. Mathematics, digital electronics, and engineering are powerful partners.

can also override the system and add an additional injection if needed. This is a lifesaving use of automation.

A pacemaker is another example of a lifesaving device designed by engineers. This small, battery-operated device helps ensure that a patient's heart beats regularly and at an appropriate rate (Figure 19-7). A pacemaker can usually sense if a heartbeat is above a certain rate, at which point it will automatically turn off. Likewise, the pacemaker can sense when the heartbeat slows down too much and automatically starts pacing again. More than 100,000 pacemakers are surgically implanted each year in the United States.

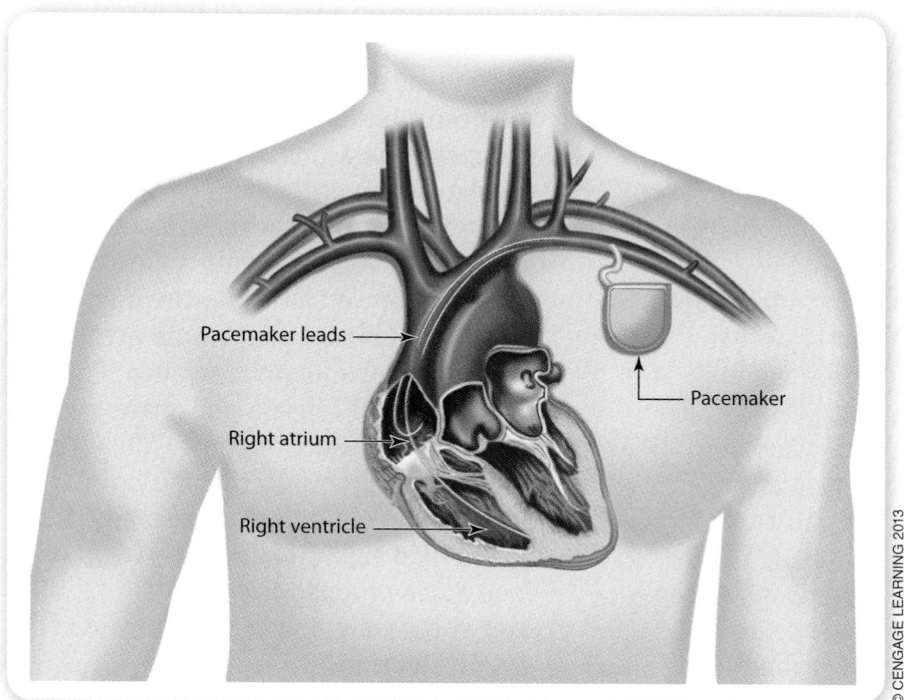

FIGURE 19-7 **Pacemakers are automated devices that help maintain a regular heartbeat.**

Engineering Challenge

The purpose of this activity is to think about the many ways in which automating routine tasks can simplify your life. Learning to think like an engineer is an important skill that will help you later in life, no matter what career path you pursue. Engineers analyze problems or situations and then develop practical solutions.

Your engineering challenge is to think about the many routine activities in your life, select one, and then design an automated solution to handle the task for you. Remember, to automate a system it must be able to operate without human intervention. You may not be able to actually develop a prototype of your design solution, but you can describe the different parts and how they would function together. Share your design idea with your class.

1. Begin by selecting a task or activity that you have to do frequently. The task you select should be one that you would be glad to have a machine do for you.
2. Precisely describe the activity that must be completed. For example, perhaps you have to feed your dog at 6 A.M. and 6 P.M. every day, and you do not like having to get up and feed him so early. (Sorry, but you cannot use this example.)
3. Describe the process that must be used to complete the task. For this example:
   a. Retrieve the bag of dog food from the cupboard.
   b. Fill a 1-cup scoop with dog food from the dog food bag. Remember, too much food or too little food is not good for your dog.
   c. Fill the dog's bowl with dog food from the 1-cup scoop.
   d. Place the scoop back in the dog food bag.
   e. Close the dog food bag and put it away.
4. Now think about different ways in which the food could be placed in the bowl. For example, you could premeasure the food into separate containers in advance. Another idea is to have a holding tank that will release exactly 1 cup of dog food.
5. Describe what input would start the process you decided on in step 4. If your dog were trained to step on a lever to dispense his food, then he would be able to eat as much as he wanted, so this would not be a good solution. A better solution might be to use an alarm clock in a circuit that would trigger the release of the dog food.
6. Describe the different technological processes involved in your solution. In the example we are using, there might be a digital circuit that would activate a programmable logic controller (PLC). The PLC would cause a pneumatic cylinder to push a door open to allow the dog food to drop into the dog bowl. There would have to be a mechanism in the food tank to allow just 1 cup of food to be dropped each time.
7. Sketch a simple drawing to show your design solution.
8. Combine the sketch and your written descriptions into a project proposal.
9. Present your proposal to the class.

SECTION 2: AUTOMATION IN MANUFACTURING

The factory of the future will have only two employees, a man and a dog. The man will be there to feed the dog. The dog will be there to keep the man from touching the equipment.

—Warren G. Bennis

This statement was written long ago and before automated manufacturing had evolved to the level it has today. Although amusing as a quotation, it is not far from the truth. Today we have factories that literally work through the night with the lights turned off and no humans present (Figures 19-8 and 19-9). Much of today's manufacturing equipment is so sophisticated that it takes highly trained specialists to program and operate it. Engineering technologists are very important people in today's automated manufacturing.

automated manufacturing
Automated manufacturing is the general term for computer-controlled machine tools and robots in manufacturing.

Automated manufacturing is the general term used for computer-controlled machine tools and robots in manufacturing. Computers are used in every part of modern manufacturing. They are an integral part of every process from design, materials procurement, production, and distribution (Figure 19-11). In this section, you will read about how computers are involved in the actual design and production of consumer products.

© ISTOCKPHOTO/RICARDO AZOURY

FIGURE 19-8 Modern manufacturing is largely an automated process.

GETTY IMAGES

FIGURE 19-9 Even marshmallow peeps are manufactured using automation.

Did You Know?

Many manufacturing facilities today are as sterile as a hospital operating room. High-tech manufacturing of products such as computer chips require a sterile, dust-free environment. Static electricity must also be controlled. Manufacturers may also take other measures to minimize product and machine contamination. People who paint Harley Davidson motorcycle frames, for example, sign a contract and agree not to wear any jewelry, deodorant, or makeup, and the brand of shampoo they can use is also regulated. These employees wear the same type of sterile protective clothing that surgeons wear (Figure 19-10).

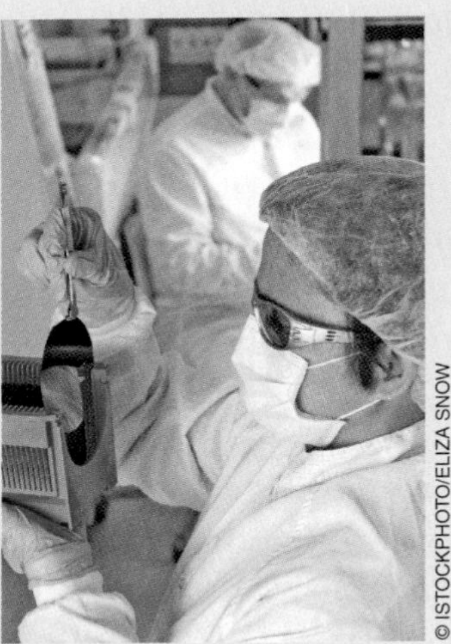

FIGURE 19-10 *Many manufacturing facilities today are as sterile as an operating room.*

FIGURE 19-11 *Computers are involved in every phase of manufacturing.*

The primary advantages of automation in manufacturing include lower costs for the consumer and better-quality products (Figure 19-12). Automation allows a company to produce more products in less time and with fewer mistakes. In the past, parts or products that were not made correctly were a large expense for a company. These parts often became scrap waste. Automation has greatly increased product quality and reduced the waste of natural resources. In this way, automation has also helped protect the environment.

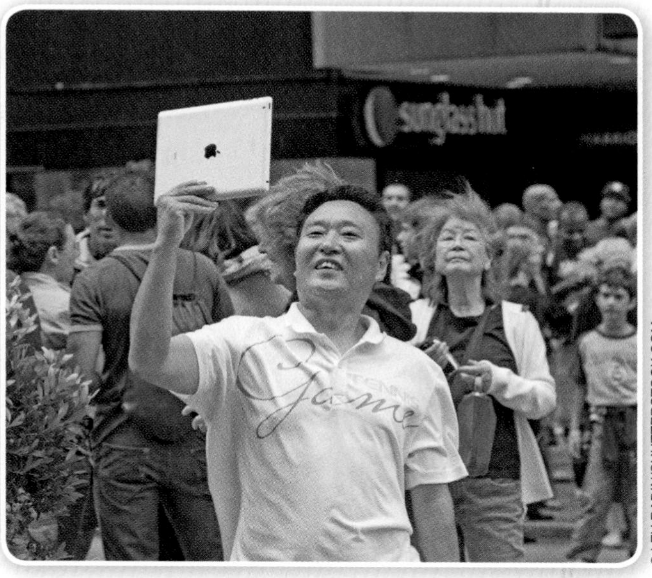

© LEV RADIN/SHUTTERSTOCK.COM

FIGURE 19-12 Consumer products such as Apple iPad® are a benefit of automation in manufacturing.

Computer-Aided Design

The manufacturing process begins with a well-designed product. Although engineers often begin by sketching their ideas (see Chapter 4), the actual design is done with sophisticated **computer-aided design (CAD)** software on powerful computers (Figure 19-13). CAD is a faster process than manual drafting. And CAD software can also generate three-dimensional (3-D) models on the computer screen to allow engineers to rotate their design to see all sides of the part.

Three-dimensional models also allow engineers to test their designs before the part or product is actually made. The CAD software can also run a stress analysis on a part to determine fracture points. Engineers are able to see what their part will look like, how it will work with other parts, what materials are best to use, and whether any changes are needed in the design.

CAD systems are also useful for estimating material costs. The CAD software has libraries of different materials that allow an engineer to compare the cost, strength, and other properties of each. With larger products made of multiple parts, the CAD software can also quickly calculate new product costs for each part that is changed.

COURTESY OF KELLI J. MCGREGOR, WESTFIELD HIGH SCHOOL

FIGURE 19-13 Computer-aided design software can generate 3-D models that allow engineers to test their designs.

FIGURE 19-14 Computer-aided manufacturing is the process of using computers to control machines in the manufacturing process.

computer-aided manufacturing (CAM)
Computer-aided manufacturing is the process of using computers to control machines used in the manufacturing process.

computer numerical control (CNC)
Computer numerical control is a programming language or code that tells machines exactly what to do and where to move.

computer-integrated manufacturing (CIM)
In computer-integrated manufacturing, the entire manufacturing process is coordinated through computers.

Computer-Aided Manufacturing

Computer-aided manufacturing (CAM) is the process of using computers to control the machines used in the manufacturing process (Figure 19-14). These machines read a code called **computer numerical control (CNC)**. CNC is a programming language that tells machines exactly what to do. In the case of a CNC milling machine, the code tells the machine how fast to spin the cutting tool, where to start the cut, how deeply to cut into the material, which direction to cut, and how fast to move the cutter through the material.

CAD software can also generate the cutting instructions for machine tools. When the CAD station is connected directly to a CAM machine, the process is called CAD/CAM. This eliminates the need for a human to program the CNC instruction into the machine. Typically, the CAD software will perform a virtual cutting process on the computer screen for the engineer to approve. Then the CAD station sends the necessary CNC codes directly to the machine through the network.

Computer-Integrated Manufacturing

Computer-integrated manufacturing (CIM) is really an expansion of CAD/CAM. In CIM, the entire manufacturing process is coordinated through computers. This includes robots to load and unload parts for the machine tools, conveyor belts to transport parts in process between workstations, and often the delivery of raw materials and pickup of finished parts by automated guided vehicles.

Engineers and managers can check on any phase of the manufacturing cycle from their offices. The CIM process will tell them when they need to order additional materials and when to arrange for shipping finished products. The CIM software will also tell engineering technologists when a machine needs servicing or if a cutting tool needs to be replaced with a sharper one.

CIM is very helpful to managers because it keeps a computer record of every phase of the manufacturing process. Managers can use this information to analyze the efficiency of their operations. Good record keeping allows for more efficient operations that will lower costs and increase profits for the company.

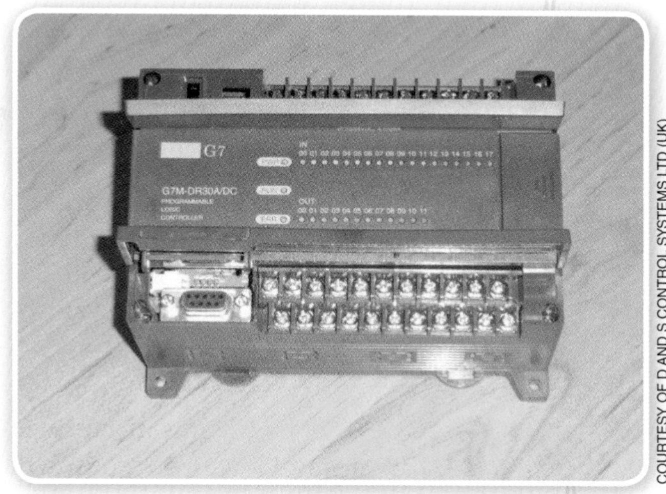

COURTESY OF D AND S CONTROL SYSTEMS LTD (UK)

FIGURE 19-15 **A programmable logic controller is a small computer that is used to control machines or the actions of automated devices.**

Programmable Logic Controllers A **programmable logic controller (PLC)** has many uses in manufacturing. A PLC is a small programmable computer that gives a machine or device specific, limited directions (Figure 19-15). PLCs are often connected to some type of a sensor. When the sensor sends a signal to the PLC, the PLC will tell the machine what to do. For example, a PLC on an assembly line could help route boxes to the appropriate location for shipping. A sensor would read a bar code label on the box as it is moves down the conveyor and sends the signal to the PLC. The PLC would then send a signal to a pneumatic piston to push the box onto a side conveyor for the appropriate destination.

programmable logic controller (PLC)

A programmable logic controller is a small programmable computer that gives a machine or device specific and limited directions.

Workcells

A workcell in manufacturing is designed to improve the quality, speed, and cost of production. This involves having the necessary resources readily available in the same area. The **workcell** might include robots and machine tools working together to accomplish several processes on the same part. The idea is to perform several processes quickly in the same location without wasting time moving parts long distances by forklift or conveyor. The adage "Time is money" is especially true in manufacturing today.

Workcells are a recent innovation in manufacturing. In the traditional manufacturing plant, each machine or station was designed to accomplish a specific task. Everything was highly specialized, including the people who operated the machines. In a workcell, the opposite is true. Each station is

workcell

A workcell is a group of machines and devices such as robots that work together to accomplish several processes on the same part.

designed to accomplish several tasks without transporting the part to another station.

Not all workcells are completely automated. Workcells that are not fully automated require teams of people who are versatile. That means they are trained in more than one skill. Managers, engineers, and skilled laborers work together to ensure that the highest-quality product is made with the least amount of wasted movement, time, or resources. Teamwork is an important skill to develop. The ability to work well with others is one of the primary traits employers look for when they hire new employees.

Flexible Manufacturing Systems

In the past, manufacturing assembly lines were often dedicated to making a specific part or product. They had dedicated machines and processes that could not easily accommodate changes as consumer demand changed. As a result, some manufacturing firms were unable to keep pace with market demands for different products and went out of business. Many employees lost their jobs.

flexible manufacturing system (FMS)

A flexible manufacturing system is a system that uses sophisticated machines that can perform more than one type of process through rapid tool changes.

A **flexible manufacturing system (FMS)** was a new approach to manufacturing. FMS is a manufacturing process that allows much greater flexibility in production. FMS uses sophisticated machine tools that can perform more than one type of task through rapid tool changes. In FMS, programmable robots and automation are used (Figure 19-16). This enables machines to easily accommodate new designs and part specifications. In the past,

REIS ROBOTICS-TWO ROBOTS IN MASTER/SLAVE OPERATION FOR AUTOMATED PRODUCTION OF SCAFFOLD COMPONENTS

FIGURE 19-16 Flexible manufacturing systems allow companies to make rapid adjustments in production.

employees often had to be retrained when there were significant changes in the production process. In FMS, you simply need a new computer program.

FMS typically relies on three systems working together. First, as mentioned previously, CNC machine tools can be programmed to perform more than one process. Second, an automated material-handling system delivers the appropriate raw materials just when needed and removes the finished stock. Third, a central computer system coordinates the CNC machine tools, robots, and material-handling processes (Figure 19-17).

The main advantage of FMS is its ability to respond quickly to changes in market demand. Other advantages include (1) increased productivity and improved product quality, largely because of automation; (2) reduced setup and preparation time to run a new product line; (3) lower labor costs, again because of automation; and (4) the ability to run small or large batches of different parts or products.

COURTESY OF FANUC ROBOTICS AMERICA, INC.

FIGURE 19-17 Automation also includes robots making robots. This robotic arm is performing tasks to manufacture robot parts.

SUMMARY

In this chapter you learned:

- Automation is a system in which some action happens as a result of another event without human intervention.
- Automation affects your personal life in many ways.
- Automated manufacturing methods lower costs and improve product quality.
- Computers play an important role in every aspect of manufacturing.
- Computer-aided design software is used to design products.
- Computer-aided manufacturing is the process of using computers to control the various machines used in the manufacturing process.
- Computer numerical control is a programming code that tells machine tools what to do.
- CAD software can generate the machine code for a CNC machine tool.
- In computer-integrated manufacturing, the entire manufacturing process is coordinated through computers.
- Flexible manufacturing system is a manufacturing system that can respond more quickly to changes in market demand.

VOCABULARY

Write a definition for each term in your own words. After you have finished, compare your answers with the definitions provided in this chapter.

Automation
Automated manufacturing
Computer-aided manufacturing (CAM)
Computer numerical control (CNC)
Computer-integrated manufacturing (CIM)
Programmable logic controller (PLC)
Workcell
Flexible manufacturing system (FMS)

STRETCH YOUR KNOWLEDGE

Provide thoughtful, written responses to the following questions and assignments.

1. In what ways does automation affect your personal life? Describe at least three examples other than those provided in the chapter.
2. Describe three advantages of automated manufacturing compared to traditional manufacturing processes.
3. Compare three advantages of computer-aided design over conventional board drafting.
4. In what ways have computers improved the manufacturing process?
5. Why is a workcell an improvement over traditional manufacturing?
6. Compare and contrast a flexible manufacturing system with traditional manufacturing.

Onward to Next Destination

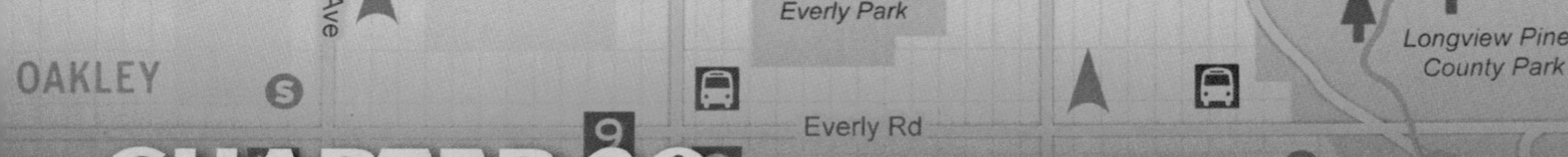

CHAPTER 20
Emerging Technologies

Menu

Before You Begin

Think about these questions as you study the concepts in this chapter:

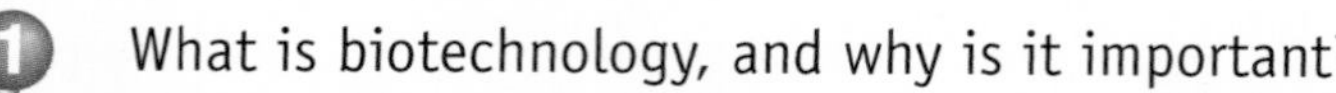

1. What is biotechnology, and why is it important?

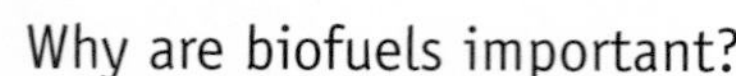

2. Why are biofuels important?

3. Can we really make electricity from garbage?

4. Can I clone my pet?
5. What are some of the benefits of nanotechnology?
6. What size is a nanorobot?
7. How is telemedicine similar to a video game?
8. What is prosthesis technology?

© ISTOCKPHOTO/JEFF NAGY

Engineering in Action

S T E M

More than 150 million laptops are sold annually around the world, more than the number of desktop computers sold. Affordable, available laptop computers are a positive technology outcome for people everywhere, but the increase in laptop sales has also produced an unintended outcome for the environment.

Many people keep their laptop computers for only about three years or so before they throw them away. This throwaway habit creates about 19,000 tons of waste each year from laptops alone. And that waste can harm the environment: laptop computers are made from plastic and contain toxic materials that can leak out into the environment.

Potential environmental damage from old laptop computers is just one example of problems today's students (tomorrow's engineers) will have to solve. Engineers are already working to reduce the harmful effects that disposable electronic products can have on the environment. To solve this growing problem, tomorrow's engineers must think ahead about their products' lifecycles and create innovative options to reuse product parts through recycling. Protecting the environment through careful product engineering will be even more important in the future.

biotechnology
Biotechnology is a technological process that uses living or once-living organisms to make or modify products.

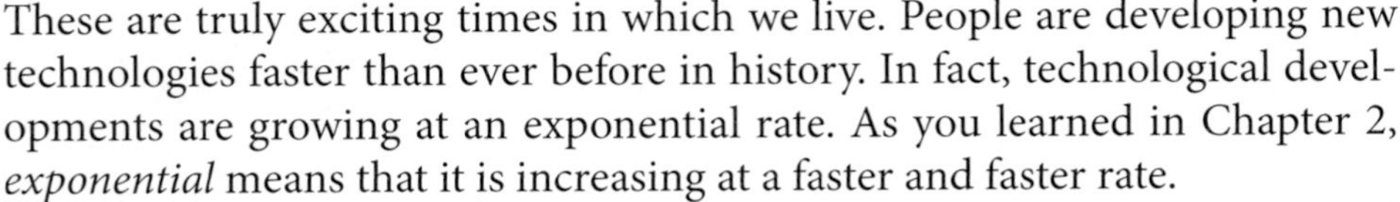

These are truly exciting times in which we live. People are developing new technologies faster than ever before in history. In fact, technological developments are growing at an exponential rate. As you learned in Chapter 2, *exponential* means that it is increasing at a faster and faster rate.

To compete successfully in the twenty-first-century global economy, and to find exciting and rewarding careers, you will need to understand new emerging technologies. Three key emerging technologies are biotechnology, nanotechnology, and medical technologies. Careers in these fields will require a solid foundation in technology, mathematics, and science.

1 SECTION 1: BIOTECHNOLOGY

Biotechnology is one of the fastest-growing fields today. **Biotechnology** is technology based on biology. In other words, it is the technology based on things that were once alive. Biotechnology processes have actually been used for centuries. However, the modern use of the term *biotechnology* is only 50 years old. In the last two decades, there have been many major developments in biotechnology. In this section, we will explore just a few of the many ways in which biotechnology can be used to improve our lives.

biofuel
Biofuel is any fuel made from biomass, such as ethanol.

2 Biofuels

Biofuel refers to any fuel made from biomass. As you may remember from Chapter 9, *biomass* is the term used to describe any organic material made from plants or animals. Biomass can be made from agricultural crops, forest by-products, and even city garbage. Ethanol is the most widely used biofuel today. Also, used cooking oil can be converted to a fuel source to power cars.

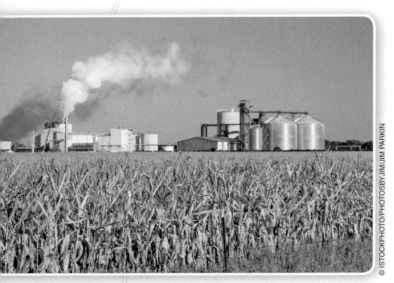

© ISTOCKPHOTO/PHOTOSBY.JIM/JIM PARKIN

FIGURE 20-1 *We can fuel our cars with biofuels such as ethanol, which can be made from renewable agricultural products instead of oil.*

Engineers are interested in biofuels because they are a renewable energy source. Oil is an exhaustible (nonrenewable) energy resource that pollutes the environment when burned. Biofuels are more environmentally friendly than fossil fuels and do not pollute the air when they burn as fossil fuels do. Also, biomass is a resource that is more evenly distributed around the world than oil.

Ethanol is an alcohol fuel made from sugars and starch found in plants. In the United States, corn is the most common source of starch for making ethanol (Figure 20-1). Wheat, corn stalks, straw,

Science in Engineering

Two chemical processes are needed to make ethanol. First, the biomass is converted to sugar through a process called hydrolysis. Then the sugar is converted to ethanol through fermentation. Fermentation is really a series of chemical reactions.

and fast-growing trees are also good sources of biomass to make ethanol. Ethanol produced from grain (such as corn) is the most common. However, this product alone cannot begin to satisfy our need for fuel. New ways to produce bioethanol need to be developed in the near future.

Biodiesel is another type of biofuel that can be made from new or used vegetable oils and animal fat. Biodiesel is a fuel that burns cleanly and can replace petroleum-based diesel. This is a good substitute because it is non-toxic and does not contain sulfur. Another advantage is that the used oil is not dumped and therefore will not seep into underground water supplies, rivers, or oceans.

biodiesel
Biodiesel is a type of biofuel that can be made from vegetable oil or animal fat.

Biopower

Biopower is the use of biomass to generate electricity. Biopower could also be called *biomass power*. There are several ways to generate electricity using biopower. The most common method is a direct-fired system: biomass (such as plant stalks) is burned to heat water and create steam, which then drives a turbine (see Chapter 9). The turbine turns a generator that produces electricity. In efficient systems, the steam could also be used to heat buildings after it passed through the turbine.

Biomass can also be used to create biopower through a process called **anaerobic digestion**. Basically a "digester" is a tank in which decomposing biomass produces a gas (not gasoline, but a vapor) (Figure 20-2). With a special type of engine, this gas can produce renewable electricity.

This anaerobic digestion process is similar to what happens in a composting pile in your backyard. Have you ever let grass and leaves sit in a pile for more than a day? They will begin to decompose, which generates heat that you can feel with your hands or see rising off the pile on a cold fall day.

Animal waste on a farm can also be used as biomass to produce electricity. This new technology holds great promise for disposing of municipal and animal wastes in a positive way. The benefits of producing electricity from biomass include:

- It requires little energy to produce electricity.
- It is more environmentally friendly than burning fossil fuels.

biopower
Biopower is the use of biomass to generate electricity.

anaerobic digestion
Anaerobic digestion is a process in which decomposing biomass in a tank produces a gas vapor that can be burned in a special engine to produce renewable electricity (one form of biopower).

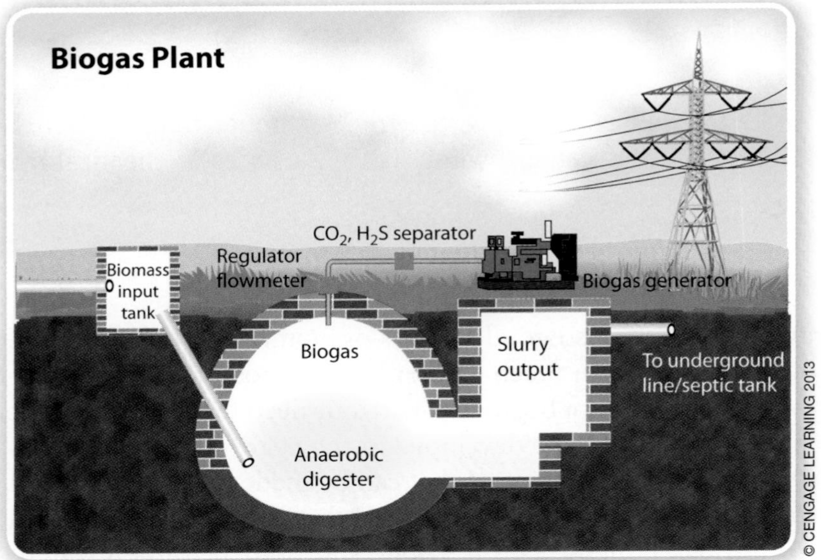

FIGURE 20-2 An anaerobic digester is an efficient and environmentally friendly way to produce electricity.

- It requires little space compared to other electricity-generating plants.
- It has low operating costs because the source is readily available.
- The process is easily automated and requires little human power.
- It is a sustainable technology.

Food Production

Genetic improvements in livestock are a practical example of how biotechnology research and engineering can directly benefit society. Livestock for food production, such as dairy cows, can be genetically altered to improve quantity and quality. The same principles can be applied to chicken, pork, and beef. Biotechnology research and genetic engineering will improve our ability to feed the world's population.

Molecular Farming

Molecular farming is another new area that poses great promise. **Molecular farming** refers to a new area of engineering and science that combines biotechnology and plants to produce a variety of products. These products might be used for agricultural crops, livestock, cosmetics, or medical purposes.

Researchers and engineers are developing plants that can be used for medical drugs. Molecular farming may enable us to grow plants that will be easily converted into medicines with a minimum of production. This activity will require society to develop new regulations to protect the public. In Chapter 2, you learned how society often controls which technologies are developed.

KAREN KASMAUSKI/SCIENCE FACTION/GETTY IMAGES

FIGURE 20-3 Dolly was the world's first cloned mammal from an adult.

Cloning

Biotechnology may also be able to help restore endangered species. This is viewed by some people as a logical extension of the cloning of the first mammal in the late 1990s. Dolly the sheep was cloned by scientists in Scotland in 1997. She was the first mammal to be cloned from an adult (Figure 20-3).

Cloning is a process that makes an exact genetic duplicate of an organism. Despite continued research, many problems with cloning still remain. Nevertheless, the potential to restore shrinking wild populations is an exciting prospect for scientists and engineers in the biotechnology field (Figure 20-4).

Farmers and others have a keen interest in animal cloning. Imagine that a dairy farmer had a cow that could naturally produce twice as much milk as other cows. If you could clone this cow, then the farmer would not need as

© TIM JENNER/SHUTTERSTOCK.COM

FIGURE 20-4 Biotechnology may help save endangered species.

cloning
Cloning is a process that makes an exact genetic duplicate of an adult in a baby animal.

© STEVE BYLAND/SHUTTERSTOCK.COM

FIGURE 20-5 **Scent engineering may increase crop production and help feed the world.**

many dairy cows. Or what if we could create an exact duplicate of a winning race horse? On a more personal level, someday it might be possible for you to clone your favorite pet.

Scent Engineering

Although on the surface it may not appear as significant as saving endangered species, another interesting prospect for biotechnology is *scent engineering*, or controlling how flowers smell. You might be thinking of perfume, but actually this engineering research could have a significant impact on improving the yield and quality of many crops.

Floral scent plays an important role in the reproductive process of many plants because it attracts the insects or birds that carry pollen to other plants (Figure 20-5). Improved pollination of crops will increase our food supply. Feeding the world's population continues to be a challenge, and it is certainly as important as saving endangered species.

Did You Know?

A nanometer is about the amount a man's beard would grow during the 2 seconds it would take him to raise the razor to his face!

Math in Engineering

One nanometer (nm) is one-billionth, or 10^{-9}, of a meter. It is not easy to comprehend anything so small. If we were to compare a nanometer to a marble, then a meter would be the size of planet Earth.

SECTION 2: NANOTECHNOLOGY

Nanotechnology is a technology that creates small materials in the size of molecules by manipulating single atoms. *Nanotechnology* also refers to making devices in that size range. The prefix *nano* comes from the size of molecules, which are measured in *nanometers*, or one-billionth of a meter (0.000000001 meter). Nanotechnology may very well be the most exciting emerging technology and will significantly affect future generations. Engineers and scientists are rushing to develop new products and gain patents. Nanotechnology is expected to be one of the fastest-growing industries of the twenty-first century.

nanotechnology
Nanotechnology is a technology that creates materials or devices at the scale of molecules by manipulating single atoms.

Potential Benefits of Nanotechnology

As you read in Chapter 2, there are both positive and negative outcomes, planned and unplanned, with all technologies. Potential benefits from nanotechnology include:

- Nanotechnology can be used to purify drinking water. Contaminated drinking water is a major cause of death and disease worldwide. From the medical and social ethics points of view, the potential of nanotechnology to purify drinking water may be its greatest contribution to humanity.

Did You Know?

The term *nanotechnology* was first defined in 1974 by Japanese Professor Norio Taniguchi as the processing, separation, consolidation, and deformation of materials by one atom or by one molecule. At that time, nanotechnology was purely theoretical and did not yet exist. Until 20 years ago, we could not even see or work with particles of that size.

- Engineered food and crops could increase agricultural productivity. Or they might allow us to grow food in poor soil that cannot support farming now.
- The so-called smart foods could be tailored to just your needs. Each person's body has slightly different nutritional needs, and these smart foods would provide just the nutrients your body needed.
- Cheap energy can be generated. Our need for energy is increasing (see Chapter 9). Nanomaterials can absorb the sun's energy and convert it into electricity. Nanowires can generate small amounts of electricity from body movements. These wires could be assembled into larger arrays to someday power small medical implants—or even power your MP3 player.
- Manufacturing can be made cleaner and more efficient. Everything we use and consume is a manufactured product, including most of our food. Increased efficiency would raise the standard of living in the world for everyone while saving our environment.

6

- Health care can be improved. Health care is one of the fastest-growing concerns in the world today. Nanotechnology shows tremendous potential in many areas of health care. Nanorobots can seek out specific cancer cells and deliver deadly medicine just to those cells (Figure 20-6). Unlike current chemotherapy or radiation treatments, this technique leaves the surrounding healthy cells untouched.
- More information can be stored, and communication capacities can be improved. Nanowires that can store data have been developed, and these data can be accessed a thousand times faster than we currently can using flash memory or other portable devices. These nanowires also use far less energy and take up less space. On a personal level, this benefit could affect entertainment by allowing you to have thousands of movies and songs at your fingertips, thus making your MP3 player a thing of the past.

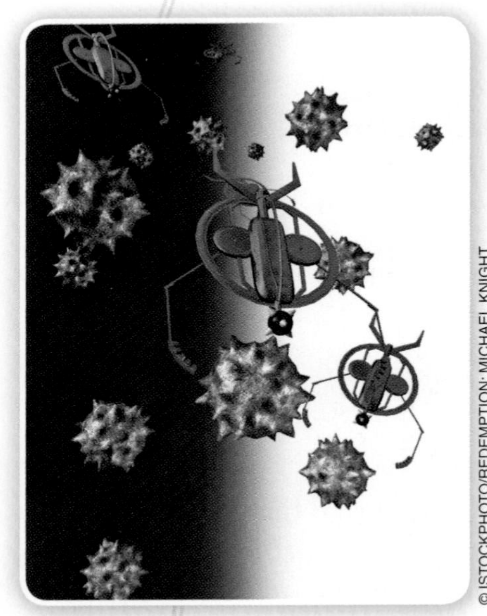

FIGURE 20-6 Nanorobots may be small enough to seek out and inject only a diseased cell.

Potential Risks of Nanotechnology

Nanotechnology also poses a number of associated risks. Risks come with anything that is so small that it can be accidentally inhaled, ingested, or passed through your skin. What would happen if the cancer-killing nanorobots described previously accidentally went off course in your body? The accidental or deliberate release of some forms of nanotechnology into our water supply could be disastrous.

Research the potential impacts of nanotechnology. Remember to get your teacher's permission before conducting Internet searches. Develop a list of both the potential benefits and possible dangers of nanotechnology. Do you think the government should take action before nanotechnology is fully developed, or should we just wait and see what happens?

Write a letter to your state's U.S. senators or representatives or both. In the letter, list three potential benefits of nanotechnology and three potential risks. Ask them what their position is on supporting nanotechnology research. You might also ask what the government is doing to establish priorities and policies for the development and use of nanotechnologies. Make sure you include your school name, class, and grade level. And, of course, get your teacher's permission before mailing the letters.

Social Impacts of Nanotechnology

As you learned in Chapter 2, all technological developments have social impacts. Some people believe we should be proactive in regulating nanotechnology to control how it is developed. They want us to develop governmental policies to make sure nanotechnology meets social objectives. Public participation is very important in this process. This is one more reason why it is important for you to study technology education in school.

Many people are making predictions about the impact of nanotechnology on society. Some people believe it will drive a "nanorevolution" that, like the Industrial Revolution, will change our economy, social structures, and even the environment. The impact may be like a tsunami. This suggests that the nanotechnology revolution will hit so fast and hard that it will be very disruptive and catch society unprepared. These critics want government to take action before that happens.

SECTION 3: MEDICAL TECHNOLOGY

Americans are enjoying longer, healthier lives than ever before, partly because of advances in medical technologies. **Medical technology** is either the diagnostic or the therapeutic applications of technology to help improve people's health. Developments in medical technologies will be closely linked to developments in biotechnology and nanotechnology.

medical technology
Medical technology is the use of therapeutic applications of technology to improve human health.

Engineers play a major role in designing new devices and products in cooperation with medical doctors. Many treatments and devices that we

Career Spotlight

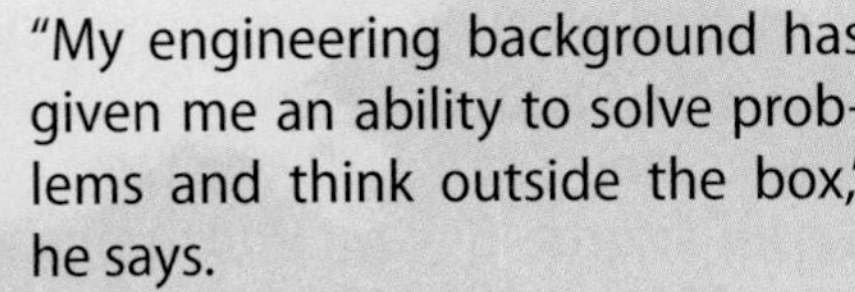

Name:
Orok Duke

Title:
Asset and Growth Support Team Leader, BP Pipelines NA, Inc.

Job Description:
Orok Duke always had a passion for science and math, and he knew he wanted to funnel that passion into his career. He studied chemical engineering in college, but now he uses his technical expertise working on business issues for BP Pipelines NA, Inc.

BP searches for oil and gas all around the world, processing it into useful materials such as gasoline and diesel and jet fuel. The company is also developing alternative sources of energy such as wind and solar.

Mr. Duke leads a team that helps develop new business opportunities for the company. His team works on acquisitions, divestments, and asset management.

Mr. Duke's technical background helps him understand the issues related to each new project. "When building a new pipeline, for example, technical issues arise," he says. "I can make connections between the technical issues and the business issues. Examples of technical issues include how temperature will affect the density of the fluid that flows through the pipe."

Duke also learned skills in engineering that help him find business solutions. "My engineering background has given me an ability to solve problems and think outside the box," he says.

Duke started with BP as an engineer and later expressed an interest in the business side, learning it on the job. His new career suits him well. "I like the idea of creating value for the company and for society at large," he says.

Education:
Duke received his bachelor's degree in chemical engineering from Loughborough University in the United Kingdom. He received a master's degree in chemical engineering from Princeton University.

"Loughborough University offered a practical program in engineering," he says. "The program offered a chance to work on industry-related projects."

Advice to Students:
Duke is dedicated to making the engineering field more diverse. He is active in the BP African American Network and the National Society of Black Engineers.

He thinks it is a good idea for high school students to shadow people doing different kinds of jobs. "That way you can get a good sense of what you enjoy and what you don't enjoy," he says. "You'll have a better sense of what path to follow in college."

take for granted now did not even exist a decade ago. It is important that we do not assume this will continue by itself. It takes dedicated engineers and scientists to keep developing new medical innovations.

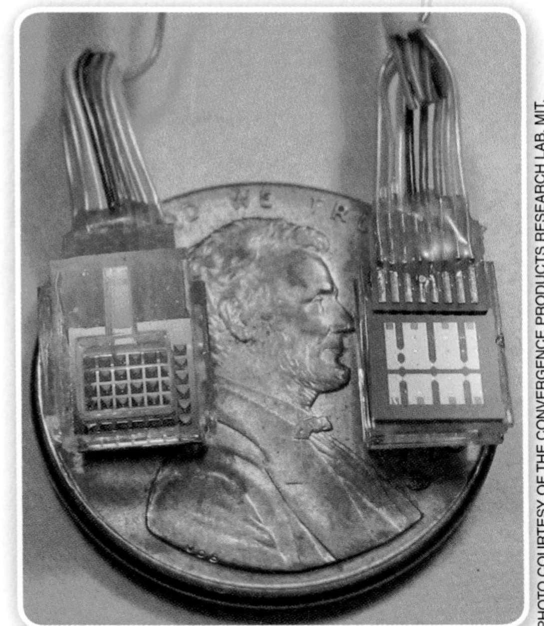
PHOTO COURTESY OF THE CONVERGENCE PRODUCTS RESEARCH LAB, MIT, CAMBRIDGE, MA

FIGURE 20-7 A micro drug-delivery system smaller than a penny can dispense medicine to a patient.

Personalized Medicine

The medical field's continuing innovation is moving us closer toward personalized medicine. **Personalized medicine** means getting the right treatment to the right patient at the right time. We need to develop technologies to help diagnose the exact treatment a patient requires. New drug-delivery systems are being developed that can dispense medicine to patients in seconds (Figure 20-7).

Telemedicine

One development in medical technology that shows great promise is telemedicine. **Telemedicine** is medical information that is shared and medical procedures that are performed from a distance. Telemedicine allows doctors to assist patients in more than one place. This is made possible by improvements in Internet connections and robotics.

telemedicine
Telemedicine is the process of sharing medical information or medical procedures from a distance.

Telemedicine gives doctors the ability to conduct surgery on patients hundreds or thousands of miles away. In some cases, a doctor can perform surgery on a patient by guiding a robot in a virtual reality environment. This is similar to some of today's video games. Did you ever think that playing a video game would be good practice for becoming a doctor? At a simpler level, a doctor in one location can be guided by an expert surgeon in another location who watches the surgery over the Internet.

Developments in robots made telemedicine surgery possible. Robots are capable of making very fine movements, in the range of thousandths of an inch—and without hands that shake. A doctor can control much more precisely the location and depth of cuts using a robotic device. These devices can also have a camera mounted on them to enable the surgeon to see much more clearly.

Endoscopy

Currently, if doctors need to see what is going on in your body, they can feed a small camera into your digestive system in a process called *endoscopy*. Although this camera can be guided into you from either end of your digestive track, neither way is pleasant. That may soon change.

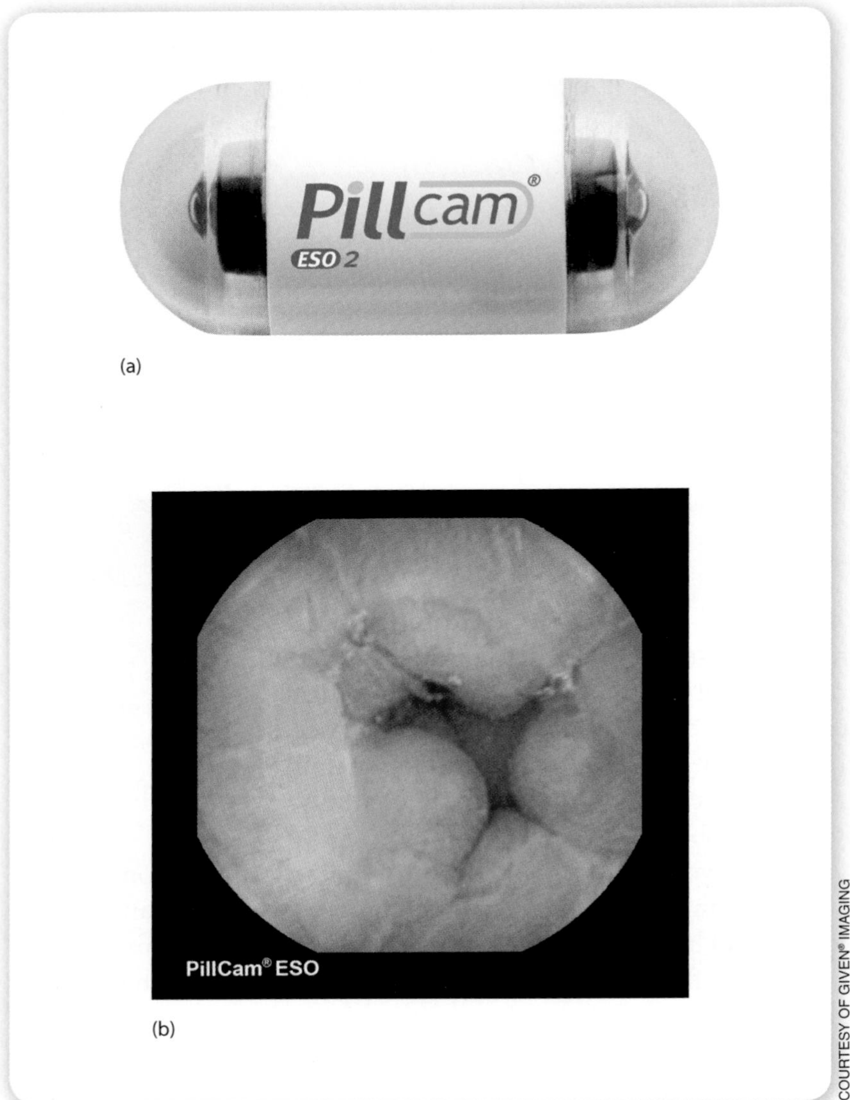

FIGURE 20-8 (a) A tiny camera may soon look inside your body from inside a swallowed pill. This tiny camera produced this image (b) of a patient's esophagus.

Engineers are developing disposable cameras that are small enough to fit inside a pill (Figure 20-8). All you have to do is swallow the single-use pill and allow doctors to see on a broadcast monitor what is going on inside your digestive track. This is made possible because cameras have become smaller and more disposable, and their images have become clearer. Your parents or grandparents, if they have had endoscopies, will indeed think this innovation an improvement.

Prosthesis Technology

A prosthesis is a device that has been designed to replace a missing part of the body or to make a part of the body work better. An artificial limb or part is called a **prosthetic device**. Doctors have had to amputate limbs to save injured patients' lives for thousands of years (Figure 20-9). Perhaps you have seen a cartoon or a movie of a pirate who had a peg leg—a wooden stump from the knee down. Or perhaps you have seen fictional movies or television shows about a bionic man or woman. Significant improvements in prosthetic devices have been made in the past decade, and the future looks even better (Figure 20-10).

prosthetic device
A prosthetic device is an artificial limb or body part such as an artificial knee or hip.

FIGURE 20-9 Early prosthetic devices were often just pieces of wood.

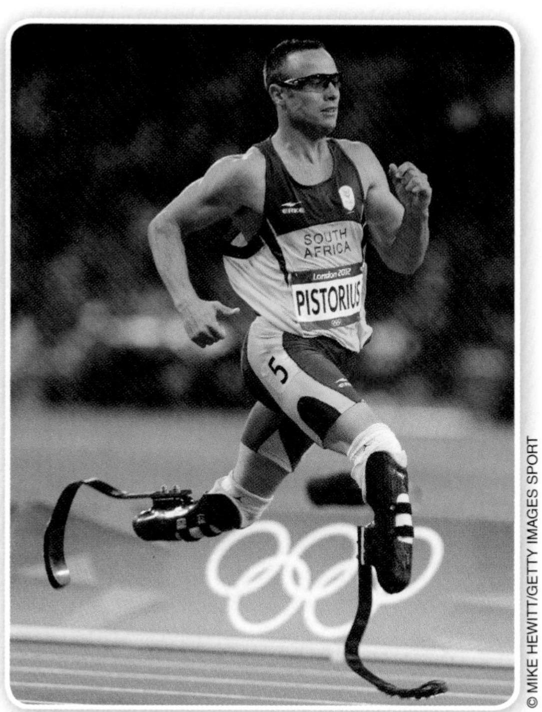

FIGURE 20-10 Prosthetic devices are helping many people lead normal lives after an injury.

Many wounded soldiers have been helped by new developments in prosthetics. The soldiers have primarily needed legs and arms. It takes a lot of rehabilitation training for people with prosthetic devices to learn how to use them, but many people with prosthetic devices have returned to fairly normal activities, including skiing.

One of the new developments in prosthetics is adding computer technology. These devices have built-in sensors that feed information to a computer. The computer can adjust for certain variables such as hills and walking speed. Or, in the case of an arm, the computer can control finger grip. Claudia Mitchell received the first thought-controlled bionic arm in 2007 (Figure 20-11).

WIN MCNAMEE/GETTY IMAGES

FIGURE 20-11 Claudia Mitchell is the first female recipient of a thought-controlled bionic arm, an advanced prosthesis, which was developed by the Rehabilitation Institute of Chicago.

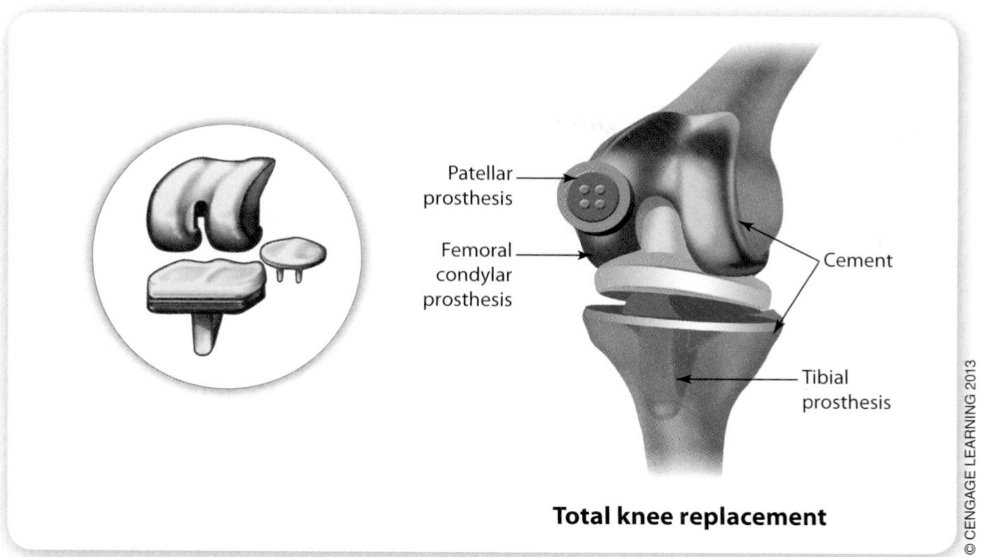

FIGURE 20-12 New materials and techniques have made knee replacement common.

Not all prosthetic devices are for limbs. Joint replacement occurs far more often. New materials and techniques have made joint replacement for knees and hips much more common surgeries (Figure 20-12).

Health care is an increasing concern because of the large number of people who are living longer. Fortunately, medical technologies are advancing rapidly. As with other technologies, there are issues about the availability of medical technologies for all those who need them. Society will need dedicated, competent engineers to solve many of the issues facing society. Are you ready for the challenge? Working as an engineer in these emerging technology fields will be an exciting and rewarding career.

SUMMARY

In this chapter you learned:

- Creative, innovative thinking and engineering are required to solve many of our most challenging problems in the twenty-first century.
- Biotechnology, nanotechnology, and medical technology are three emerging fields that hold great promise for the future.
- Biotechnology is technology based on biology.
- *Biomass* is the term used for any organic material made from plants or animals.
- *Biofuel* is any fuel made from biomass.
- Biofuels are more environmentally friendly and are an important resource to reduce our dependency on fossil fuels such as oil.
- Biopower is the use of biomass to generate electricity.
- Nanotechnology is technology that creates small materials or devices at the scale of molecules by manipulating single atoms.
- Nanotechnology has great potential, but it also has many risks.
- *Medical technology* is a term for either diagnostic or therapeutic applications of technology to help improve people's health.
- New developments in medical technologies will be closely linked to developments in biotechnology and nanotechnology.
- Through telemedicine, a doctor can perform surgery hundreds of miles away by guiding a robot in a virtual reality environment.

VOCABULARY

Write a definition for each term in your own words. After you have finished, compare your answers with the definitions provided in the chapter.

Biotechnology
Biofuel
Biodiesel
Biopower
Anaerobic digestion
Cloning
Nanotechnology
Medical technology
Telemedicine
Prosthetic device

STRETCH YOUR KNOWLEDGE

Provide thoughtful, written responses to the following questions.

1. Why do emerging technologies hold great promise for the future?
2. Develop a time line of significant technological innovations that have been developed since you were born.
3. Describe three advantages of biofuels over oil.
4. Explain two ways to generate biopower.
5. Compare the benefits of biopower with coal-fired electrical generation (see Chapter 9 to refresh your memory).
6. Research the latest prosthetic developments on the Web. Remember to obtain your teacher's permission before conducting Web searches. Write a brief report on two developments that interest you. Include information about the technology and how it was different than what was previously available.

Glossary

A

absolute temperature: Absolute temperature is the measurement system in which zero (0) is the point at which all molecular movement stops. It is also known as the Kelvin (K) scale; 0 K = −463 degrees Fahrenheit.

active solar heating: Active solar heating is energy captured by liquid circulating through tubes exposed to sunlight.

actuator: An actuator transfers fluid power into mechanical power.

adhesives: Adhesives are materials such as glue that are used for combining items.

aerodynamics: Aerodynamics is (1) the science that deals with the motion of air and the forces acting on objects as a result of the motion between the air and the object, and (2) the study of forces and the resulting motion of objects through the air.

aerospace engineer: An aerospace engineer designs machines that fly.

airfoils: Airfoils are wings.

aileron: An aileron is the controlling surface that regulates an aircraft's roll.

alloy: An alloy is a combination of two or more metals into a single compound.

alphabet of lines: The alphabet of lines shows the various line types that, when used together, create an engineering drawing of an object.

altitude: Altitude is the highest point of a rocket's flight.

amperes (amps): Amperes are the unit measure of electric current.

anaerobic digestion: Anaerobic digestion is a process in which decomposing biomass in a tank produces a gas vapor that can be burned in a special engine to produce renewable electricity (one form of biopower).

AND gate: AND gate is a type of logic gate that requires two "on" (1) inputs to produce an "on" (1) output.

angle of attack: Angle of attack is the measurement in degrees between a wing's chord and the movement of wind against the leading edge.

annealing: Annealing is a conditioning process that makes a material softer and easier to work.

annotations: Annotations are notes placed on an engineering sketch to clarify the viewer's understanding of the object or objects drawn.

anode: An anode is the positively charged structure within a voltaic cell.

architect: An architect is a person who designs and supervises the construction of buildings. The architect is responsible for the building design, including how the space is laid out and how the building relates to its environment.

architecture: Architecture is the art and science of designing and erecting buildings.

armature: An armature is the rotating part of an electrical generator.

array: An array is a group of connected photovoltaic cells.

arrowhead: An arrowhead contacts a line that has been extended for a drawing.

articulated robot: An articulated robot is a robot that has two or more joints that swivel.

ASCII code: ASCII is short for American Standard Code for Information Interchange; the code regulates the use of bits and bytes.

aspect ratio: The aspect ratio is the wingspan length divided by the cord.

assembly drawing: An assembly drawing is a pictorial drawing that shows the parts of an object placed in their correct locations.

atoms: Atoms are small particles that are the building blocks of our universe.

automated guided vehicle (AGV): An automated guided vehicle is a robot that can be programmed to pull carts with raw materials or parts from one location in a factory to another.

automated manufacturing: Automated manufacturing is the general term for computer-controlled machine tools and robots in manufacturing.

automation: Automation is a system in which some action happens as a result of another event without human intervention.

autonomous robot: An autonomous robot is able to react to its environment to complete its task.

B

balance: Balance is the principle of design dealing with the various areas of a structure as they relate to an imaginary centerline; sometimes referred to as *symmetry*.

base: The part of a transistor that receives the signal in the form of either a negative or a positive electrical charge.

battery: A battery is two or more voltaic cells connected together.

beam: A beam is a horizontal structural member that supports roof or wall loads.

Bernoulli's principle: Bernoulli's principle states that as the speed of a fluid increases, its pressure decreases.

bevel gear: A bevel gear is a gear in which the axis of the drive gear forms an angle with the axis of the driven gear.

binary system: The binary system is the simplest number system, using only 0 and 1.

biodiesel: Biodiesel is a type of biofuel that can be made from vegetable oil or animal fat.

biofuel: Biofuel is any fuel made from biomass, such as ethanol.

biomass: Biomass is (1) a type of energy and (2) living or recently dead biological material that can be used as fuel.

biopower: Biopower is the use of biomass to generate electricity.

biotechnology: Biotechnology is a technological process that uses living or once-living organisms to make or modify products.

bit: In the binary number system, *bit* (short for *b*inary dig*it*) notes either off or on (0 or 1).

blow molding: Blow molding is a molding technique for plastic used to make containers, bottles, and hard-sided plastic cases.

Boyle's law: Boyle's law states that for a fixed amount of gas kept at a fixed temperature, pressure and volume are inversely proportional.

brainstorming: Brainstorming is a method of shared problem solving in which all members of a group engage in unrestrained discussion and spontaneously generate ideas.

brittleness: Brittleness is the opposite of plasticity; brittle material is not flexible and is easily broken.

brushes: Brushes are parts of an electrical generator.

byte: A byte is a combination of eight bits.

C

CAD: CAD is the acronym for *computer-aided drafting* (or *design*).

cam and follower: A cam and follower is a common mechanism used to convert rotary motion into reciprocating motion.

capacitor: A capacitor is an electronic device that stores electrical energy but can release it quickly.

Cartesian coordinate system: The Cartesian coordinate system is the graphical mathematical system used to locate points in two- and three-dimensional space.

casting: Casting is a forming process in which the material is first made into a liquid and then poured into a mold.

cathode: A cathode is the negatively charged structure within a voltaic cell.

center lines: Center lines are lines that cross at a 90° angle in the center of the circle or hole on an engineering drawing.

Charles's law: Charles's law states that the volume of a confined gas is proportional to its temperature, provided its pressure remains constant.

chemical energy: Chemical energy is stored energy.

chord: The chord is the distance between the leading and trailing edges of a wing.

circuit: A circuit is an electrical pathway in which electrons travel.

circuit breaker: A circuit breaker is an electrical device designed to open when a current flow exceeds a limit.

civil engineer: A civil engineer is a person who designs and supervises the construction of public works projects (such as highways, bridges, sanitation facilities, and water-treatment plants).

cloning: Cloning is a process that makes an exact genetic duplicate of an adult in a baby animal.

closed-loop system: A closed-loop system includes an automatic feedback loop to regulate the system.

collector: A collector is the part of a transistor from which the electrons flow when the base is signaled.

column: A column is a vertical structural member that transfers loads to a structure's foundation.

combination circuit: A combination circuit is an electrical circuit that contains both series and parallel circuits.

combining: Combining is the process of joining two or more materials together.

combustion: Combustion is the process of burning.

communication system: A communication system is a technological system in which information is transferred between people and other people or machines.

commutator: A commutator is part of a generator through which electricity flows to the generator's brushes.

compound machine: A compound machine is a combination of two or more simple machines.

compound pulley/block and tackle: A compound pulley/block and tackle is the combination of two or more pulleys to gain a mechanical advantage typically used for lifting heavy objects.

compression: Compression is a force that pushes or squeezes material.

compressor: A compressor is a device that increases the pressure of air and pumps it into a tank.

computer-aided design (CAD): Computer-aided design is design done using powerful computers and software.

computer-aided manufacturing (CAM): Computer-aided manufacturing is the process of using computers to control machines used in the manufacturing process.

computer-integrated manufacturing (CIM): In computer-integrated manufacturing, the entire manufacturing process is coordinated through computers.

computer numerical control (CNC): Computer numerical control is a programming language or code that tells machines exactly what to do and where to move.

concurrent engineering: Concurrent engineering is the process of involving everyone responsible for the design and manufacturing of a product right from the beginning of the process.

conditioning: Conditioning is a set of processes that change the internal characteristics of a material.

conduction: Conduction is the way in which heat travels through a solid material.

conductor: A conductor is a material composed of elements with atoms that contain one, two, or three valence electrons, which allows the flow of electrons.

constraint: A constraint is a limit, such as appearance, budget, space, materials, or human capital, in the design process.

construction-management engineers: Construction-management engineers oversee the actual building of a structure, including reviewing plans, ordering materials, and scheduling subcontractors. They are responsible for quality control and ensuring that each structure is built as designed.

consumer products: Consumer products are manufactured products made for consumers.

contours: Contours are lines on a map that show topographical outlines or elevations of land.

control valves: Control valves regulate the flow of gases.

convection: Convection is the way in which heat travels by movement or circulation within liquids or gases.

conveyor belt: A conveyor belt is a continuous looped system that is used to move parts.

crank: A crank is a device that changes rotary motion into reciprocating motion.

criteria: Criteria are special requirements that a product has met or must meet.

crown-and-pinion gear: A crown-and-pinion gear changes the direction of rotary motion, typically decreasing the speed of rotation while increasing the torque.

current: Current is the rate of flow of the electrons along an electrical pathway; measured in amperes (A).

cylinder: In pneumatic systems, a cylinder is a container.

D

design: Design is a process that turns a concept into a product that can be produced.

design brief: A design brief is a written plan that identifies a problem to be solved, its criteria, and its constraints.

design process: The design process is a systematic problem-solving strategy used to satisfy human wants or needs.

detailed sketch: A detailed sketch is a type of freehand technical sketch that provides detailed information about the object such as annotations (notes), dimensions, and shading.

details: Details are drawings that enlarge a specific part to help a builder understand how the particular pieces fit together.

digital electronics: Digital electronics is the use of digital logic to operate an electronic system.

dimension lines: Dimension lines in a drawing show an object's physical dimensions.

dimensions: Dimensions are the sizes and notes placed on a mechanical drawing that record an object's linear measurements such as width, height, and length, as well as the location of the object's features.

diode: A diode is an electronic device that allows current to flow in only one direction.

documentation: Documentation is the organized collection of records and documents that describe a project's purpose, processes, and related activities for future reference.

doping: Doping is the addition of impurities to a semiconductor to modify its electronegativity.

drag: Drag is (1) a force that causes resistance to moving through the air, (2) resistance of the air (technically a fluid) against the forward movement of an airplane, (3) the force that acts opposite to the direction of motion (caused by friction and differences in air pressure), and (4) the resistance of the motion of an object through a fluid.

drill bit: A drill bit is the cutting tool used in drilling.

drilling: Drilling is a separating process that produces a round hole.

drill press: A drill press is a machine used for drilling.

drive gear: A drive gear is the first gear in a gear train.

driven gear: A driven gear is the last gear in a gear train.

drones: Drones are robotic aircraft; they are used extensively by the military.

ductility: Ductility is a material's ability to bend, stretch, or twist without breaking.

E

elasticity: Elasticity is a material's ability to return to its original shape once the force is gone.

electrical energy: Electrical energy is produced by the flow of electrons.

electrical engineer: An electrical engineer designs electronic systems and products.

electricity: Electricity is the flow of electrons through a pathway.

electrolyte: An electrolyte is usually a liquid in a voltaic cell through which electrons flow from a cathode to an anode.

electromagnet: An electromagnet is a metal core that is rendered magnetic by the passage of an electric current through a surrounding coil.

electromotive force: Electromotive force is the pressure on electrons to move them through a conductor that is supplied by the power source.

electron: An electron is part of an atom; it holds a negative charge.

electronegativity: Electronegativity is the measure of the attraction that an atom has for the electrons orbiting about its nucleus.

electronics: Electronics is the study and control of the flow of electrons, typically using low voltage.

elevations: Elevations are drawings that show the exterior of a building or facility.

elevator: An elevator is the controlling surface that regulates an aircraft's pitch.

ellipse: An ellipse is a flattened circle.

emitter: An emitter is the part of a transistor where the electrons exit to complete the circuit.

end effector: An end effector is the attachment at the end of a robot's arm that is used for a specific task such as holding a tool, painting, or welding.

energy: Energy is the ability to do work.

energy conversion: Energy conversion is a change in energy from one form to another.

engineer: An engineer is a person who designs products, structures, or systems to improve people's lives.

engineering: Engineering is the process of designing solutions.

engineering analysis: Engineering analysis is the process used by engineers to test the strength of their design or the mechanical movement of parts within the design before making the product.

engineering design process: The engineering design process applies mathematics, science, and engineering principles to help in decision making.

engineering drawing: An engineering drawing is made using instruments.

engineering notebook: An engineering notebook is the documentation of all of the steps and calculations for and an evaluation of the engineering design process for a particular item.

engineering technologist: An engineering technologist is a person who works in a field closely related to engineering. The technologist's work is usually more applied or practical, whereas the engineer's work is more theoretical.

environmental engineer: An environmental engineer designs solutions to protect and maintain the environment.

environmental impact study: An environmental impact study assesses the impact of a proposed project on the environment.

ethanol: Ethanol is an alcohol fuel made from sugars and starch found in plants.

evaluation: Evaluation is the process of collecting and evaluating information and data to determine how well a design meets requirements.

exclusive gate: The exclusive gate is one type of logic gate. It has two inputs (A and B) to generate its output (Q).

exhaustible energy: Exhaustible energy is any source of energy that is limited and cannot be replaced when it is used, such as oil, coal, and natural gas.

experimentation: Experimentation is the act of conducting a controlled test on a prototype.

exploded view: An exploded view is part of a pictorial drawing that shows the parts of an object disassembled but in relation to each other.

exponential: Exponential is the quality of something increasing at a set pattern.

extension line: An extension line is a line that has been extended for a drawing.

extruding: In parametric modeling, extruding is the process of adding depth to the geometric shape.

extrusion: Extrusion is a forming process that forces material through an opening in a die with a desired shape.

F

fabrication: Fabrication is the process of making or creating something.

features: Features are objects or parts of an object such as holes, pins, slots, braces.

feedback: Feedback is information about the output of a system that can be used to adjust it.

ferrous metals: Ferrous metals contain iron.

fission: Fission is the process of an atom's nucleus splitting into its many smaller subparticles.

fixtures: Fixtures are drill press accessories that are used to align material to ensure accurate drilling.

flexible manufacturing system (FMS): A flexible manufacturing system is a system that uses sophisticated machines that can perform more than one type of process through rapid tool changes.

floor plans: Floor plans show the space and size relationships of rooms and how people will move through a proposed facility.

fluid power: Fluid power is the use of a fluid under pressure to transmit power.

force: Force is the transferring of energy to an object, typically by pushing or pulling on that object.

forging: Forging is a forming process that uses heat and force to shape metal.

form: Form is the principle of design that is described by lines and geometric shapes.

forming: Forming is a family of processes that change the shape and size of a material without cutting it.

forms: Forms of energy can be converted from one to another.

fossil fuels: Fossil fuels are fuels produced by deposits of ancient plants and animals.

foundation: A foundation transfers a structure's loads to the ground. It is the substructure of the building and is usually made from concrete.

freehand technical sketching: Freehand technical sketching is the process of drawing technical images without the use of drafting tools such as a T square, a triangle, or a compass.

front view: A front view shows an object's height and width.

fulcrum: A fulcrum is the support about which a lever turns.

fuse: A fuse is a safety device for an electronic circuit that breaks a circuit.

fusion: Fusion is the combination of two or more atomic particles to form a heavier nucleus.

G

gear: A circular device that transfers rotary motion using interlocking teeth.

gear train: A gear train is a set of gears designed to increase or decrease speed or to increase or decrease torque.

generator: A generator is a device that converts rotational motion into electricity.

geometric constraint: A geometric constraint establishes a fixed relationship between the features of a drawing such as lines and shapes to each other—for example, constraining two lines to be parallel or perpendicular.

geometry: Geometry is a branch of mathematics concerned with the properties of space. This includes points, lines, curves, planes, and surfaces in space, as well as figures bounded by them.

geotechnical engineers: Geotechnical engineers make sure that the ground for a designated site can support the weight of a proposed structure.

geothermal energy: Geothermal energy is energy stored in the Earth in the form of heat.

grading: Grading is the moving and leveling of earth as part of site preparation for a project.

green building: Green building is both the process of building in an environmentally responsible way and the resulting structure.

H

hardening: Hardening is a conditioning process that makes a material harder.

hardness: Hardness is a material's ability to resist permanent indentation by another object.

hardwoods: Hardwoods come from deciduous trees; examples are oak, maple, and walnut. They are typically used in furniture construction because of their appearance and strength.

heat: Heat is energy in transit.

horsepower: Horsepower is a common unit for measuring mechanical power in the U.S. customary system of measurement.

hydraulic fluid: Hydraulic fluid is an oil-type liquid used in hydraulic systems.

hydraulics: Hydraulics is the transfer of power through a liquid.

hydroelectricity: Hydroelectricity is electricity generated from water.

hydropower: Hydropower is electricity generated from water.

I

idler gear: An idler gear is the gear used in the middle of a gear train.

inclined plane: An inclined plane is a sloping surface used to alter the force required to move the load in a perpendicular direction.

industrial engineer: An industrial engineer designs the most efficient means of producing products.

industrial products: Industrial products are manufactured products that are used as inputs for other manufacturing operations.

industrial robot: An industrial robot is a computer-controlled, reprogrammable, and multipurpose unit that can move in three or more directions.

inexhaustible energy: Inexhaustible energy is any energy source that cannot be used up, such as solar, wind, water, and geothermal.

injection molding: Injection molding is one of the most common processes for molding plastics. Injection molding forces molten plastic into a mold under pressure.

innovation: An innovation is an improvement of an existing product, process, or system.

insulator: An insulator is a material made of elements that do not have free electrons.

integrated circuit (IC): An integrated circuit uses a single crystal chip of silicon that has all the components needed to make a functioning electrical circuit.

interchangeable parts: Interchangeable parts are standardized parts that are uniform and can be assembled with other types of interchangeable parts in manufactured products.

internal-combustion engine: An internal-combustion engine is a mechanical device that converts chemical energy to heat energy and then to mechanical energy.

International System (SI): The International System (SI), or metric system, is the system of measurement used in much of the world. (The abbreviation is from the French term, Systéme International d'units.)

invention: An invention is a new product, system, or process that has never existed before, created by study and experimentation.

isometric sketching: Isometric sketching is a method of representing three-dimensional objects with three axes—*x*, *y*, and *z*—with the *x*- and *y*-axis 120° apart.

J

jet engines: A jet engine is a heat engine that converts chemical energy to thrust through the use of compressors and turbines.

joule: A joule is a unit measure of work, one newton-meter.

K

kerf: A kerf is the groove left by the sawing process.

kinetic energy: Kinetic energy is energy that is in motion or what happens when potential energy is released.

L

lamination: Lamination is the layering of materials; plywood is an example.

lathe: A lathe is a machine used in a separating process called turning.

law of conservation of energy: The law of conservation of energy states that energy can be neither created nor destroyed. There is a fixed amount of energy present in different forms that can be converted from one form to another, but energy does not go away.

laws of magnetism: The laws of magnetism are the scientific principles that like magnetic poles repel and unlike magnetic poles attract.

leader: A leader is a line that points to the center of a circle but extends only to the circle's outer edge.

leading edge: The leading edge is the front surface of a wing.

lead screw: A lead screw converts rotary motion to linear motion.

lever: A lever is an arm that pivots on a fulcrum.

lift: Lift is (1) a component of aerodynamic forces acting on an object in flight, (2) a force produced by an airfoil shape that works against gravity, (3) the force that acts at a right angle to the direction of motion through the air (created by differences in air pressure), and (4) the force that directly opposes the weight of an airplane and holds the airplane in the air.

light-emitting diode (LED): A light-emitting diode is an electronic device much like a small lightbulb that allows current to flow in only one direction.

light energy: Light energy is a form of radiant energy.

linear motion: Linear motion is movement in a straight line.

load: Load is the weight a structure must support.

load device: A load device is a component of an electrical circuit that consumes power.

logic gate: A logic gate is a digital electronic device that uses inputs of 0 or 1 to signal outputs of 0 or 1.

M

machine screw: A machine screw is a small bolt that must be used with a nut.

magnet: A magnet is a device that attracts iron and has a surrounding force field.

magnetic field: A magnetic field is the force field that surrounds a magnet.

magnetic induction: Magnetic induction is the generation of electron flow by a magnetic field.

manufacturing: Manufacturing is the use of tools and machines to make things. Manufacturing converts materials into usable products.

market research: Market research is a survey of potential product users to find out their likes and dislikes about a product.

mass production: Mass production is the manufacture of large numbers of identical pieces (interchangeable parts).

matrix: A matrix is an arrangement of mathematical elements to help solve problems.

mean chord: The mean chord is an imaginary line halfway between a wing's top and bottom surfaces.

mechanical advantage: Mechanical advantage is gaining increased force or motion by using a machine.

mechanical energy: Mechanical energy is the energy of moving objects.

mechanical engineer: A mechanical engineer designs products ranging from simple toys to very large and complex machines.

mechanical fasteners: Mechanical fasteners are common materials such as nails, staples, screws, bolts, and rivets used in a combining process.

mechanical properties: Mechanical properties are descriptions of a material's characteristics and include hardness, toughness, elasticity, plasticity, brittleness, ductility, and strength.

mechanism: A mechanism is a device that transmits movements so that the output movement is different than the input movement. It can be used to change the direction, speed, or type of movement.

medical technology: Medical technology is the use of therapeutic applications of technology to improve human health.

metal: Metal is a durable material that is opaque, is a conductor, is reflective, is heavier than water, and can be melted, cast, and formed.

microprocessor: A microprocessor is a preprogrammed chip used in an integrated circuit.

mock-up: A mock-up is a model of a design that does not actually work.

model: A model is a three-dimensional representation of an existing object.

molding: Molding is a process that forces plastic into a mold or cavity of the desired shape.

molecular farming: Molecular farming refers to a new area of engineering and science that combines biotechnology and plants to produce a variety of products.

motor: A motor is the opposite of an electrical generator.

multiview drawing: A multiview drawing contains more than one view of a product.

N

nail: A nail is a wedge made of metal that is used to join lumber together.

nanotechnology: Nanotechnology is a technology that creates materials or devices at the scale of molecules by manipulating single atoms.

neutron: A neutron is part of an atom; it holds no charge.

newton-meter: A newton-meter is a unit measure of work.

Newton's third law of motion: Newton's third law of motion states that for every action, there is an equal and opposite reaction.

nonferrous metals: Nonferrous metals do not contain iron.

nonisometric planes: Nonisometric planes are inclined surfaces that do not lie within an isometric plane.

NOT gate: A type of logic gate that reverses the input signal. Sometimes called an inverter gate.

NPN-type transistor: An NPN-type transistor is a transistor that is signaled by a positive charge.

N-type semiconductor: An N-type semiconductor has been doped with phosphorus to produce a material with extra electrons.

nuclear energy: Nuclear energy is energy that results from combining (fusing) or splitting atoms (fission).

O

ohm: The ohm (Ω) is the unit measure of electrical resistance.

Ohm's law: Ohm's law is a mathematical formula that describes the relationship among voltage, current, and resistance: $E = I \times P$.

open-loop system: An open-loop system is the simplest type of system and requires human action to be regulated.

optics: Optics is the study of light, including how humans see the world around them.

optimization: Optimization is an act, process, or methodology that is used to make a design as effective or as functional as possible within the given criteria and constraints.

OR gate: An OR gate is a type of logic gate where either input of "on" (1) signals an output of "on" (1).

origin: The origin point (O or 0,0) is where the *x*- and *y*-axes intersect in the Cartesian coordinate system.

orthographic projection: An orthographic projection is a method of transferring the views of an object onto the planes that are perpendicular to each surface of the object.

P

parallel circuit: A parallel circuit is an electrical circuit that contains multiple pathways for current to flow.

parameter: A parameter is a physical property whose values determine the characteristic of something.

parametric modeling: Parametric modeling is a 3-D computer drawing program; also known as *feature-based solid modeling*.

Pascal's law: Pascal's law states that when there is an increase in pressure at any point in a confined fluid, there is an equal increase at every other point in the container.

passive solar heating: Passive solar heating is solar energy absorbed by stone, slate, or other materials that radiates back into the air.

patent: A patent is a unique number assigned to an invention by the U.S. Patent and Trademark Office that protects the inventor's idea; a contract between the federal government and the inventor that gives the inventor exclusive rights to make, use, and sell a specific product for a period of 17 years.

pathway: A pathway is the path electrons take in an electric circuit.

periodic table of the elements: The periodic table of the elements is a chart that displays all the elements and their corresponding numbers of neutrons, protons, and electrons.

personalized medicine: Personalized medicine is the use of technology to deliver the right medical treatment to the right patient at the right time.

perspective: Perspective is the way we see distance and depth in the things in our world.

perspective drawing or sketching: In perspective drawing or sketching, objects appear to become shorter and closer together as they move back and away from the viewer.

photoresistor: A photoresistor is one type of variable resistor. Its resistance depends on light.

photovoltaic cell: A photovoltaic cell is a device that converts the sun's energy into electricity.

pictorial sketching: Pictorial sketching is the process of graphically representing objects in three-dimensional form.

pitch: Pitch is (1) the up or down movement of an aircraft and (2) the slope of a roof.

plane: A plane is a flat surface that has no thickness, sort of like a piece of paper that extends forever.

plasticity: Plasticity is a material's ability to be deformed and then remain in the new shape.

plastics: Plastics are synthetic materials that have large, heavy molecules, can be formed by heat, and are primarily produced from petroleum products.

pneumatics: Pneumatics is the transfer of power through a gas, usually air.

PNP-type transistor: A PNP-type transistor is signaled by a negative charge.

potential energy: Potential energy is stored energy that is readily available under certain conditions.

potentiometer: A potentiometer is a variable resistor that can be regulated by the operator or engineer.

power: Power is how much work has been or could be done in a certain amount of time. Power is also the mathematical description of how much work is done.

power source: A power source provides the electron flow for a circuit.

presentation sketch: A presentation sketch is very detailed and made to look realistic, usually as a three-dimensional pictorial view such as an isometric or perspective drawing.

pressure: Pressure is force on a surface area.

problem: A problem is a challenge that needs to be solved. It is not necessarily a negative situation.

production systems: Production systems produce physical goods efficiently and at the lowest cost possible.

profile: A profile is the two-dimensional outline of one side of an object.

programmable logic controller (PLC): A programmable logic controller is a small programmable computer that gives a machine or device specific and limited directions.

propellants: Propellants are the chemical mixtures used in rockets that produce thrust.

propeller: A propeller is an airfoil mounted on a revolving shaft. It creates low pressure in front of it, thereby moving an aircraft forward because of high pressure area behind the propeller.

proportion: A proportion is the relationship between one object and another, or between one size and another size.

proportional: Proportional is the quality that, given two values in a ratio, a change in one value will cause a corresponding change in the other value, keeping the same ratio. If the first value increases, then the second value increases in the same ratio.

propulsion: Propulsion is the means by which aircraft and spacecraft are moved forward. It is a combination of factors such as thrust (forward push), lift (upward push), drag (backward pull), and weight (downward pull).

prosthetic device: A prosthetic device is an artificial limb or body part such as an artificial knee or hip.

proton: A proton is part of an atom; it holds a positive charge.

prototype: A prototype is a full-scale working model used to test a design concept by making actual observations.

P-type semiconductor: A P-type semiconductor has been doped with boron to produce a material seeking electrons.

pulley: A pulley is a rope or a cable wrapped over a grooved wheel.

pump: A pump is a device that creates flow in a system.

Q

quality assurance: Quality assurance is the process of making sure the entire manufacturing process is the best it can be.

quality circles: Quality circles are teams of managers and skilled and unskilled labor.

quality control: Quality control is the process of making sure each individual part and manufacturing process is of the highest possible quality.

R

rack-and-pinion gear: A rack-and-pinion gear is a mechanism that converts rotary motion to linear motion.

radiation: Radiation is the way in which heat travels through air in the form of waves.

rafters: Rafters are the structural members of a roof on a house.

ratio: For a gear train, the ratio is determined by the number of teeth on the various gears in the gear train.

reciprocating motion: Reciprocating motion is linear movement in a back-and-forth motion.

rendering: Rendering is the use of color, surface texture, shading, and shadow to give a sketched object a realistic look.

renewable energy: Renewable energy is any source of energy that can be replaced, such as ethanol and biomass.

reservoir: A component that stores either hydraulic fluid prior to its being pressurized by the pump or air after it has been compressed.

resistance: Resistance is the ability of a material to resist the flow of electrons; it is measured in ohms.

resistors: Resistors are used to place opposition to the flow of electrons in an electrical circuit.

reverse engineering: Reverse engineering is the process of measuring and analyzing an existing item and then re-creating technical drawings of that item.

rhythm: Rhythm is the illusion of flow or movement created by having a regularly repeated pattern of lines, planes, or surface treatments.

robot: A robot is a mechanical device that acts in a seemingly human way and that is most often guided by automatic (usually programmable) controls.

robotics: Robotics is the science of designing and using robots.

rocket: A rocket is a vehicle, missile, or aircraft that obtains thrust by the reaction to the ejection of fast moving exhaust from within a rocket engine

roll: Roll is the clockwise or counterclockwise rotating motion of an aircraft.

roof system: A roof system is the top of a structure and the primary protection for a building's interior from weather.

roof trusses: Roof trusses are structural supports common in residential and commercial construction and are premade at factories and then shipped to a building site.

rotary motion: Rotary motion is circular movement.

rudder: A rudder is a controlling surface on an aircraft's tail that regulates yaw.

S

sanding: Sanding is a separating process that uses small abrasive grain particles to remove material from a softer surface.

satellite: A satellite is an object placed into Earth's orbit by humans.

sawing: Sawing is a common separating process that uses a set of teeth to remove material.

scale: Scale is the mathematical relationship of an object to its full size; it is expressed in a ratio such as 1:2 (read as "1 to 2"), which represents one-half scale; or 1:1 (read as "1 to 1"), which is full scale.

schedules: Schedules are tables that list specific construction details, such as information about windows, doors, and light fixtures. They inform the builder about which kinds of materials to order and contain specifics about the sizes, models, and manufacturers of construction parts.

schematic: A schematic is a map of an electronic circuit. It uses symbols to represent electrical components along the pathway.

science: Science is the study of the natural world and the laws that govern it.

screw: A screw is an inclined plane wrapped around a cylinder.

sections: Sections show details of a building through a cutting plane across the proposed design.

semiconductor: A semiconductor is a material that is neither a good conductor nor a good insulator; its electrical conductivity can be precisely altered by a manufacturing process.

separating: Separating is a family of processes that remove excess material to change the size, shape, or finish of a product.

series circuit: A series circuit is an electrical circuit that contains only one path for current to flow.

servomotor: A servomotor is a special type of motor that provides very precise position feedback used in robotics and automation. The motor moves in small increments to certain positions.

shear: Shear is a splitting set of forces, one on each side of a material or object.

shearing: Shearing is a process that separates material without producing waste using such tools as scissors.

short circuit: A short circuit is a circuit that does not contain a load device or has a "shortcut" past the load device and thus no resistance in the circuit.

side view: The side view is the orthographically projected view of an object that shows its height and depth.

simple machine: A simple machine is a basic device that makes work easier.

sketch geometry: Sketch geometry is the process of drawing geometric shapes in a parametric modeling program.

sketch plane: In parametric modeling software, a sketch plane is the surface on which a geometric feature is drawn.

softwoods: Softwoods are the products of conifer or evergreen trees that have needles and bear seed cones; they typically do not have the strength of hardwoods.

solar energy: Solar energy is energy from the sun.

soldering: Soldering is a process in which two pieces are joined by a molten metal that adheres to the base metal but the base metal is not melted as in welding.

spacecraft: A spacecraft is a vehicle designed for spaceflight.

static electricity: Static electricity is the flow of electrons found in nature—for example, lightning.

strength: Strength is a material's ability to withstand a force without changing shape or breaking.

structural engineer: A structural engineer is a person who designs structures such as sports stadiums, buildings, bridges, towers, and dams.

structural engineering: Structural engineering is the analysis of forces and loads and their effects on a structure.

subsystem: A subsystem is a system that operates as part of another system.

supply chain management: Supply chain management is the process of making sure materials and products are delivered on time from raw material to finished product.

survey: A survey is a method used to establish legal boundaries of land.

sustainable architecture: Sustainable architecture, sometimes termed green architecture, refers to designing a structure with the environment in mind.

sustainability: Sustainability implies that we design engineering solutions with future needs and the environment in mind.

switch: In an electronic device, a switch controls the flow of electrons along an electronic pathway.

system: A system is a set or group of parts that all work together in a systematic, organized way to accomplish a task.

T

technological resource: A resource is something that has value and that can be used to satisfy human needs and wants.

technology: Technology is (1) the processes humans use to develop new products to meet their needs and wants, and (2) the products or artifacts actually made.

telemedicine: Telemedicine is the process of sharing medical information or medical procedures from a distance.

tempering: Tempering is a conditioning process that makes a material more flexible.

tension: Tension is a force that pulls material from each end of an object.

testing: Testing is the process of analyzing or assessing the performance of the design solution based on the established design criteria.

texture: Texture refers to the roughness or smoothness, including reflective properties, of a material or building.

thermistor: A thermistor is a variable resistor that is regulated by temperature.

three-dimensional (3-D) sketching: In three-dimensional sketching, sketched objects show width, height, and depth.

three-view drawings: A three-view drawing is a multiview drawing with three views.

thrust: Thrust is (1) a force applied to a body to propel it in a desired direction, (2) the force that propels an aircraft in the direction of motion and is produced by engines, and (3) the force that moves an aircraft through the air. Thrust is generated by an aircraft's engines.

thumbnail sketch: A thumbnail sketch is usually small and simple and has just enough detail to convey a concept. It is often used in brainstorming to record ideas quickly.

tolerance: Tolerance is the amount of variation that is allowable from the established value.

top view: A top view shows an object's width and depth.

torque: Torque is the force of rotary motion.

torsion: Torsion is the force that twists material.

toughness: Toughness is the ability of a material to withstand a force applied to it.

trade-off: A trade-off is an exchange of one option in order to gain a better option.

trailing edge: The trailing edge is the rear edge of a wing.

trajectory: Trajectory is the path an object takes during flight.

transistor: A transistor is a device made by combining three semiconductors together. It uses a small amount of current to control a larger amount of current flow.

transmission lines: In pneumatics, transmission lines are tubing through which compressed air flows.

transportation engineers: Transportation engineers are involved with the design and analysis of roads and highways, traffic-control signals, railways, and airports.

transportation system: A transportation system is a system designed to move people or goods from one place to another as efficiently as possible.

truth table: A truth table is a chart that represents the inputs and outputs for a desired outcome or electronic component.

turbojet engine: A turbojet engine is a heat engine that converts chemical energy to thrust through the use of compressors and turbines. It is based on Newton's third law of motion.

turning: Turning is a separation process used with a lathe.

two-point perspective: A two-point perspective view allows the viewer to see two sides of an object "vanishing" to two different points on the horizon line.

U

unity: Unity is the principle of design that ties the various elements of a structure's design together.

universal joint: A universal joint is the simplest means of transferring rotary motion. It allows two axles to vary their alignments and also move slightly during operation.

V

vacuum forming: Vacuum forming is the use of vacuum suction to pull a softened plastic sheet down into a mold.

valence electron: A valence electron is an electron on an atom's outermost orbit.

vanishing point: The vanishing point is the location on the horizon line where the viewer's line of sight merges.

V-belts and pulleys: V-belts and pulleys change rotary motion; the belts change the direction of their rotation (clockwise to counterclockwise) while the pulleys continue to turn in the same direction.

visualizing: Visualizing is the ability to form a mental picture of an object. Freehand technical sketching records that mental picture on paper.

voltage: Voltage is the measurement of the force or pressure of the electrons in a pathway (measured in volts); also known as *electromotive force.*

voltage drop: A voltage drop is a decrease in voltage along a series circuit.

volts: Volts are the measure of voltage.

W

waste-treatment engineers: Waste-treatment engineers design and analyze water-treatment facilities.

water-management engineers: Water-management engineers design dams and structures to control the flow of water, including drainage systems and waterways for transportation.

wedge: A wedge is a device that converts motion in one direction into a splitting motion acting at right angles.

welding: Welding is a process in which materials are heated and "melted" together.

wheel and axle: A wheel and axle is a rotating lever (wheel) that moves around a fulcrum (axle).

wind tunnel: A wind tunnel is a facility used by engineers to measure the effect of airflow across a device.

wind turbine generators: Wind turbine generators are wind-turned devices that produce electricity from rotary motion.

wingspan: A wingspan is distance between two ends of an airfoil.

wood screw: A wood screw is a tapered screw that is used to join lumber but not metal.

work: Work is (1) how much force it takes to move some object a set distance, and (2) the transfer of energy from one physical system to another.

workcell: A workcell is a group of machines and devices such as robots that work together to accomplish several processes on the same part.

work envelope: A work envelope is the area of normal reach for an industrial robot.

working drawing: A working drawing is a 2-D drawing of individual parts that shows different views (e.g., front, side, and top views) with dimensions and annotations (notes). This type of drawing is used to fabricate an object.

worm gear: A worm gear is a gear that changes the direction of rotation and increases the torque of movement.

X

***x*-axis:** The *x*-axis is the horizontal axis in the Cartesian coordinate system.

XOR gate: An XOR gate is a type of logic gate where only one input can be "on" (1) to produce an output of "on" (1). Sometimes called an exclusive gate.

Y

yaw: Yaw is an aircraft's left or right turning motion.

***y*-axis:** The *y*-axis is the vertical axis in the Cartesian coordinate system.

Index

Note: Page numbers followed by *f* indicate figures.

A

B

D

E

F

G

P